Mass Spectrometry of Inorganic, Coordination and Organometallic Compounds

Inorganic Chemistry

A Wiley Series of Advanced Textbooks

Editorial Board

Derek Woollins, *University of St Andrews, UK*
Bob Crabtree, *Yale University, USA*
David Atwood, *University of Kentucky, USA*
Gerd Meyer, *University of Hannover, Germany*

Previously Published Books In This Series

Chemical Bonds: A Dialog
Author: J. K. Burdett

Bioinorganic Chemistry: Inorganic Elements in the Chemistry of Life –
An Introduction and Guide
Author: W. Kaim

Synthesis of Organometallic Compounds: A Practical Guide
Edited by: S. Komiya

Main Group Chemistry (Second Edition)
Author: A. G. Massey

Inorganic Structural Chemistry
Author: U. Muller

Stereochemistry of Coordination Compounds
Author: A. Von Zelewsky

Forthcoming Books In This Series

Lanthanides and Actinides
Author: S. Cotton

Mass Spectrometry of Inorganic, Coordination and Organometallic Compounds

Tools – Techniques – Tips

William Henderson

University of Waikato, Hamilton, New Zealand

and

J. Scott McIndoe

University of Victoria, Canada

John Wiley & Sons, Ltd

Other Wiley Editorial Offices

John Wiley & Sons Inc., 111 River Street, Hoboken, NJ 07030, USA

Jossey-Bass, 989 Market Street, San Francisco, CA 94103-1741, USA

Wiley-VCH Verlag GmbH, Boschstr. 12, D-69469 Weinheim, Germany

John Wiley & Sons Australia Ltd, 33 Park Road, Milton, Queensland 4064, Australia

John Wiley & Sons (Asia) Pte Ltd, 2 Clementi Loop #02-01, Jin Xing Distripark, Singapore 129809

John Wiley & Sons Canada Ltd, 22 Worcester Road, Etobicoke, Ontario, Canada M9W 1L1

Wiley also publishes its books in a variety of electronic formats. Some content that appears in print may not
be available in electronic books.

Library of Congress Cataloging-in-Publication Data

Henderson, William.
    Mass spectrometry of inorganic, coordination and organometallic compounds
/ William Henderson & J. Scott McIndoe.
        p. cm.
    Includes bibliographical references.
    ISBN 0-470-85015-9 (cloth : alk. paper) – ISBN 0-470-85016-7 (pbk. : alk. paper)
    1. Mass spectrometry.   2. Chemistry, Inorganic.   3. Organometallic
compounds.   I. McIndoe, J. Scott.   II. Title.
    QD96 .M3H46 2005
    543'.65–dc22   2004023728

British Library Cataloguing in Publication Data

A catalogue record for this book is available from the British Library

ISBN 978-0-470-85016-9

This book is dedicated to our families:

Angela, Laura and Liam (WH)

Angela, Seth and Grace (JSM)

Bill Henderson was born in Darlington, County Durham, and grew up in Stockton-on-Tees, both in the North-East of England. He studied chemistry and geochemistry at the University of Leicester, and stayed at Leicester for his PhD in organometallic chemistry under the supervision of Dr. Ray Kemmitt, studying metallacyclic complexes of platinum, palladium and nickel. An NSF-supported postdoctoral fellowship at Northwestern University, Evanston, Illinois, USA with Professor Du Shriver followed. This included a period of collaborative research involving metal clusters as catalyst precursors at Hokkaido University in Sapporo, Japan with Professor Masaru Ichikawa. Bill then returned to England, with a period spent in industry with Albright & Wilson Ltd in the West Midlands, where he carried out research and development work in organophosphorus chemistry and surfactants. In 1992, a lectureship in New Zealand beckoned, at the University of Waikato in Hamilton, where he has been ever since. Since 2000 he has been an Associate Professor, and has been Head of Department since 2002.

Research interests cover a range of areas, with the characterisation of inorganic compounds using mass spectrometry being one of the central themes. Other research areas include the chemistry of the platinum group metals and gold, and applications of organophosphorus chemistry to the synthesis of novel ligands and the immobilisation of enzymes. He has published over 150 articles in refereed journals, together with three textbooks.

Bill is married to Angela, a high school teacher, and they have two children, Laura and Liam. In his spare time, other interests include music, gardening and English mediaeval history.

Born in Rotorua, New Zealand, Scott McIndoe completed all his degrees at the University of Waikato in Hamilton. His DPhil in organometallic chemistry was supervised by Professor Brian Nicholson. The New Zealand Foundation for Research, Science & Technology (FRST) awarded him a postdoctoral fellowship in 1998 to work in the group of Professor Brian Johnson FRS at the University of Cambridge, England. In 2000 he took up the post of college lecturer at Trinity and Newnham Colleges, also at Cambridge. After three years in this position, he moved to an assistant professorship at the University of Victoria in British Columbia, Canada. Scott's research interests focus around using mass spectrometry as a first-resort discovery tool in organometallic chemistry and catalysis.

In curious symmetry with Bill, Scott is also married to an Angela who is a high school teacher, and they have two children, Seth and Grace. His other interests include cricket, windsurfing and finding excuses to add to his power tool collection.

Contents

Preface

Mass spectrometry (MS) is just one of many powerful instrumental techniques that is available to the inorganic, coordination or organometallic chemist. When we set out to write this book, our principal aim was to make it understandable by a typical inorganic or organometallic chemist who might use mass spectrometry, but who is by no means an expert in the field. Our own scientific backgrounds – as synthetic chemists who have discovered the power of mass spectrometry techniques for studying inorganic systems– are in accord with this philosophy.

Mass spectrometry applied to the analysis of inorganic substances has a long and fruitful history. However, relatively recent developments in ionisation techniques have placed two of these – MALDI (Matrix Assisted Laser Desorption Ionisation) and especially ESI (Electrospray Ionisation) – at the forefront of the pack. These are extremely powerful, soft ionisation methods that provide valuable mass spectrometric information to the chemist. Furthermore, the gentle nature of these ionisation techniques often results in spectra that can be easily analysed by chemists, as opposed to experts in mass spectrometry, as was often the case with harsher ionisation methods. Coupled with major advances in instrument robustness, automation, computer hardware, operating software and ease of operation and maintenance, such instrumentation is becoming widely used by inorganic chemists worldwide. We therefore felt that a textbook describing the mass spectrometric characterisation of inorganic and organometallic compounds was timely.

Many excellent textbooks and review articles cover the principles behind the various ionisation techniques and their applications, which are dominated by organic and biochemical systems. Readers wanting more detailed expositions on the finer points of mass spectrometry are encouraged to consult these texts.

This book is roughly divided into two main sections. In the first half of the book (Chapters 1 to 3), the basic principles of operation of various types of mass spectrometry systems are included, with an emphasis on mass analysers and ionisation techniques. Again, this has been written with the chemist in mind, so the treatment is primarily descriptive rather than mathematical. Also included are fundamental aspects such as resolution, data presentation methods and the use of isotope information. We have tried, where possible, to provide helpful suggestions for practical use, in the form of end-of-section summaries.

The second half of the book (Chapters 4 to 7) describes the applications of just one ionisation technique – electrospray – that without doubt is the most versatile and widely used mass spectrometry technique for the characterisation of inorganic and organometallic compounds today. The material is divided into chapters according to the type of compound, for example, coordination compounds (Chapter 5) and transition metal organometallic compounds (Chapter 7). In these chapters we have endeavoured to discuss the behaviour patterns of the various classes of compounds, such that the reader will be able to successfully apply modern mass spectrometry techniques to their own area of chemistry. Finally, Chapter 8 discusses some 'Special Topics' involving the application of modern mass spectrometry techniques in imaginative ways to particular inorganic and organometallic systems.

William Henderson
and
J. Scott McIndoe

Acknowledgements

We are grateful to our publishers, John Wiley & Sons, for the opportunity to write this book, and to the other publishers who have generously allowed reproduction of some of the figures.

JSM

Many thanks to David McGillivray for running the LSIMS and EI mass spectra, and to Orissa Forest for assisting in collating, tabulating and graphing the data in the Appendices. I greatly appreciate the discussions I've had with many chemists and mass spectrometrists who have shown me and described in detail their laboratories and instrumentation. Also those I've met at conferences, many of whom made extremely useful comments and suggestions on a wide variety of topics. Brian Fowler of Waters Canada went well beyond the call of duty in installing my mass spectrometer by answering an incessant stream of questions about componentry. The Canada Foundation for Innovation, the British Columbia Knowledge Development Fund and the University of Victoria are thanked for their support for purchasing and maintaining this instrument. Also thanks to Brian Nicholson (Waikato) and Brian Johnson (Cambridge), inspiring mentors and educators, and to Paul Dyson (EPFL), for posing thoughtful problems that led to many of our collaborative ventures. Pat Langridge-Smith (Edinburgh) gave me a unique introduction to some of the more esoteric aspects of mass spectrometry. Members of my research group – notably Nicky Farrer, Sarah Luettgen and Colin Butcher – and numerous undergraduates have asked good questions requiring clear answers that have helped clarify my own thinking.

WH

I am indebted to Pat Gread and Wendy Jackson, for their dedication in maintaining the Waikato mass spectrometry instrumentation, and to the University of Waikato for their generous investment in mass spectrometry. I would also like to thank Brian Nicholson for numerous fruitful discussions concerning all aspects of chemistry, including mass spectrometry, and my students, past and present, who have each made distinctive contributions. Through mass spectrometry I have been able to develop a number of productive and enjoyable collaborations with other chemists around the world, and especially acknowledge Professor Andy Hor and his coworkers at the National University of Singapore.

List of commonly-used abbreviations

Mass spectrometric

API	Atmospheric Pressure Ionisation
APCI	Atmospheric Pressure Chemical Ionisation
CE	Capillary Electrophoresis
CI	Chemical Ionisation
CIS	Coordination Ionspray
CID	Collision Induced Dissociation
DFT	Density Functional Theory
EDESI	Energy-Dependent Electrospray Ionisation
EI	Electron Impact (ionisation)
ESI	Electrospray Ionisation
ES, ESMS	See ESI
FAB	Fast Atom Bombardment
FD/FI	Field Desorption/Field Ionisation
FTICR	Fourier Transform Ion Cyclotron Resonance
FTMS	Fourier Transform Mass Spectrometry
FWHM	Full Width at Half Maximum
GCMS	Gas Chromatography Mass Spectrometry
HPLC	High Pressure Liquid Chromatography
ICP-MS	Inductively Coupled Plasma Mass Spectrometry
LCMS	Liquid Chromatography Mass Spectrometry
LDI	Laser Desorption Ionisation
LSIMS	Liquid Secondary Ion Mass Spectrometry
MALDI	Matrix Assisted Laser Desorption Ionisation
MCA	Multi Channel Analysis
MS	Mass Spectrometry
MS/MS	Mass Spectrometry/Mass Spectrometry
MS^n	n^{th} generation Mass Spectrometry
m/z	Mass-to-charge ratio
NMR	Nuclear magnetic resonance
PDMS	Plasma Desorption Mass Spectrometry
PSD	Post Source Decay
Q	Quadrupole
rf	Radiofrequency
SIMS	Secondary Ion Mass Spectrometry
TIC	Total Ion Current
TOF	Time-of-Flight
oa-TOF	orthogonal Time-of-Flight
UV	Ultraviolet

Non-SI units often encountered in mass spectrometry

Atm	Atmospheric pressure, defined as 101 325 Pa.
Bar	10^5 Pa. Unit of pressure approximating to one atmosphere. Low pressures often listed in millibar, 1 mbar = 100 Pa.
Da	Dalton, the *unified atomic mass unit* = $1.660\ 540 \times 10^{-27}$ kg. ^{12}C = 12 Da exactly.
eV	electron volt, the kinetic energy acquired by an electron upon acceleration through a potential difference of 1 V = $1.602\ 177 \times 10^{-19}$ J.
mm Hg	one atmosphere of pressure will force a column of mercury to a height of 760 mm, so 1 mm Hg = 1/760 atm = 133.32 Pa. Equivalent to Torr.
Torr	see mm Hg.
Th	Thomson, unit for mass-to-charge ratio i.e. 1 Th = 1 *m/z*.

Chemical

acac	Acetylacetonate anion (2,4-pentanedionate)
An	Actinide metal
bipy	2,2′-bipyridyl (also 2,2′-bipyridine)
Cp	η^5-cyclopentadienyl (C_5H_5)
Cp*	η^5-pentamethylcyclopentadienyl (C_5Me_5)
Cy	Cyclohexyl
DMSO	Dimethylsulfoxide, $S(O)Me_2$
EDTA	Ethylenediamine tetra-acetic acid or anion thereof
en	Ethylene-1,2-diamine (1,2-diaminoethane), $NH_2CH_2CH_2NH_2$
Fc	Ferrocenyl, $(\eta^5\text{-}C_5H_4)Fe(\eta^5\text{-}C_5H_5)$
GSH	Glutathione (Glu-Cys-Gly)
HMPA	Hexamethylphosphoric triamide $OP(NMe_2)_3$
L	Ligand coordinated to a metal centre
Ln	Lanthanide metal
M	Metal centre
MeCN	Acetonitrile
py	Pyridine, C_5H_5N
PPh$_3$	Triphenylphosphine
PPN	Bis(triphenylphosphine)iminium, $[Ph_3P=N=PPh_3]^+$
pta	Phosphatriaza-adamantane
R	Alkyl or aryl group
THF	Tetrahydrofuran
TM	Transition Metal
X	Halogen

1 Fundamentals

Introduction

A mass spectrometer is an instrument for generating gas-phase ions, separating them according to their mass-to-charge ratio using electric fields (sometimes magnetic fields as well) in an evacuated volume, and counting the number of ions. A computer system controls the operation and stores, manipulates and presents the data. The features of the mass spectra so produced relate to the properties of the original sample in a well understood way. This chapter deals with some of the fundamental aspects of mass spectrometry: how samples are introduced to the instrument (inlets); how the ions are fragmented; how ions are counted (detectors) and the type of output and how it is manipulated (data systems and data processing) and interpreted (isotope patterns). How the ions are separated (mass analysers) is dealt with in Chapter 2 and the how the ions are formed (ionisation techniques) in Chapter 3.

Figure 1.1 is a schematic drawing of a mass spectrometer. The sample is introduced through an inlet to the ionisation source. The source generates gas-phase ions, which are transferred to the mass analyser for separation according to their mass-to-charge ratio. A detector registers and counts the arriving ions. The data system controls the various components of the mass spectrometer electronically, and stores and manipulates the data. All mass spectrometers have a vacuum system to maintain the low pressure (high vacuum) required for operation. High vacuum minimises ion-molecule reactions as well as scattering and neutralisation of the ions.

Modern instruments often have the facility to perform more than one mass analysis on a single sample, i.e. MS/MS or even MS^n (where n = number of stages of mass spectrometry). Such machines require more than one mass analyser, or alternatively, have the facility to trap ions in a small volume of space and carry out repeated experiments on them. Both types of mass spectrometer require the ability to fragment ions, and this is usually achieved by collision-induced dissociation, another topic covered in this chapter.

Inlets

The way in which a sample is introduced to the mass spectrometer is very dependent on its phase (gas, liquid, solid or solution) and the means by which ionisation is induced. Gaseous samples are easily transferred to a mass spectrometer, as the gas may simply be allowed to leak into the low pressure source region. The effluent from the capillary

Mass Spectrometry of Inorganic, Coordination and Organometallic Compounds   W. Henderson and J. S. McIndoe

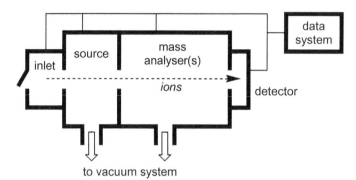

Figure 1.1
Schematic of a
mass spectrometer

column of a gas chromatograph (GC) may be conveniently plumbed directly into the source of a mass spectrometer. Condensed phase analytes (liquid or solid) are placed on a sample holder and passed through a door into the instrument. The door is closed, sealed and the inlet/source region is evacuated, after which time whatever ionisation technique being used is applied. Analytes dissolved in a solvent are usually introduced to the mass spectrometer *via* a combined inlet/ionisation source, in which sample introduction, desolvation and ionisation are intimately related. The solution is commonly the effluent from a liquid chromatograph (LC), or it may be injected directly into the instrument by means of a syringe pump.

Collision-Induced Dissociation[1]

Collision-induced dissociation (CID, sometimes known as collision-*activated* decomposition, or CAD) of ions occurs when some of the translational energy of an accelerated ion is converted into internal energy upon collision with a residual gas (typically nitrogen or one of the noble gases helium, argon, xenon). The increase in internal energy can induce decomposition (fragmentation) of the ion. CID was of limited importance in mass spectrometry – indeed, some instruments were fitted with 'metastable suppressors' designed to eliminate this troublesome effect – until the advent of soft ionisation techniques. The ability of these techniques to obtain practically intact molecular ions for many classes of compound was enormously useful in itself, but obtaining structural information through characteristic fragmentation patterns is also highly desirable and CID proved to be the ideal answer to this problem.

The first step in the CID process is the actual collision between a fast-moving ion and an immobile neutral target, resulting in an increase in the internal energy of the ion. The ion then rapidly redistributes this extra energy amongst its vibrational modes, which number $3N - 6$ for an ion with N non-linear atoms. The much slower second step is the unimolecular decomposition of the excited ion to generate product ions and neutral fragments. Because the timescale of the first step is very much shorter than the second, large ions are more difficult to fragment using CID as they have more vibrational modes in which to deposit the extra energy, making decomposition of the ion less likely. Two collision regimes for CID may be defined, low energy (tens of electron volts (eV)) and high energy (thousands of eV).

In practice, low-energy CID is carried out by allowing an accelerated beam of ions to traverse a volume occupied by gas molecules or atoms as the target. In MS/MS instruments in which the mass analysers are separated in space, such as the triple quadrupole (QqQ)[2] or hybrid quadrupole-Time-of-flight (QqTOF), an rf-only quadrupole (the 'q' in QqQ) encloses this volume, called a **collision cell**. The directional focusing abilities of the rf-only quadrupole are used to good effect here, redirecting ions back on to the right axis after collisions drive them off-course. However, the potential well created by a rf-only quadrupole field is not particularly steep-sided and ion losses do occur. Better ion guides are rf-only hexapoles or octapoles and the recently introduced **ion tunnels**. The latter are a series of ring shaped, alternately charged electrodes, 60 or more of which describe a hollow cylinder inside of which the ions are tightly confined. Whatever its configuration, the collision cell is separated from the mass analysers either side by narrow apertures and is filled with an inert gas. Ions emerging from the first mass analyser are fragmented (and often scattered) upon collision with the gas, strongly refocused back on to the ion optical axis by the rf-only field, transmitted to the second mass analyser and then detected. A large number of collisions is allowed to occur in the collision cell, so collision yields (the percentage of fragmented ions that reach the detector) are frequently very high for this form of CID. In MS/MS instruments that rely on each stage of MS being carried out sequentially (*in time*) in the same space, such as ion traps or Fourier Transform Ion Cyclotron Resonance (FTICR) analysers, the collision gas is simply introduced to the chamber. The ions are energised and fragmented by CID. The process is especially simple for ion traps, which typically contain a background pressure of helium gas at about 10^{-3} mbar during operation, so the trap does not even need to be filled and emptied between stages of MS/MS.

The nature of the target gas is important in low-energy CID. A large proportion of the translational energy of the ion is transformed into internal energy upon collision with an effectively stationary target, the mass of which has a significant effect on the spectra (so the extent of dissociation increases He < Ar < Xe). Atomic gases are more efficient than polyatomic gases in causing CID, because the latter can be vibrationally excited themselves upon collision and hence reduce the amount of energy transferred to the ion. The chemical effects of the target are also important due to the possibility of ion/molecule reactions, so if dissociation of the precursor ion only is sought, an inert target gas is desirable (making the noble gases doubly appropriate). However, there are some circumstances in which ion/molecule reactions are of great interest.

Low-energy CID spectra are very sensitive to small absolute changes in the collision energy, to collision gas pressure and to the mass of the neutral target. These factors conspire to make the reproducibility of low-energy CID spectra between instruments poor compared to electron ionisation mass spectra, for which searchable libraries of spectra are very well established.

Instruments with an atmospheric pressure source have another region in which low-energy CID can occur, located just before the ions enter the high-vacuum region of the mass spectrometer. Here, the pressure is low enough that the mean free path length of the accelerating ions is sufficiently long that they can attain a high enough velocity for collisions with residual solvent molecules and/or desolvation gas to cause fragmentation. This process is called **in-source** CID, and is an especially important facility for instruments with a single mass analyser. The ions are accelerated by application of a variable voltage between the sampling cone and the skimmer cone (which separate differentially pumped regions of the instrument; Chapter 3, Section 8 on electrospray ionisation gives more details), and this 'cone voltage' generally has the most profound effect on the mass spectrum of any of the parameters used to tune the instrument.

High-energy CID is the preserve of sector instruments (Chapter 2, Section 2), which accelerate and analyse ions with energies of thousands of eV. rf-only multipoles are useless as collision cells under these circumstances, as they are unable to refocus such energetic ions after a collision. A simple reaction region containing the collision gas is quite sufficient; ions deflected more than a few tenths of a degree upon collision are lost. The lack of means by which to refocus errant ions and a peak-broadening effect due to kinetic energy release upon collision conspire to make high-energy CID markedly less efficient in terms of conversion of precursor ion to *detected* product ion than its low-energy cousin. The distribution of energies transferred at collision energies of thousands of eV is broad, and high-energy processes result in some product ions that do not appear at all in low-energy CID spectra.

1.3.1 Bond Dissociation Energies from CID Studies

Bond dissociation energies may be obtained from low-energy ('threshold') CID studies, by analysing the kinetic energy dependence of the reactions of metal complexes with an inert collision gas,[3] and ion thermochemistry remains an active research field.[4] Threshold CID experiments are carried out using guided ion beam mass spectrometers, custom-made instruments that allow the sequential generation, thermalisation (cooling), mass selection, fragmentation and mass analysis of ions.[5] To obtain precise data, multiple ion-neutral collisions are eliminated, careful consideration is taken of internal energies of the complexes and their dissociation lifetimes, and the experiments are backed up by Density Functional Theory (DFT) calculations. Fundamental information such as the stepwise energies for dissociation of $[Pt(NH_3)_x]^+$ ($x = 1 - 4$) or $[Cr(CO)_x]^+$ ($x = 1 - 6$) complexes can be obtained using this approach.[6] The main limitation for wider applicability of this technique is that experiments cannot yet be implemented on commercially available instruments. Metal-ligand bond dissociation energies have also been established using FTICR experiments under single-collision conditions.[7]

1.3.2 Presentation of CID Data

Detailed CID investigation of a compound can generate huge quantities of data – in a typical low-energy CID experiment, the collision energy can be varied from $0 - 200$ eV, and the analyst must decide which spectra are most representative and informative. This is traditionally carried out by means of a stacked plot, selecting values for the collision energy so that all product ions show up in at least one of the spectra chosen. Numerous examples of this approach can be seen in Chapters 4 to 7 (*e.g.* Figures 4.3, 4.6, 5.6, 5.8 etc.).

If the appearance/disappearance potentials of a particular ion are of special interest, the breakdown graph is an effective way of presenting this data.[8] A breakdown graph plots the intensity of a given ion against the fragmentation energy, represented by the cone voltage (for in-source CID) or collision voltage (for CID in a collision cell). Multiple ions may be presented on a single breakdown graph (Figure 1.2).

In more complicated cases, where there are many fragment ions, and/or a mixture of ions, it may be beneficial to collect spectra across the entire energy range and present all the information simultaneously. This approach is encapsulated in energy-dependent electrospray ionisation mass spectrometry (EDESI MS), which uses a presentation style reminiscent of two-dimensional NMR spectra.[9] The precursor and all product ions appear as cross-peaks in a contour map, where the contours represent ion intensity. The approach is best illustrated with an example (Figure 1.3).

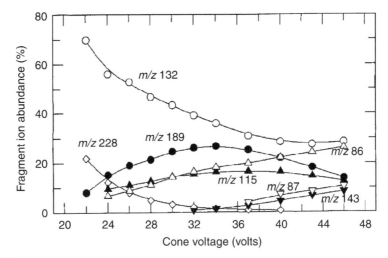

Figure 1.2
Breakdown graphs obtained by CID of protonated H-Gly-Gly-Leu-OH. From Harrison. Reproduced by permission of Wiley Interscience

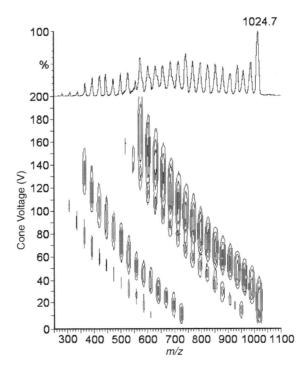

Figure 1.3
EDESI mass spectrum of a mixture of four anionic metal carbonyl clusters, $[Ru_5CoC(CO)_{16}]^-$, $[HRu_4Co_2C(CO)_{15}]^-$, $[Ru_3Co(CO)_{13}]^-$ and $[RuCo_3(CO)_{12}]^-$.[10] Note how each component of the mixture is clearly discriminated in the map, but the summed spectrum at the top is uninformative

Detectors

The abundance of ions at each mass-to-charge ratio (*m/z*) value must be measured, and this is the role of the **detector**. The ideal detector will have a wide dynamic range (able to detect a few ions arriving just as well as tens of thousands) and a response as linear as possible (provide a peak 100 × larger for 1000 ions than that produced for 10). In the earliest days of mass spectrometry detectors were simply photographic paper but this method was essentially made obsolete by the introduction of electron multipliers. These devices convert the kinetic energy of the arriving particles into electrical signals. The incoming ions strike a surface called a dynode, which is capable of releasing one or more electrons when struck by a particle having an energy above a certain level. Usually, there are a series of dynodes and the released electron is accelerated towards the second dynode, which releases further electrons (Figure 1.4). By repeating this input and release process many times, the number of electrons increases in a geometrical progression (10^6 to 10^8 ×).

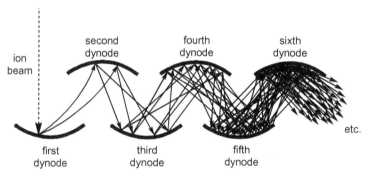

Figure 1.4
An electron multiplier. An ion travelling at high speed causes secondary electrons to be ejected from a metal surface (a dynode) upon impact. These electrons are accelerated through an electric potential towards a second dynode, releasing more electrons, and so on until a blizzard of electrons strikes the final dynode, producing a detectable current which may be amplified further

A scintillator or 'Daly detector' accelerates the secondary electrons (generated when the incoming ions strike the first dynode) towards a dynode made of a substance that emits photons (a phosphor). A photomultiplier tube enhances the signal which is ultimately converted into an electric current. This arrangement has some advantages over the electron multiplier as the photomultiplier may be sealed from the vacuum of the mass spectrometer and does not suffer ill effects from the presence of residual gas or discharged ions, significantly increasing the lifetime of the detector.

In some applications it is advantageous to collect ions over an area using an **array detector**, rather than a point detector which relies on ions arriving sequentially at a single location. Array detectors can detect ions arriving simultaneously at different points in space. This property is particularly useful in sector instruments, which disperse ions in space, so a number of detectors arranged in a line are capable of measuring a section of the mass spectrum in the same amount of time that a single detector can measure a single *m/z* value. For example, an array detector containing ten collectors could simultaneously measure ten times the mass range that a single collector could in the same time. The efficiency of detection is thus greatly improved, important in applications requiring high sensitivity.

TOF instruments generate a pulse of ions of a wide range of *m/z* values, all of which arrive at the detector within a few microseconds, and ions of adjacent *m/z* value are separated in time by less than a nanosecond. A detector is required that has a very fast recovery time. Furthermore, orthogonal-TOF mass analysers pulse a whole section of an ion beam at once, and this spatial dispersion in the original direction of travel is preserved as the ions progress down the flight tube. The ions arrive at the detector across a broad front, demanding a detector able to accept ions over an equally wide area. Both of these obstacles are solved by the use of a **microchannel plate** (MCP), which consists of a large number (thousands) of tightly packed individual detection elements all connected to the same backing plate. Each of these 'microchannels' is a tiny electron multiplier tube, and an ion arriving in any of them sets off a cascade of electrons to provide a detectable signal. A **time-to-digital converter** (TDC) sets up timing increments separated by intervals of less than a nanosecond, and a signal detected in any of these intervals is recorded as an arrival time. It does not, however, record the intensity of the signal, so two ions arriving with the same time interval on different parts of the MCP are still recorded as a single arrival time. Generally, this does not pose a problem; a TOF analyser is typically recording 30 000 spectra per second so the number of ions arriving in any one individual spectrum is low. However, it becomes an issue when recording particularly high ion currents, and is exacerbated by the fact that the TDC itself has a 'dead time', in which it takes some time to recover before it can record a new event. These effects conspire to affect the quantitative response of TOF detectors, and high ion currents tend to distort peak shape and underestimate intense signals, though computer processing does mitigate the detrimental effects to a large degree.

Mass Resolution

The resolution of a mass spectrometer represents its ability to separate ions of different *m/z*. It is manifested in the sharpness of the peaks seen in the mass spectrum. An instrument with high resolving power will be able to distinguish two peaks very close in mass. Calculating the resolution is done in one of two ways. Magnetic instruments tend to give peaks which are essentially Gaussian in shape, and the usual definition is R = $m/\Delta m$, where *m* is the mass of an ion peak and Δm is the distance to another peak overlapping such that there is a 10 % valley between the two peaks (Figure 1.5).

In the figure, $m = 1000$ and $\Delta m = 0.208$, so the 10 % valley definition gives a resolution of $1000/0.208 = 4800$. It is generally more convenient to conduct the calculation on a single ion, in which case Δm is the full width of the peak at 5 % of its maximum intensity. Another common resolution calculation uses the full width of the peak at half maximum intensity (FWHM). This definition is commonly used for TOF and ion trap instruments, which typically have relatively broad-based peak profiles and as such the 5 % definition exaggerates the peak width and hence gives an unreasonably low value for the resolution. In the figure shown, FWHM = $1000/0.1 = 10\,000$. Clearly, when comparing resolution performance between instruments it is important to apply the same definition in each case. Generally in this text, 'resolution' will correspond to the FWHM definition unless otherwise stated.

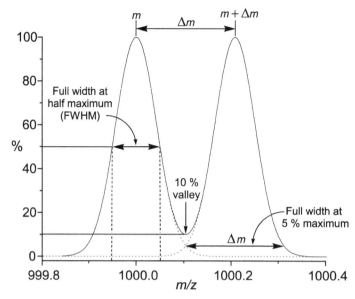

Figure 1.5
Measurements used in calculations of resolution. The FWHM definition will be used in this book

The ability of instruments with different resolution to differentiate between low-mass ions of the same nominal mass is illustrated in Figure 1.6. At low resolution (1000, e.g. quadrupole/ion trap in low resolution mode or linear TOF), the three ions are not discriminated at all and just a single peak is observed. At slightly higher resolution (2500, e.g. quadrupole/ion trap in maximum resolution mode) the higher m/z ion is differentiated, but the remaining two ions appear as just a single peak at an m/z value intermediate between the two real values. Three peaks can be clearly observed at a resolution of 5000 (e.g. reflectron TOF), and the signals are baseline resolved at 10 000 resolution (e.g. magnetic sector, high performance reflectron TOF, FTICR).

However, the major criterion for an inorganic/organometallic chemist should be the ability to provide good baseline-resolved isotope patterns in the m/z range of most interest. The need for high resolution becomes less stringent when it is isotope pattern information that is required. Baseline resolution of the individual members of the isotopomer envelope is the most important criterion for satisfactory data. The majority of coordination complexes and organometallic compounds are below 1000 Da, at which a resolution of 2500 is generally sufficient for good data (Figure 1.7).

However, a resolution much below 2500 will drastically reduce confidence in assignment, as can be clearly seen in the lumpy, indistinct and unsatisfactory profile observed for the spectrum collected at a resolution of 1000. Higher resolution than 2500 is always desirable, especially when collecting data on ions of mass > 1000 Da and for multiply charged ions, and the higher quality the data the correspondingly higher confidence can be had in assignment. A resolution of 2500 can be achieved for practically all modern research level instruments, regardless of type – even relatively inexpensive ion trap and quadrupole machines can be scanned slowly over the isotope envelope region (usually not

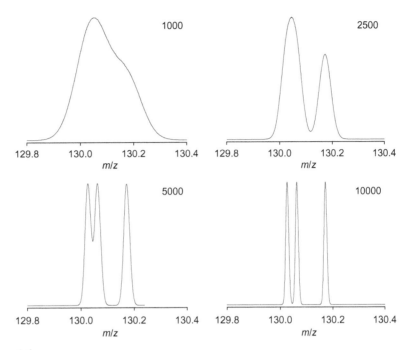

Figure 1.6
Effect of increasing resolution in differentiating the ions $[C_5H_6O_4]^+$, $[C_6H_{10}O_3]^+$ and $[C_9H_{22}]^+$, all with a nominal mass of 130 Da and present in equal amounts. Monoisotopic masses are 130.0266, 130.0630 and 130.1722 m/z respectively

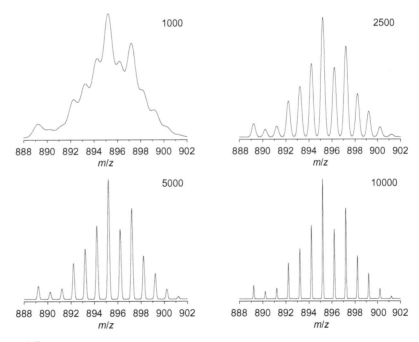

Figure 1.7
Effect of increasing resolution on the isotope pattern for $[C_{51}H_{53}ClP_3Ru]^+$

more than 20 *m/z* wide) to push the resolution up (though there is always a trade-off between resolution and intensity).

1.5.1 Mass Accuracy

The mass accuracy of a spectrometer is the difference observed between the calculated mass of an ion and its observed mass, $\Delta m = m_{\text{calculated}} - m_{\text{observed}}$, expressed relative to the observed mass. It is usually reported in parts per million (ppm):

$$\text{Mass accuracy (ppm)} = 10^6 \times \Delta m / m_{\text{observed}}$$

Careful calibration in conjunction with a reference compound enables high resolution mass spectrometers to provide a mass accuracy of 5 ppm or better. Instruments capable of this are said to provide **accurate mass** data. The requirement for high resolution becomes obvious if we consider again our four hypothetical mass spectrometers (Figure 1.8).

For low resolution mass spectrometers, the peak width is so broad that reliably picking the maximum value to within the required limits is fraught with error, though sound experimental protocols can allow surprisingly good results.[11] Resolution is must be at least 5000 for an instrument to realistically claim the ability to collect accurate mass data, and resolution of 10 000 is desirable. The rate of digitisation is also an issue (Figure 1.9).[12]

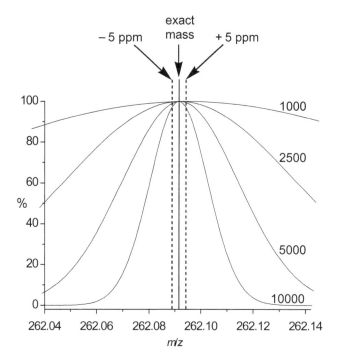

Figure 1.8
Peak profiles for $[C_{18}H_{15}P]^+$, exact mass 262.0911, at resolutions of 1000, 2500, 5000 and 10 000. The dotted lines correspond to the 5 ppm error limits

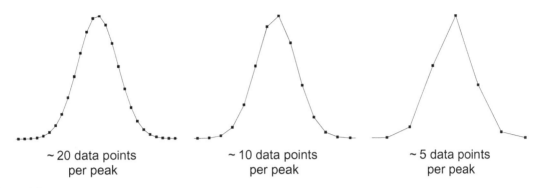

Figure 1.9
The effect of low rates of digitisation on reconstruction of a peak – at least 12 data points across the width of the peak is recommended

A distorted peak shape due to poor tuning or any contribution from an overlapping (isobaric) ion will also detrimentally affect the ability of the instrument to generate an accurate centroid.

Accurate mass data can unambiguously determine the elemental composition of an ion, but this statement comes with some important caveats. First, the number of possible combinations of elements that can fit within a given ppm range increases exponentially with *m/z*. Second, the elements chosen must consist of a severely reduced subset (often just carbon, hydrogen, oxygen and nitrogen plus any other possibilities based on the history of the sample). The elemental composition predicted is reasonably reliable for organic compounds with *m/z* < 500, but becomes increasingly suspect as the mass increases, as a search[13] for matching elemental formulae (limited to C,H,O,N) for increasing *m/z* values demonstrates:

m/z	100.1	200.2	300.3	400.4	500.5	600.6	700.7	800.8	900.9	1001.0
Matches within 5 ppm	1	1	2	2	2	3	4	10	50	139

Historically, the low mass/limited composition restriction posed few problems, as the vast majority of compounds that could be successfully transferred into the gas phase fitted this description. However, with the advent of new ionisation techniques, ions of higher mass and of almost limitless elemental composition can be analysed with ease. Blind faith in the reliability of a match between experimental data and theoretical composition is ill-advised, and in most cases, a well-matched isotope pattern provides more compelling evidence for a correct match.

For example, a minor product isolated from the reaction of $Na[Fe(CO)_2Cp]$ with an R_3SiCl compound provided the isotope pattern reproduced in Figure 1.10. No sensible match for the exact mass of 555.9030 Da using the elements carbon (C), hydrogen (H), oxgyen (O), chlorine (Cl), iron (Fe), and sodium (Na) can be obtained, and extending the search to the whole periodic table resulted in an enormous number of hits. However, comparison of the isotope pattern with various possible elements quickly led to the identification of mercury (Hg) as the likely culprit.

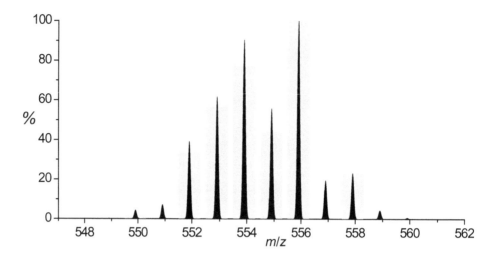

Figure 1.10
The isotope pattern for $[Hg\{Fe(CO)_2Cp\}_2]^+$ superimposed on isotopic signature for Hg. Note the dominant effect the seven isotopes of Hg have on the pattern. Inclusion of the remaining elements Fe, C, H and O in the calculation gave an exact match

The mercury came from inadvertent introduction of some sodium/mercury amalgam used to prepare $Na[Fe(CO)_2Cp]$ from $[Fe(CO)_2Cp]_2$. The usefulness of the isotope pattern over the exact mass measurement is clear in this case, and most inorganic and coordination chemists will find this to be true in nearly all instances.

Data Processing

The software used for controlling a modern mass spectrometer has powerful abilities to alter the appearance of the raw data collected from an experiment. The raw data may display much noise, an elevated baseline, a curved baseline, and contain a relatively low ratio of useful information to total information. The various data manipulation functions of the software are designed to address one or more of these 'problems'. Three main steps can be applied which affect the appearance of the raw data: smoothing, subtracting and centering (Figure 1.11).

Smoothing

Raw data may be smoothed easily using the mass spectrometry software. One type of algorithm used is the 'moving average' method, which converts each point to a new value generated by averaging it and the n points either side of it. The greater the value of n ($2n + 1$ is called the 'filter width'), the more intense the smoothing effect. This approach is deceptively impressive, and information is lost or distorted because too much weight is given to points far removed from the central point. An improvement is the **Savitsky-Golay** algorithm,[15] which is a computationally efficient way to perform a least-squares fit of the data to a polynomial function. The smoothing effect is less aggressive than the

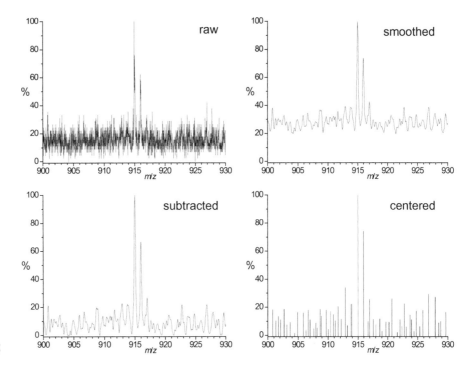

Figure 1.11
Raw data, and its cumulative treatment through smoothing, subtracting the baseline and centering

moving average method, and distortion of the data is less pronounced. However, all smoothing functions inherently lose some of the original information, and are generally applied for cosmetic effect. The *best* way to smooth data is not through mathematical treatment of a single data set, but rather to collect more sets (Figure 1.12).

Repetitive additions of noisy signals tend to emphasise their systematic characteristics and to cancel out random noise, which will average to zero. The signal-to-noise ratio will diminish by a factor of $\sqrt{N}$, where N is the number of repeat scans.

Subtracting

Subtraction essentially adjusts the baseline of the spectrum to equal zero. A variety of polynomial functions can be applied if the baseline is curved. Little information is lost during this process and the resulting mass spectrum is generally easier to interpret, especially when comparing the relative intensities of peaks.

Centering

The process of centering the data involves reducing a peak profile to a single line, indicating the peak centroid and intensity. This has the advantage of greatly reducing the amount of data required to display the spectrum – even in the rather noisy spectrum shown in Figure 1.11, less than 1/30th of the amount of information is needed to display the centered spectrum compared to the others. The data compression advantage is less important than it once was, due to tremendous improvements in computer hardware and

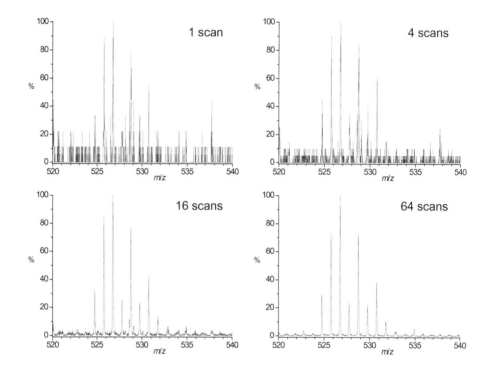

Figure 1.12
The effect of collecting and averaging multiple spectra

data storage. However, centering remains an important precursor to applying computational analysis of the data, such as library searching or instrument calibration.

In general, it is desirable that the raw data that make up a mass spectrum are manipulated as little as possible. The signal-to-noise ratio tells the analyst something about the strength of the signal and hence the quality of the data. Smoothing disguises this indicator, and artificially broadens peaks. Centered data are highly compressed but much information is thrown out, and with today's practically limitless electronic storage this saving may represent a false economy.

Isotopes

Perhaps the most immediately obvious difference between the mass spectra of organic compounds and those of inorganic and organometallic compounds is the wide occurrence of polyisotopic elements (Appendix 1). A mass spectrometer separates individual ions, so an ion of a given elemental composition containing one or more polyisotopic elements will give rise to a number of **isotope peaks**. These peaks have a characteristic pattern of relative intensities and spacing, which depends on both the masses and the relative abundances of the isotopes in the ion, and this envelope of peaks is known as the **isotope pattern** of an ion.

Spectra of organic compounds generally show rather simple isotope patterns. The reason for this becomes obvious when the isotopic abundances of the elements commonly encountered in organic chemistry are inspected:

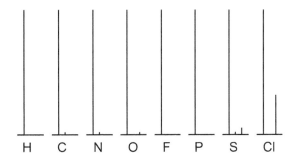

Choosing a part of a row of the heavier elements, the difference is quite striking. Below are the isotopes of elements with atomic numbers 44 to 51:

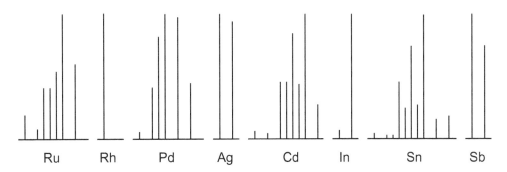

Isotope patterns of polyatomic ions are calculated using the binomial expansion:

$$(a_A.^xA + b_A.^yA + c_A.^zA + \cdots)^n \times (a_B.^xB + b_B.^yB + c_B.^zB + \cdots)^m \times \cdots$$

where a_A, b_A, c_A etc. are the fractional abundances of the isotopes x^A, y^A, z^A of element A and n the number of atoms of A present in the ion; similarly for m atoms of B, and so on. These calculations are accomplished extremely quickly using modern computers, but a simple example is illustrative of the process that goes on.

Example: $[IrCl_2(CO)_2]^-$. Iridium has two isotopes (^{191}Ir 37 %, ^{193}Ir 63 %), as does chlorine (^{35}Cl 76 %, ^{37}Cl 24 %). The contributions of the minor isotopes of carbon (^{13}C, 1 %) and oxygen (^{17}O 0.00038 % and ^{18}O 0.002 %) are negligible in this case. Table 1.1 shows how to calculate the relative abundance of each of the peaks in the isotope pattern and Figure 1.13 plots the resulting pattern.

Table 1.1 Calculating the relative abundance of peaks

Ion composition	*m/z*	Fractional abundance		Relative abundance
^{191}Ir ^{35}Cl$_2$ (CO)$_2$	317	$(0.37)(0.76)^2$	= 0.214	43 %
^{193}Ir ^{35}Cl$_2$ (CO)$_2$	319	$(0.63)(0.76)^2$	= 0.364	100 %
^{191}Ir ^{35}Cl ^{37}Cl (CO)$_2$		$(0.37)(0.76)(0.24) \times 2*$	= 0.135	
^{193}Ir ^{35}Cl ^{37}Cl (CO)$_2$	321	$(0.63)(0.76)(0.24) \times 2*$	= 0.230	50 %
^{191}Ir ^{37}Cl$_2$ (CO)$_2$		$(0.37)(0.24)^2$	= 0.021	
^{193}Ir ^{37}Cl$_2$ (CO)$_2$	323	$(0.63)(0.24)^2$	= 0.036	7 %

* $\times$ 2 because the ion intensity is made up of contributions from ^{35}Cl ^{37}Cl and ^{37}Cl ^{35}Cl.

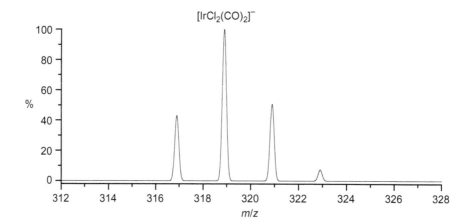

Figure 1.13
Calculated
isotope pattern
of $[IrCl_2(CO)_2]^-$

Table 1.2 Number of carbon atoms in molecule

	C	C_{10}	C_{25}	C_{50}	C_{90}
M	100	100	100	100	99
M+1	1	11	27	55	100
M+2			3	15	49
M+3				2	16
M+4					3

While the presence of ^{13}C has little effect on the isotope patterns of small ions, in high molecular weight compounds the multiplying effect makes the influence from carbon significant. The effect can be plainly seen in Table 1.2, where the contribution to the $M + n$ peaks increases with the number of carbon atoms, to the point where the $M + 1$ peak has greater intensity than that for M for C_{90}. This effect is particularly significant for large molecules of biological origin such as proteins.

However, the presence of multiple polyisotopic metal atoms has a much more dramatic effect. Compounds with high molecular weights that contain many polyisotopic metal atoms have extremely broad isotope patterns; for example, the hexamer of $SnPh_2$, Sn_6Ph_{12} has a pattern that stretches across more than 30 Da, and has a near-perfect Gaussian distribution of isotope peaks (Figure 1.14).

Most proprietary mass spectrometric software packages come with an isotope pattern calculator. However, there are resources available on the world-wide web to perform these calculations. Online examples (as of 2004) include:

- Mark Winter's[†] isotope patterns calculator at
  http://www.shef.ac.uk/chemistry/chemputer/isotopes.html
- Jonathan Goodman's molecular weight calculator at
  http://www.ch.cam.ac.uk/magnus/MolWeight.html

[†] Mark Winter is also the author of the compendious and reliable WebElements (*http://www. webelements.com/*), *the* online reference for all things related to the periodic table.

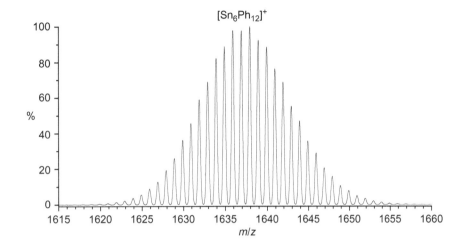

Figure 1.14 Calculated isotope pattern for $[C_{72}H_{60}Sn_6]^+$

- More sophisticated downloadable programs for offline determination of patterns are also available, such as Matthew Monroe's excellent (and free) Molecular Weight Calculator at

  http://jjorg.chem.unc.edu/personal/monroe/mwtwin.html.

1.7.1 Isotopic Abundances of the Elements[16]

The periodic table of the elements colored by number of isotopes (Figure 1.15) demonstrates a distinct alternation, whereby elements with a even atomic number have more isotopes than neighboring odd atomic number elements. Furthermore, cosmic abundances of the elements show the same alternation.

Why? Spin-pairing is an important factor for protons and neutrons, and of the 273 stable nuclei, just four have odd numbers of both protons and neutrons. Elements with even numbers of protons tend to have large numbers of stable isotopes, whereas those

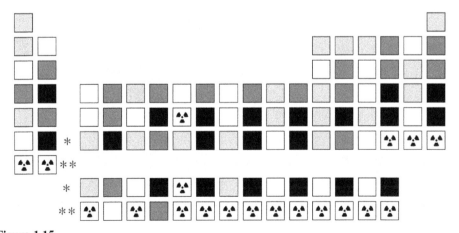

Figure 1.15
The periodic table, coloured by number of stable isotopes. White = 1 isotope, grey = 2, dark grey = 3–5, black = 6+; symbol = all isotopes are radioactive. See Appendix 2 for a more detailed version

with odd numbers of protons tend to have one, or at most two, stable isotopes. For example, $_{52}$Te has eight stable isotopes and $_{54}$Xe has nine but $_{53}$I has just one.

An additional feature is the existence of especially stable numbers of nucleons of one kind, and these '**magic numbers**' are 2, 8, 20, 28, 50, 82 and 126. These numbers correspond to completed quantum levels for the nuclei, and just like for electrons these confer particular stability (though the order in which the levels are filled differ for nuclei). Patterns among the stable isotopes bear this extra stability out. For example, tin (50 protons) has the greatest number of stable isotopes (10); lead (82 protons) isotopes are the end result of all decay pathways of the naturally occurring radioactive elements beyond lead; seven elements include isotopes with 82 neutrons (**isotones** ^{136}Xe, ^{138}Ba, ^{139}La, ^{140}Ce, ^{141}Pr, ^{142}Nd and ^{144}Sm) and six elements have isotopes with 50 neutrons (^{86}Kr, ^{87}Rb, ^{88}Sr, ^{89}Y, ^{90}Zr and ^{92}Mo). Similarly, 'doubly-magic' nuclei are strongly favoured: ^{4}He (2 p, 2 n) is the second-most common isotope in the universe, and **α-particles** (helium nuclei) are frequently ejected in nuclear reactions; ^{16}O (8 p, 8 n) makes up 99.8 % of all oxygen, ^{40}Ca (20 p, 20 n) is 97 % abundant, and ^{208}Pb (82 p, 126 n) is the most common isotope of lead. As the atomic number increases, the number of neutrons required to provide stability to the nucleus increases at a greater rate (Figure 1.16). The plotted points represent the naturally occurring isotopes; careful inspection reveals the extra stability of spin-paired nuclei.

The precision to which the **atomic weight** is known for any given element is related to the number of isotopes an element has. Generally, the atomic weight of any given isotope can be determined experimentally using mass spectrometry to an extremely high level. Gold, for example, has a single stable isotope (^{197}Au) whose relative atomic mass is 196.96655(2). However, the relative abundances of the various isotopes of a polyisotopic element are known to a lower level of precision. Mercury has seven stable isotopes, and its relative atomic mass is 200.59(2) (i.e. five significant figures compared to eight

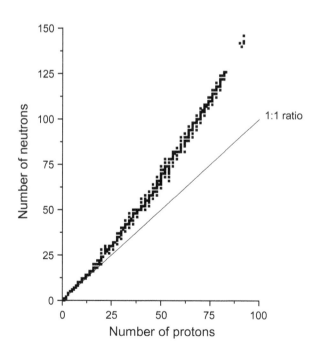

Figure 1.16
Plot of neutron number vs. proton number for the naturally occurring isotopes

significant figures for gold). Furthermore, relative isotopic abundances can depend on the source of the sample, and this phenomenon is the basis of stable isotope geochemistry. Mass spectrometry can, for example, easily detect the difference between carbon dioxide exhaled by humans from that generated by burning coal by analysing the $^{12}C/^{13}C$ ratio. Similar studies are done on $^{14}N/^{15}N$ and $^{16}O/^{18}O$ ratios.[17]

Another complication can arise from **isotopic enrichment** of samples. Most notably, this is performed on uranium ore to obtain ^{235}U for use as a nuclear fuel, leaving **depleted** uranium behind as predominantly ^{238}U, and it is this isotope that generally finds itself in chemistry laboratories. In the middle of the last century, 'strategic lithium', ^{6}Li, was stockpiled for use in nuclear weapons and as a raw material for the production of tritium (itself used in nuclear weapon production). The depleted lithium, predominantly ^{7}Li, was sold on and this had a noticeable effect on the atomic weight of some supplies of lithium. The International Union for Pure and Applied Chemistry (IUPAC) publishes periodic reports on the history, assessment and continuing significance of atomic weight determinations.[18]

1.7.2 Isotope Pattern Matching

Comparisons between theoretical and experimental isotope patterns are often done by eye, but a direct comparison is highly recommended. This approach enables direct matching between each member of the isotope pattern. Using the proprietary mass spectrometer software provides an easy way of achieving this, simply by presenting the two spectra overlaid (Figure 1.17).

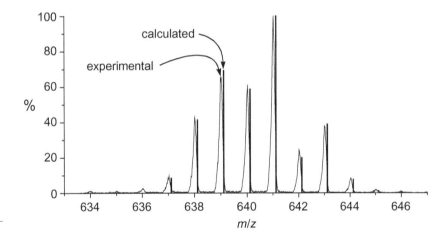

Figure 1.17 Experimental (line) and calculated (bar) isotope patterns for [Yb(EDTA)(HMPA)]$^{-}$

Immediately apparent in the example is the fact that the calibration is off by approximately 0.1 *m/z*, but the match between experimental and calculated patterns is good. In cases as clear-cut as this one, assignment made be made with confidence. Difficulties tend to arise in the following examples:

(a) Signal-to-noise ratio is low. This can cause the lower intensity peaks of the isotope pattern to disappear into the noise. The best remedy is to sum many scans over a small window, and if the signal is very weak this may take minutes.

(b) Resolution is low. This can cause the peaks to overlap, and is particularly problematic for multiply charged species of high molecular weight. Many instruments can trade-off sensitivity for resolution, and this approach is always recommended especially if the signal is a strong one.

(c) Calibration is poor. Easily remedied by recalibrating the instrument; there is generally no need to rerun the sample as most software allows spectra to be calibrated retrospectively.

(d) Two patterns overlap when two ionisation pathways are competing, for example between oxidation to form $[M]^{\bullet+}$ (for example, as occurs with certain ferrocene derivatives, see Section 7.4.1) and protonation to form $[M + H]^+$ or the appearance of $[M + NH_4]^+$ and $[M + H + H_2O]^+$. Rather than trying to deconvolute the overlapping isotope patterns, the best approach is promote one ionisation pathway at the expense of the other (e.g. add extra H^+) or to look for another ionisation route altogether. If solvent adducts are present, application of gentle in-source CID should remove this complication.

(e) Two patterns overlap from different compounds. In organometallic and coordination chemistry this case is certainly rarer than in organic chemistry, but the latter has the advantage that chromatographic separation is always an option (LCMS or GCMS). However, it may happen; for example in a mixture of lanthanide complexes (which

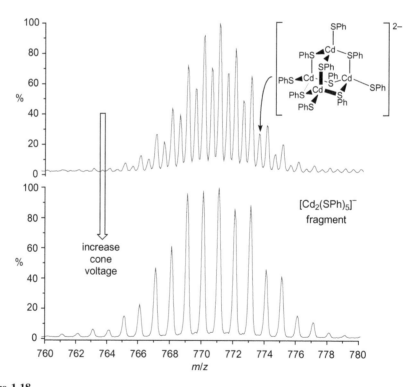

Figure 1.18

Top spectrum (ESI-MS, negative-ion mode) shows the overlapping isotope patterns of $[Cd_4(SPh)_{10}]^{2-}$ and its symmetrical fragment ion $[Cd_2(SPh)_5]^-$.[19] Increasing the cone voltage (in-source CID) removes the intact parent ion completely (bottom spectrum). The same strategy can be used to produce $[M]^{+/-}$ from its dimer, $[2M]^{2+/-}$

have very similar chemistry and whose isotope patterns frequently overlap with their neighbours). If the two components can be identified separately the isotope patterns can be calculated and combined in the appropriate proportion to model the experimentally observed pattern. Overlap of patterns when both components are unknown can be very difficult to unravel, and sample purification before reanalysis is probably the best way to proceed.

(f) Two patterns overlap where one is from $[M]^{+/-}$ and another from the dimer, $[2M]^{2+/-}$. For polyisotopic species, the presence of the dimer is obvious from the isotope pattern, because there will be peaks separated by 0.5 *m/z*. Symmetrical fragmentation of a doubly-charged ion has the same effect. Again, in-source CID can simplify the picture (Figure 1.18).

References

1. K. R. Jennings, *Int. J. Mass Spectrom.*, 2000, **200**, 479; A. K. Shukla and J. H. Futrell, *J. Mass Spectrom.*, 2000, **35**, 1069; E. de Hoffmann, *J. Mass Spectrom.*, 1996, **31**, 129; K. L. Busch, G. L. Glish and S. A. McLuckey, *Mass Spectrometry/Mass Spectrometry: Techniques and Applications of Tandem Mass Spectrometry*, VCH, New York, 1988.

2. R. A. Yost and C. G. Enke, *J. Am. Chem. Soc.*, 1978, **100**, 2274.

3. K. M. Ervin, *Chem. Rev.*, 2001, **101**, 391; P. B. Armentrout, *Acc. Chem. Res.*, 1995, **28**, 430; B. S. Freiser, *Acc. Chem. Res.*, 1994, **27**, 353.

4. see P. B. Armentrout, *Int. J. Mass Spectrom.*, 2003, **227**, 289, and 80+ references to this group's work therein; R. Amunugama and M. T. Rodgers, *Int. J. Mass Spectrom.*, 2003, **227**, 1.

5. F. Muntean and P. B. Armentrout, *J. Chem. Phys.*, 2001, **115**, 1213.

6. R. Liyanage, M. L. Styles, R. A. J. O'Hair and P. B. Armentrout, *Int. J. Mass Spectrom.*, 2003, **227**, 47.

7. R. A. Forbes, L. Lech and B. S. Freiser, *Int. J. Mass Spectrom. Ion Proc.*, 1987, **77**, 107; C. E. C. A. Hop, T. B. McMahon and G. D. Willett, *Int. J. Mass Spectrom. Ion Proc.*, 1990, **101**, 191; J. B. Westmore, L. Rosenberg, T. S. Hooper, G. D. Willett and K. J. Fisher, *Organometallics*, 2002, **21**, 5688.

8. A. G. Harrison, *Rapid Commun. Mass Spectrom.*, 1999, **13**, 1663.

9. P. J. Dyson, B. F. G. Johnson, J. S. McIndoe and P. R. R. Langridge-Smith, *Rapid Commun. Mass Spectrom.*, 2000, **14**, 311; P. J. Dyson, A. K. Hearley, B. F. G. Johnson, J. S. McIndoe, P. R. R. Langridge-Smith and C. Whyte, *Rapid Commun. Mass Spectrom.*, 2001, **15**, 895; C. P. G. Butcher, P. J. Dyson, B. F. G. Johnson, P. R. R. Langridge-Smith, J. S. McIndoe and C. Whyte, *Rapid Commun. Mass Spectrom.*, 2002, **16**, 1595.

10. P. J. Dyson, A. K. Hearley, B. F. G. Johnson, T. Khimyak, J. S. McIndoe and P. R. R. Langridge-Smith, *Organometallics*, 2001, **20**, 3970.

11. A. W. T. Bristow and K. S. Webb, *J. Am. Soc. Mass Spectrom.*, 2003, **14**, 1086.

12. A. N. Tyler, E. Clayton and B. N. Green, *Anal. Chem.*, 1996, **68**, 3561.

13. J. E. Deline, *Molecular Fragment Calculator 1.0*, 1995.

14. J. S. McIndoe and B. K. Nicholson, *Acta Cryst. Sect. E*, 2002, **E58**, m53.

15. A. Savitsky and M. J. E. Golay, *Anal. Chem.*, 1964, **36**, 1627.

16. P. A. Cox, *The Elements*, Oxford University Press, Oxford, 1989.

17. R. Corfield, *Chem. Brit.*, 2003, **39**, 23.

18. J. R. de Laeter, J. K. Böhlke, P. de Bièvre, H. Hidaka, H. S. Peiser, K. J. R. Rosman and P. D. P. Taylor, *Pure Appl. Chem.*, 2003, **73**, 683.

19. T. Løver, W. Henderson, G. A. Bowmaker, J. Seakins and R. P. Cooney, *Inorg. Chem.*, 1997, **36**, 3711.

2 Mass Analysers

Introduction

Separating ions according to their mass-to-charge ratio is the job of the mass analyser and numerous ingenious methods have been developed. All employ electric fields, sometimes in conjunction with magnetic fields, to enable discrimination between ions of different mass-to-charge ratio (m/z). Table 2.1 lists the five main types of mass analyser (sector, quadrupole, ion trap, TOF, and FTICR) and includes two popular instruments that incorporate two mass analysers, the triple quadrupole and the hybrid quadrupole-TOF. All have slightly different strengths and weaknesses, which will be expanded upon in this chapter.

Sectors

Perhaps the most familiar type of mass spectrometer is the **sector** instrument, thanks to its long history and faithful description in many textbooks. The action takes place inside a curved region between the poles of a magnet, whose purpose is to disperse the ions (which have been accelerated out of the ion source) into curved trajectories that depend on the m/z of the ion (Figure 2.1). Low-mass ions are deflected most, and the heaviest least. The radius of deflection necessary for an ion to impinge on the detector is determined by the curvature of the flight tube, so in order to obtain a mass spectrum either the accelerating voltage or (usually) the field strength of the magnet is varied.

In a **magnetic sector** mass spectrometer, ions leaving the ion source are accelerated to a high velocity. The ions then pass through a magnetic sector in which the magnetic field is applied in a direction perpendicular to the direction of ion motion. From physics, it is known that when acceleration is applied perpendicular to the direction of motion of an object, the object's velocity remains constant, but the object travels in a circular path. Therefore, the magnetic sector follows an arc; the radius and angle of the arc vary with different instrument designs. The dependence of mass-to-charge ratio on the magnetic field, accelerating voltage and arc radius may be easily derived. Ions of mass m and charge z are accelerated by a potential V, and enter the magnetic sector with a kinetic energy, KE, given by:

$$\mathrm{KE} = zV = \tfrac{1}{2}mv^2$$

Mass Spectrometry of Inorganic, Coordination and Organometallic Compounds   W. Henderson and J. S. McIndoe
© 2005 John Wiley & Sons, Ltd   ISBNs: 0-470-85015-9 (HB); 0-470-85016-7 (PB)

Table 2.1   Strengths and weaknesses of mass spectrometers

	Price	Size	Resolution	Mass range	MS/MS
Sector	+	+	+ +	+ +	+
Quadrupole	+ + +	+ + +	+	+	−
Triple quad	+ +	+ + +	+	+ +	+ +
Ion trap	+ +	+ + +	+	+ +	+ + +
TOF	+ +	+ +	+ +	+ + +	+
Q-TOF	+	+ +	+ +	+ + +	+ +
FTICR	−	+	+ + +	+ +	+ + +

Low price, small size, high resolution, large mass range and good MS/MS abilities are all seen as desirable and rate '+ + +'. Poorer performance on any of these rate '+ +', '+', or '−'.

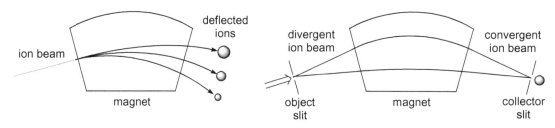

Figure 2.1
The effect of a magnetic field on the trajectory of ions. The heaviest ions are deflected least, the lightest the most (left). A divergent beam entering a magnetic field exits with the beam converging (right), a property known as directional focusing

The centrifugal force experienced by the ions must balance the Lorentz force exerted by the magnet field (strength B, radius r) for them to be focused:

$$mv^2/r = \mathrm{B}zv$$

Combining the two equations leads to the expression for focusing ions of a given *m/z* ratio:

$$m/z = \mathrm{B}^2\mathrm{r}^2/2\mathrm{V}$$

A magnetic sector is rarely operated on its own, because the ions emerging from the source have a range of kinetic energies, and the magnetic sector will focus these ions in slightly different locations. The resolution suffers as a result of this effect, but it can be improved by first focusing the energy of the ion beam. Once accelerated away from the ion source, the ions pass through slits then between a pair of smooth, curved metal plates called an **electrostatic analyser**. Like the magnetic sector, the electric sector applies a force perpendicular to the direction of ion motion, and therefore has the form of an arc (Figure 2.2).

Combination of electrostatic and magnetic analysers constitutes the double-focusing mass spectrometer, the geometry of which is configured in such a way that ions of the same mass but of different energies converge at the collector (Figure 2.3).

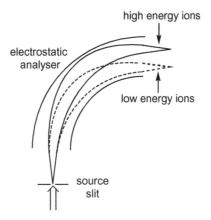

Figure 2.2
The effect of an electric field on the trajectory of ions. Ions having different kinetic energies follow different trajectories. As is the case for magnetic sectors, electrostatic analysers also focus ions in a directional sense

The simplest mode of operation of a magnetic sector mass spectrometer keeps the accelerating potential and the electric sector at a constant potential and varies the magnetic field. Ions that have a constant kinetic energy but different mass-to-charge ratio are brought into focus at the detector slit (called the 'collector slit') at different magnetic field strengths. The magnetic field is usually scanned exponentially or linearly to obtain the mass spectrum. A magnetic field scan can be used to cover a wide range of mass-to-charge ratios with a sensitivity that is essentially independent of the mass-to-charge ratio. An alternative is to hold field strength B constant and scan potential V.

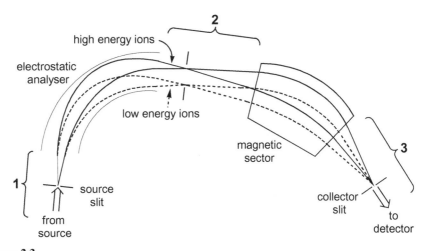

Figure 2.3
A double-focusing electrostatic/magnetic analyser configuration (referred to as EB or Niers-Johnson). The reverse arrangement (magnet first, electrostatic analyser second or BE) is an alternative and equally valid geometry. The zones labelled with the bold numerals **1**, **2** and **3** correspond to the first, second and third field-free regions

The electric sector potential tracks the accelerating voltage. This has the advantage that the electric field is not subject to hysteresis, so the relationship between mass-to-charge ratio and accelerating voltage is a simple linear relationship. The disadvantage of an accelerating voltage (electric field) scan is that the sensitivity is roughly proportional to the mass-to-charge ratio.

The cross sectional shape and dimensions of the ion beam is controlled by a series of slits and electric focusing devices and lenses. The resolving power of a sector mass analyser is determined by the slit widths. Higher resolution is obtained by decreasing the slit widths, and thereby decreasing the number of ions that reach the detector. Double-focusing instruments are capable of very high resolution, generally 10 000 or well above. The maximum ion transmission and sensitivity occur at the maximum working accelerating voltage for a given magnetic sector mass spectrometer. The effective mass range of the mass spectrometer can be increased by decreasing the accelerating voltage, with a sensitivity that is roughly proportional to the accelerating voltage.

2.2.1 MS/MS

A double-sector instrument can not really be thought of as two mass spectrometers linked in series, so MS/MS studies can not be carried out by the simple expedient of placing a collision cell between the two sectors. Nonetheless, interesting structural information can be gathered on ions that undergo decomposition in the field-free regions (as indicated on Figure 2.3; these regions have no electric or magnetic fields to perturb the trajectory of the ions).[1] These **metastable** ions are generally low in abundance, because they must have just enough energy to decompose but not so much energy that the decomposition occurs in the source. The low abundance makes the product ions of the decomposition of metastable ions more difficult to detect. Furthermore, generally only electron ionisation imparts enough initial energy for metastable ions to have any appreciable existence; the softer ionisation techniques tend to produce abundant (pseudo)molecular ions and little or no fragmentation. A range of ingenious **linked scanning** techniques were developed where two of the three main variables in a double-focusing instrument (electric field, magnetic field, accelerating voltage) are varied in a fixed ratio.[2] Linked scans enable, for example, all the product ions from a given precursor to be identified, or in a different mode, all the precursor ions that break down to form a given fragment ion. These connections provide some useful structural information, but metastable transitions are seen mainly as a method of supporting inferences made from more direct observations.

Three- and four-sector instruments, with configurations such as BEB, EBE or BEEB have been constructed, as well as hybrid instruments coupling sectors with quadrupole or time-of-flight mass analysers. However, these are large, expensive and increasingly rare as commercial production of MS/MS instruments is now focused on more convenient and smaller (and usually cheaper) machines.

2.2.2 Summary

Strengths

- High resolution, sensitivity, and dynamic range.

Weaknesses

- Not well-suited for pulsed ionisation methods (e.g. Matrix Assisted Laser Desorption Ionisation, MALDI);
- Coupling with atmospheric pressure sources (e.g. Electrospray Ionisation, ESI) complicated;
- Very large;
- Expensive.

Mass range and resolution

- Up to 4000 *m/z*;
- Maximum resolution of 100 000 + for double-focusing instruments, more typically 25 000.

Applications

- All organic MS analysis methods;
- Accurate mass measurements;
- Quantitation;
- Isotope ratio measurements.

Popularity

- Waning. Once a mainstay, the size and cost of sector instruments has diminished their popularity, especially in the face of stiff and ever-improving competition from TOF mass analysers. The latter are also better suited to the 'modern' ion sources, ESI and MALDI.

Quadrupoles[3]

Quadrupole mass analysers consist of four parallel rods arranged as in Figure 2.4. Applied between each pair of opposite and electrically connected rods are a DC voltage

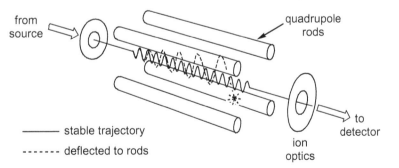

Figure 2.4

Exploded diagram of a quadrupole mass analyser showing ion paths. The distance between opposite rods is actually less than the diameter of the rods themselves. The rods are typically 5–15 mm in diameter and between 50 and 250 mm long, and must be perfectly parallel and machined with very high precision (to the micron level) in order to achieve satisfactory resolution. Ideally, the rods should have a hyperbolic cross-section but in practice they are usually cylindrical

and a superimposed radio-frequency potential. Ions are accelerated before entering the quadrupole assembly because the applied electric fields provide no forward impetus. A positive ion entering the quadrupole will be drawn towards a negatively charged rod but if the field changes polarity before the ion reaches it, it will change direction. Under the influence of the combination of fields the ions undergo complex trajectories. Within certain limits these trajectories are stable and so ions of a certain *m/z* are transmitted by the device, whereas ions with different *m/z* values will have an unstable trajectory and be lost by collision with the rods.

A derivation of the working equations for a quadrupole mass analyser is beyond the scope of this discussion, but it is based upon a second-order differential equation known as the Mathieu equation. The operation of a quadrupole mass analyser is usually treated in terms of a **stability diagram** that relates the applied DC potential (*U*), the applied rf potential (*V*) and the radio frequency (ω) to a stable *vs* unstable ion trajectory through the quadrupole rods. A qualitative representation of a stability diagram for a given mass m is shown in Figure 2.5; a and q are parameters that are proportional to *U/m* and *V/m* respectively:

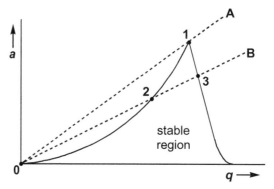

Figure 2.5
Stability diagram. The parameters (U, V, ω) must be chosen so a/q fits a line that passes as close to point 1 as possible but still lies within the stable region (i.e. just below line **0A**). Lowering the slope of the line (e.g. line **0B**) allows more ions to pass through the quadrupole assembly but also lowers the resolution

Changing the slope of the scan line will change the resolution. Increasing the resolution decreases the number of ions that reach the detector (the region at the apex of the stable region that is bounded by the scan line). Good resolution is also critically dependent on the quality of the machining for the quadrupole rods.

Quadrupoles have other functions besides their use as a mass filter. An rf-only quadrupole will act as an ion guide for ions within a broad mass range (in terms of the stability diagram, *U* – and hence a – is now zero, so the slope of the line drops to zero and all ions are passed, regardless of *m/z* value). In such applications, hexapoles or even octapoles are often employed. While hexapoles and octapoles cannot act as mass filters, they make excellent ion guides as they are able to efficiently contain, direct and constrain ions into a narrow beam. This ability is especially important in mass spectrometers with atmospheric pressure sources, where the ion beam from the source tends to be easily dispersed by collisions with residual solvent molecules. Ion guides are also used to collimate ions prior to their introduction to an orthogonal TOF mass analyser (Section 2.5.2).

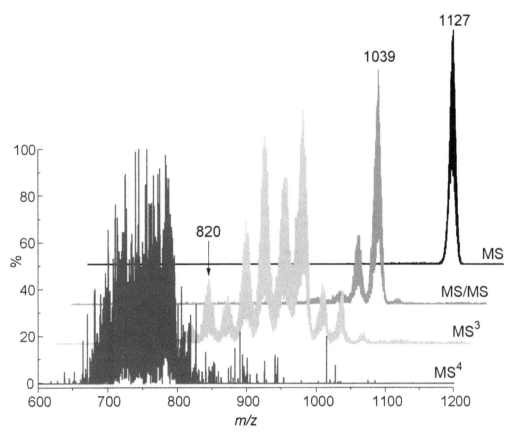

Plate 1 (Figure 2.9) ESI-MS^n spectra of [Ru$_6$C(CO)$_{17}$] in MeOH/NaOMe solution. From the top: the standard MS spectrum carried out at low temperature (50 °C) with no in-source fragmentation; MS/MS spectrum, 50% collision energy on the peak at 1127 m/z; MS^3 spectrum, with the 1127 m/z peak selected, fragmented at 50%, followed by selection of the 1039 m/z peak and further fragmentation at 50%; MS^4 spectrum, selection of the 820 m/z after MS^3 and further fragmentation at 50% collision energy

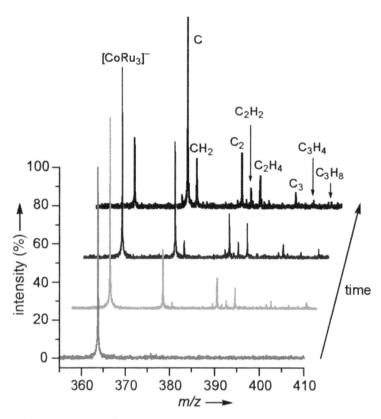

Plate 2 (Figure 2.18) Reaction of [CoRu$_3$]$^-$ with CH$_4$. Note the build-up of product ions over time

2.3.1 MS/MS

Tandem mass spectrometry with a quadrupole instrument can only be carried out with multiple mass analysers separated in space. The most popular configuration is the **triple quadrupole**, in which two mass analysers are separated by an rf-only quadrupole (or hexapole) used as a collision cell. To discriminate between the two types of quadrupole, the shorthand for this type of arrangement is QqQ, where Q is a quadrupole mass analyser and q an rf-only quadrupole. QhQ (h = hexapole) instruments are also common. MS/MS experiments are easy to set up on triple quadrupoles, and are especially valuable when examining complex mixtures of ions. For the coordination/organometallic chemist this is an enormously valuable facility; to a large extent it obviates the need for chromatography and allows analysis of, for example, a reaction mixture directly. Each ion may be selected and individually fragmented to provide structural information exclusively on that ion, without any need for consideration of interference from other ions in the mixture. Examples of this are catalyst screening and the study of fundamental organometallic reactions in the gas phase (see Section 8.5).

The **product ion scan** (once called 'fragment ion scan' or 'daughter ion scan') is probably the most important of the MS/MS experiments. The first quadrupole (MS1) is set so that it passes only ions of a certain *m/z* value. The rf-only quadrupole contains an inert collision gas (typically argon) to assist in the creation of fragment ions. The third quadrupole (MS2) scans the mass range of interest and generates a **daughter ion** spectrum, in which the fragments observed are derived from the **parent ion** originally selected in MS1 (Figure 2.6).

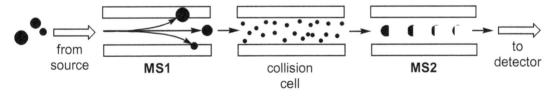

from source · · MS1 · collision cell · MS2 · to detector

Figure 2.6
A product ion scan. Ions of different *m/z* enter MS1 but only ions of a specific *m/z* value are allowed to enter the collision cell. A range of fragment ions are generated which are scanned in MS2

The **precursor ion scan** (once called 'parent ion scan') reverses the arrangement. Now, MS1 scans through all ions, which are fragmented in the collision cell, and MS2 is set to detect only product ions of a certain mass. This scanning technique is of special importance where detection of a certain class of compound is desired, all members of the class fragmenting in a characteristic way to generate a particular product ion. A **neutral loss scan** involves setting up MS1 and MS2 to scan in parallel but offset by a set mass difference, Δm. Ions passing through MS1 are fragmented in the collision cell but pass through MS2 only if the neutral fragment eliminated has mass Δm. Again, this scan is important for plucking a certain class of compound from a morass of other peaks (for example, protonated alcohols lose water, so a constant mass difference scan set at $\Delta m = 18$ will selectively detect this class of compound). The last two types of scan are enormously important in analytical chemistry but have yet to find real application in coordination and organometallic chemistry.

2.3.2   Summary

Strengths

- Compact, simple, easy to clean;
- Fast scanning (1000 m/z s^{-1} or above) well suited to chromatographic coupling;
- Good reproducibility;
- Relatively inexpensive;
- Affordable MS/MS when set up as triple quadrupole;
- Conversion of precursor to product ion in MS/MS highly efficient;
- Doesn't operate at high voltages so well-suited for coupling to atmospheric pressure sources.

Weaknesses

- Limited resolution;
- Not well-suited for pulsed ionisation methods;
- MS/MS spectra depend strongly on a host of instrumental parameters – energy, collision gas, pressure, etc.

Mass range and resolution

- Up to 4000 m/z; more typically 2000 m/z;
- Maximum resolution of little more than 3000, but a typical maximum is 2500 and for smaller, cheaper models perhaps 1000. Many are run at 'unit mass resolution', sufficient to discriminate peaks one mass unit apart in the m/z range of interest.

Applications

- Majority of benchtop GC/MS and LC/MS systems;
- Triple quadrupole MS/MS systems;
- Hybrid MS/MS systems e.g. quadrupole/TOF or sector/quadrupole.

Popularity

- High and steady. Well established and incrementally improving means of mass analysis, quadrupoles remain important in many applications.

Quadrupole Ion Trap[4]

Nearly 50 years ago Wolfgang Paul described a device for trapping gas-phase ions using electric fields,[5] a discovery for which in 1989 he was awarded a Nobel Prize for Physics.[6] The **quadrupole ion trap** (QIT, often referred to simply as an ion trap) consists of a ring shaped electrode with curved caps on the top and bottom (Figure 2.7). Ions are injected from the source through one of the caps, and by applying a combination of voltages to the ring and capping electrodes, the ions can be trapped in a complicated three-dimensional orbit. The electric field is constructed in such a way that the force on an ion is proportional to its distance from the centre of the trap. A constant pressure of helium (ca. 0.1 Pa) is usually maintained in the cell to remove excess energy from the ions, which

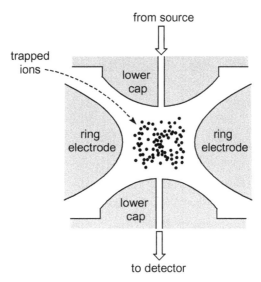

Figure 2.7
A quadrupole ion trap. Ions are injected into the trap through the upper cap electrode, and are stored in a complicated three-dimensional trajectory. The end caps and ring electrode have a hyperbolic cross-section, allowing ions of a wide mass range to have stable trajectories. Application of a resonant frequency allows selective expulsion of ions of a given *m/z* ratio, generating a mass spectrum

would otherwise repel each other to the extent that their trajectories became unstable, causing loss of ions from the trap. The relatively low operating pressure of the ion trap has a favourable impact on the cost of the instrument.

The effluent from a gas chromatograph (GC), entrained in helium, can be injected directly into an ion trap and ionised *in situ*. This provides some advantages in terms of simplicity of design and increased sensitivity compared to ion traps that form ions externally, due to ion transmission and trapping losses in the latter case, and a GC-ion trap was the first commercialised instrument. Nonetheless, the many advantages of ion traps have made them increasingly popular in conjunction with all types of ionisation method and in combination with other mass analysers (in hybrid instruments).

Only of the order of 300–1000 ions or so are trapped at a given time, as more ions than this decreases attainable resolution and fewer reduces sensitivity. A short pre-scan is used to assess the length of time needed to fill the cell with the optimum number of ions. The time taken will be dependent on supply from the source, but typically ion injection takes about half a second. Once the ions are trapped, the electrode potentials can be manipulated so that the ions' motion becomes unstable in order of their mass-to-charge ratios and they are ejected from the trap into an external detector, a process called **mass-selective ejection**. Mass analysis in this way takes less than a tenth of a second, and the resolution obtained is comparable to that of a quadrupole. However, it is possible to scan very slowly over a small mass window (a process known as the 'zoomscan'), and this approach allows high resolution (up to 5000) to be obtained. By increasing the time which ions of a particular mass spend in the ion trap the resolution can be considerably improved as the relative energies of ions with the same *m/z* become closer in value. Figure 2.8 shows the effect of collecting data around a window of interest while extending

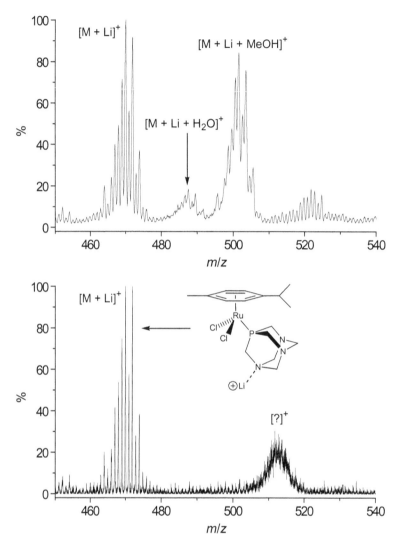

Figure 2.8
Ru(cymene)(pta)Cl$_2$ in methanol in presence of Li$^+$: (top) normal scan, resolution ~1000; note the broader peaks for the weakly bound solvent adducts: (bottom) zoomscan, the resolution of the [M + Li]$^+$ peak has improved to ~ 3000 and the weakly bound solvent adducts have disappeared, to be replaced by an unresolved peak. Inset shows the structure of the lithiated ion

the time the ions spend in the trap. The ESI mass spectrum of the neutral compound Ru(η^6-cymene)(pta)Cl$_2$ in methanol in presence of lithium ions was recorded in normal mode (top) in the range 450–540 *m/z*. The pta ligand contains basic nitrogen centres which readily associate with protons and metal cations allowing its facile ionisation, see also Section 5.3. The dominant peaks comprise the complex with an associated lithium ion, [M + Li]$^+$ at 470 *m/z*, water solvate [M + Li + H$_2$O]$^+$ at 488 *m/z*, and methanol solvate [M + Li + MeOH]$^+$ at 502 *m/z*. By selecting a slower scanning mode and prolonging the time the ions spend in the ion trap (by a factor of about ten), the spectrum at the bottom of Figure 2.8 was obtained.

Only one dominant peak is present, corresponding to the $[M + Li]^+$ ion and the resolution is superior to the normal speed scan. The second lower intensity peak is unresolved and appears at a mass that prevents assignment. This unresolved peak arises because unstable ions formed in the instrument tend to show reduced resolution and also a slight shift in mass to lower values (0.1–0.2 amu). While the ions are being scanned out of the trap, unstable species undergo CID by collisions with the helium in the trap. It should be noted that this phenomenon only occurs with very unstable ions, since the ion kinetic energy at this part of the scanning cycle is very low. If an ion fragments while scanning is underway, it will be ejected immediately (at slightly lower rf potential than the intact ion), and it will appear at slightly lower mass.

The ions contained within the ion trap can react with any neutral species present through ion-molecule reactions. Most frequently, the neutral species will be water. Chemists should be alert to this possibility. Intentional introduction of a reactive gas with the aim of probing ion/molecule chemistry is rarely successful, however, because the instruments are carefully tuned for helium as the bath gas and its displacement tends to result in drastic deterioration in performance.

2.4.1 MS/MS and MS^n

Ion trap instruments are capable of repeated cycles of MS/MS, so-called **MS^n**, where n is the number of MS operations performed. MS/MS experiments in ion trap instruments take place in the same location (in the trap) and so are separated **in time** rather than **in space**. MS/MS is achieved by keeping only ions of a certain m/z ratio in the trap by selectively ejecting all others. The remaining ions are energised and fragmented by collision-induced dissociation (CID) with the ever-present helium gas. The fragment ions are then scanned normally to generate a daughter ion mass spectrum. MS^3 is achieved by ejecting all fragment ions except the one of interest from the trap, and repeating the cycle of collision and scanning to generate granddaughter ions.

Remarkably, modern commercial instruments are quite capable of progressively fragmenting ions through ten cycles (MS^{10}), though ion loss in each stage of MS means that the effective usefulness of the approach rarely extends much beyond MS^3 or MS^4. An example is shown in Figure 2.9 (Plate 1) for the derivatised transition metal carbonyl cluster $[Ru_6C(CO)_{17} + OMe]^-$; the MS and MS/MS experiments provide clean, well resolved and intense data, the MS^3 experiment is somewhat less intense and MS^4 produced only a weak, incoherent spectrum containing no useful information.[7]

The modern ion trap also has the facility to fragment ions externally to the ion trap if the source is at atmospheric pressure via **in-source CID**. MS^n may then be performed on the fragments, and this can be an effective way to break the target down further than would otherwise be possible.

Ion traps have found application in hybrid instruments, as they represent a relatively inexpensive way to greatly enhance MS/MS capabilities.[8] The Q/QIT combination resembles a traditional triple quad in many ways, but replacing MS2 with an ion trap allows extra stages of MS. It is the cheapest hybrid instrument currently available. QIT/FTICR machines can store, fragment and deliver controlled quantities of ions for more detailed study in the FTICR cell, and similar advantages are inherent in the QIT/TOF design.

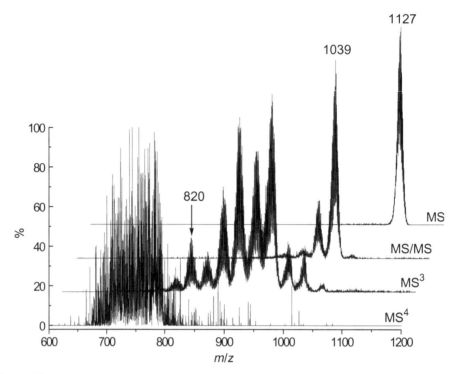

Figure 2.9
(Plate 1) ESI-MS^n spectra of [Ru$_6$C(CO)$_{17}$] in MeOH/NaOMe solution. From the top: the standard MS spectrum carried out at low temperature (50 °C) with no in-source fragmentation; MS/MS spectrum, 50 % collision energy on the peak at 1127 m/z; MS^3 spectrum, with the 1127 m/z peak selected, fragmented at 50 %, followed by selection of the 1039 m/z peak and further fragmentation at 50 %; MS^4 spectrum, selection of the 820 m/z after MS^3 and further fragmentation at 50 % collision energy

2.4.2   Summary

Strengths

- High sensitivity;
- Multi-stage mass spectrometry (analogous to FTICR experiments);
- Compact – the smallest mass analysers[9];
- Relatively inexpensive;
- Robust;
- Operate at much higher pressure than most mass analysers, so the vacuum system does not need to be as efficient;
- Versatile – can be configured with any other mass analyser in useful hybrid instruments.

Weaknesses

- Poor quantitation;
- Subject to space charge effects and ion molecule reactions;
- Collision energy not well-defined in CID MS/MS.

Mass range and resolution

- 4000 *m/z* for the best commercial instruments; more typically 2000 *m/z*;
- Maximum resolution of maybe 4000, but a typical maximum is 2500 and for smaller, cheaper models perhaps 1000.

Applications

- Benchtop GC/MS, LC/MS;
- Atmospheric pressure MALDI;
- MS^n systems;
- Hybrid systems.

Popularity

- Rapidly increasing. The small size, rapidly improving technology and ability to perform MS^n make ion traps an attractive package, and they are well suited to miniaturisation. A technique set for further growth.

Time-of-Flight[10]

A **Time-of-Flight** (TOF) mass analyser is perhaps the simplest means of mass analysis, and was first described over 50 years ago.[11] However, it is only relatively recently (1990 onwards) with advances in high speed digital timing electronics and improvements in instrument design that TOF has really taken off, as it is now possible to obtain excellent resolution along with the traditional advantages associated with TOF of extraordinary mass range, high duty factor (percentage of ions formed that are detected) and compatibility with pulsed ionisation techniques. TOF mass analysers have also be adapted for use in continuous ion sources, the so-called orthogonal-TOF, and it seems entirely likely that the popularity of this means of ion separation is set to increase further.

A pulse of ions is generated in the source, and accelerated through an electric field into the TOF analyser. Effectively, this provides all the ions with the same kinetic energy. Since $KE = \frac{1}{2}mv^2$, the resulting velocity of the ion is inversely dependent on the square root of its mass. Subsequent to the initial acceleration, the ions are allowed to travel in a straight line at constant velocity along a drift tube, typically about a meter long. Because the lightest ions will reach the detector first (and heaviest last), the arrival times of the ions can be transformed into a mass spectrum (Figure 2.10).

An entire mass spectrum is generated in a single pulse, and up to 30 000 individual spectra are collected per second. These spectra are summed to obtain a much improved signal to noise ratio.

The mathematics of TOF mass analysis are quite accessible. The ions formed by the laser pulse are extracted into the TOF analyser using an electric field. The kinetic energy given to each ion by the electric field is the product of the charge of the ion and the electric field strength:

$$KE = zeV$$

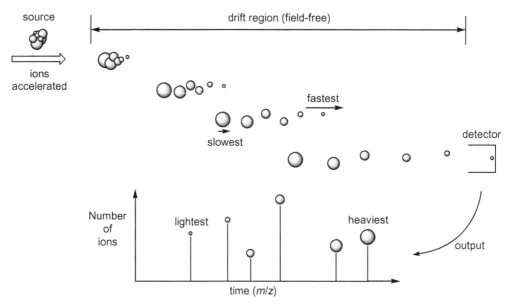

Figure 2.10
The essentials of TOF optics. A pulse of ions is formed and accelerated out of the source into the drift region (which is free from electric or magnetic fields). The ions separate along the drift region according to *m/z* value. The ions with smallest *m/z* value (fastest moving) begin arriving first at the detector, to be followed by ions of gradually increasing *m/z*. The detector counts the ions and this information in conjunction with the arrival time generates a mass spectrum (bottom)

where z is the charge on the ion, e is the charge of an electron in coulombs, and V is the strength of the electric field in volts. All ions (of the same charge), regardless of mass, enter the field-free region with the same kinetic energy. Since kinetic energy can also be represented by the Newtonian equation $KE = \frac{1}{2}mv^2 = \frac{1}{2}m(dx/dt)^2$, then:

$$zeV = \tfrac{1}{2}m(dx/dt)^2$$

By rearranging this equation, m/z for a given ion can be determined:

$$m/z = 2eV\Delta t^2/\Delta x^2$$

where m is the mass in kilogrammes, Δt is the flight time and Δx is the flight path length in meters. Rationalising the constants ($1\,\mathrm{Da} = 1.660 \times 10^{-27}\,\mathrm{kg}$, $e = 1.60 \times 10^{-19}\,\mathrm{C}$) allows the equation to be written more conveniently:

$$m/z = (1.928 \times 10^8)V\Delta t^2/\Delta x^2$$

For example, if the spectrometer has a 1 m flight path, the ions are extracted at 5 kV, and a particular ion arrives at 100 µs, the mass-to-charge ratio of the ion can be calculated to be 9640 *m/z*.

All time-of-flight mass analysers rely on very low pressures for good performance, because in order to discriminate ions at high resolution on the basis of their flight times collisions must be eliminated. The required mean free path length of the ions must be in excess of the length of the flight tube, and this generally means that a turbomolecular pump is required in addition to earlier pumping stages using a roughing pump. Despite the simplicity of the technique, this requirement of very high vacuum means that TOF analysers are generally more expensive than instruments employing quadrupoles or quadrupole ion traps.

The necessity for a pulsed ionisation source for TOF analysis stems from the fact that all ions must leave the source simultaneously so that their start time is synchronised. As such, MALDI and TOF are perfectly complementary, as MALDI produces mostly singly-charged ions, often of very high mass, upon nanosecond-length bursts of a radiation from a pulsed laser. The desorbed ions do have a range of velocities, which has a negative effect on the attainable resolution. This problem has been largely overcome by a technique called **delayed extraction**, where after a short delay a pulsed electric field accelerates the ions furthest from the desorption surface least (because these ions must have the highest velocity) and those closest most. As a result, the initial conditions of all ions can be tuned to be as identical as possible, and this has a very favourable effect on the resolution achieved.[12]

2.5.1 Reflectron Instruments

The ions leaving the ion source of a time-of-flight mass spectrometer have neither exactly the same starting times nor exactly the same kinetic energies. Improvements in time-of-flight mass spectrometer design have been introduced to compensate for these differences, and the most dramatic improvements in performance come with the use of a **reflectron**. A reflectron is an ion optic device in which ions in a time-of-flight mass spectrometer interact with an electronic ion mirror and their flight is reversed (Figure 2.11).

Ions with greater kinetic energies penetrate deeper into the reflectron than ions with smaller kinetic energies. The ions that penetrate deeper will take correspondingly longer

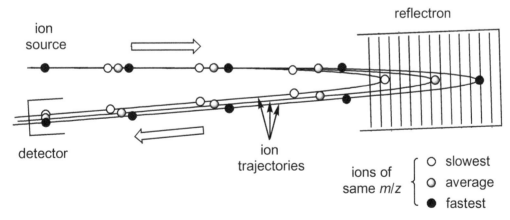

Figure 2.11
Schematic of a reflectron. Ions of the same *m/z* have a range of kinetic energies, as their formation conditions were not identical. The ions spread out after acceleration but the fastest moving ions penetrate the reflectron field further than do the slowest moving ions. By the time they reach at the detector, the ions have spent the same time in the flight tube and arrive in an ensemble. The overall effect is a considerable enhancement in resolution over that attainable in linear TOF mode

to return to the detector. If a packet of ions of a given mass-to-charge ratio contains ions with varying kinetic energies, then the reflectron will decrease the spread in the ion flight times, and therefore improve the resolution of the time-of-flight mass spectrometer. The main downside of a reflectron is loss of ions during transit and many instruments retain a detector sited for analysis in linear mode so the operator can choose between resolution or sensitivity.

2.5.2   Orthogonal TOF (oa-TOF)[13]

Orthogonal TOF analysers are employed in conjunction with continuous ion sources, especially ESI,[14] and in hybrid instruments. The continuous ion beam is subjected to a pulsed electric field gradient at right angles (orthogonal) to the direction of the ion beam. A section of the ion beam is thus pulsed away instantaneously and can be measured using a TOF analyser (Figure 2.12). Because a whole section of the beam is pulsed away at

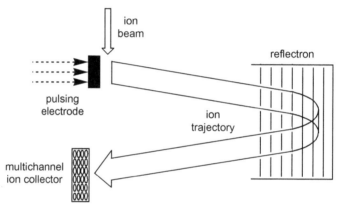

Figure 2.12
Principle of an orthogonal TOF analyser. Note that a whole section of the beam is instantaneously pulsed away, requiring the use of an array detector to efficiently collect all ions

once, the ions are dispersed in space and it is advantageous to use an array detector in order to ensure all ions are collected. A **microchannel plate** detector is typically used. It consists of an array of tiny electron multipliers all connected to the same electrified plate. An ion hitting any one of the microchannels induces a cascade of electrons and produces a detectable current. Orthogonal TOF analysers can accumulate tens of thousands of spectra per second and these are summed to provide spectra with high signal-to-noise ratios. Attainable resolution is high and modern orthogonal TOF instruments are quite capable of acquiring accurate mass data.

2.5.3   MS/MS

During the brief time (ca. tens of microseconds) the ion spends in the flight tube, particularly energetic ions may undergo dissociation and just as is the case for sector instruments (Section 2.2), this process can result in the appearance of post-source decay (PSD) peaks in the mass spectrum. In the case of linear TOF analysers, the fragments reach the detector at the same time so PSD products are not observed. However, when a reflectron is in use, if an ion decays while in the flight tube but before reaching the reflectron, the fragment ions of the decay will appear in the mass spectrum. The PSD fragment has reduced kinetic energy (it has the same velocity as the ion that underwent PSD but a lower mass) and therefore penetrates the reflectron field less, spends less time in flight and appears in the spectrum at lower *m/z*. Identification of PSD products is straightforward; a peak can be assigned as originating from a PSD process if it appears in a reflectron experiment but not one carried out in linear mode. Once the precursor ion undergoing PSD is identified, the relative flight times can be used to calculate the mass of the fragment lost during decomposition.

However, like all PSD processes, the criteria for observation are stringent and ions attributable to PSD tend to be of low relative intensity. Increasing in popularity are hybrid mass spectrometers, where fragment ions are generated in a collision cell placed between the two analysers. The number of different permutations of hybrid mass spectrometer possible with an TOF analyser is large, but certainly the most popular currently (2004) is the quadrupole/oa-TOF configuration,[15] first released commercially in the mid-1990s. In these instruments (Figure 2.13), the first mass analyser (MS1) is a quadrupole, which acts

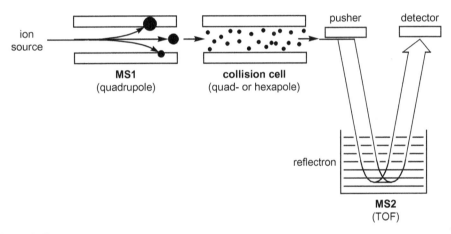

Figure 2.13
A Q-TOF instrument. MS1 is set in rf-only mode for MS studies (i.e. all ions are passed through, regardless of mass). For MS/MS studies, MS1 is used to select a precursor ion, which is fragmented in the collision cell, and the daughter ion spectrum collected using MS2

simply as an rf-only ion guide in MS mode. Next is a collision cell, followed by the oa-TOF (MS2). To collect MS/MS data, MS1 is used to select a single ion, which is then fragmented in the collision cell, and MS2 is used to collect the daughter ion spectrum.

Fast, highly efficient, sensitive and capable of high resolution, the Q/TOF provides higher quality data than the popular triple quadrupole, though it is less suitable for quantitative work. An even more recent innovation is the TOF-TOF instrument, where two reflectron TOF mass analysers are separated by a collision cell. This configuration is designed primarily for MALDI work, where the lower mass range of a quadrupole mass filter makes it poorly suited for selection of the high mass, singly-charged ions typically produced in the MALDI experiment.

2.5.4 Summary

Strengths

- Simplicity;
- Unlimited mass range;
- High transmission;
- High speed of analysis (the fastest MS analyser);
- Detection of all species simultaneously (i.e. not scanning);
- Compatibility with pulsed ionisation sources;
- MS/MS information from post-source decay.

Weaknesses

- Requires pulsed ionisation method or ion beam switching (duty cycle is a factor);
- High vacuum conditions required;
- Fast digitisers used in TOF can have limited dynamic range.

Mass range and resolution

- No theoretical upper limit to mass range. MALDI-TOF instruments typically extend well beyond 200 000 m/z; oaTOF applications rarely require $> 30\,000$ m/z;
- Resolution of the best reflectron TOF instruments is now beyond 30 000, but 10 000 is more typical. Resolution is severely compromised in linear TOF mode, and resolution much above 1000 can not be expected.

Applications

- MALDI instruments, including MALDI-TOF/TOF;
- Hybrid spectrometers with orthogonal TOF, especially quadrupole-TOF (Q-TOF);
- Very fast GC/MS systems.

Popularity

- High and climbing. The two main driving forces are the advent of MALDI, a perfect match for TOF mass analysis, and the development of oa-TOF. The latter has given TOF the ability to compete in practically every marketplace, as it is compatible with most ion sources.

Fourier Transform Ion Cyclotron Resonance[16]

Fourier transform ion cyclotron resonance (FTICR) mass spectrometers, first demonstrated in 1973 by Comisarow and Marshall,[17] offer 10–100 × higher resolution, resolving power and mass accuracy than any other analyser. They are responsible for performance records for mass analysis in practically every category.[18] Additionally, they can trap ions for extended periods and measure them non-destructively, giving the instruments powerful MS/MS abilities. Commensurate with their powers of analysis is the price of FTICR mass spectrometers, and this is the main factor preventing wider utilisation of the technique.

Ions are deflected into a circular path when moving perpendicular to a magnetic field, and this property is utilised in magnetic sector instruments (Section 2.2). If the field is very strong and the velocity of the ion is slow, the ion may be trapped in a circular orbit. FTICR instruments operate at much higher fields (above 3 Tesla), provided by large superconducting magnets similar to those used in NMR spectrometers. Under the influence of this field, ions may be trapped in a circular orbit. This orbit represents **cyclotron** motion and the frequency of the orbiting ion is characteristic of its mass. The balancing forces involved are the centripetal force (mv^2/r) and the Lorentz force (evB) (Figure 2.14).

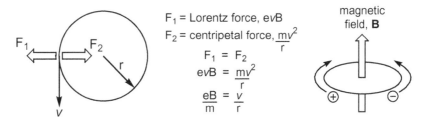

Figure 2.14
Forces involved in cyclotron motion (left), the formulae used to derive the working equation for cyclotron frequency (middle), and the direction of rotation of ions with respect to a magnetic field (right)

The v/r term is equivalent to the angular frequency, ω (omega, in radians s^{-1}), so $\omega = eB/m$, and this is the working equation used to obtain the **cyclotron frequency** of an ion. The cyclotron frequency of an ion of mass m (Da) in a field of strength B (Tesla) can be written to incorporate all constants as $f = (1.53567 \times 10^7)B/m$, so an ion of mass 577.41 Da in a 9.4 T field will have a frequency of 250 kHz for example. Note that ω is independent of the velocity of the ion, so ions of a given mass will have the same frequency regardless of their velocity. FTICR instruments are not susceptible to the peak-broadening effects suffered by other forms of mass analysis and are inherently capable of the highest resolution because, uniquely, the mass-to-charge ratio is manifested as a frequency, the experimental parameter that can be measured more accurately than any other.

The frequencies of the trapped ions must be determined in order to obtain a mass spectrum. In FTICR, all ions are excited using a radio frequency 'chirp' of short duration ($\sim$1 μs) that rapidly sweeps across a large frequency range. This has the effect of increasing the orbital radius so the ion trajectory now passes close to the walls of the cell. Two of the walls of the cubic cell (all sorts of other shapes, e.g. cylindrical, hyperbolic etc, are also known)[19] are designated receiver plates (electrodes; these are orthogonal to the pair of electrodes used to excite the ions) and as the ions approach one of the walls electrons are induced to collect on that electrode. The path of the ions takes them back towards the other electrode and the electrons flow back. This migration of electrons is amplified and detected as a sinusoidal **image current** (Figure 2.15).

Because all ions are excited by the rf chirp, the image current represents a composite of the cyclotron frequencies of all the ions in the cell. The overall signal is very complicated and dies off over time as the ions relax and return to their stable circular orbits near the

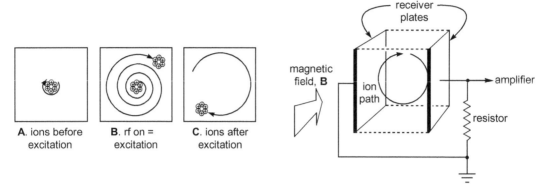

Figure 2.15
Thermal ions trapped in the cell rotate in a tight, incoherent motion (A) but upon excitation with a radio frequency burst (B) the packet of ions is excited in phase into a higher, coherent (and therefore detectable) orbit (C). Schematic of a cubic FTICR cell (right). The magnetic field points into the page. The ions rotate in the plane of the page in a cyclotronic motion, inducing image currents in the receiver plates. The ions must rotate in a coherent packet, or the charge induced in one plate will be cancelled out by an ion on the other side of the orbit

centre of the cell. To ensure that the signal is sustained over a sufficient period to collect good data, the cell must be pumped down to very high vacuum, of the order of 10^{-5} Pa (one ten-billionth of an atmosphere). Converting the image current signal into a mass spectrum is performed using a fast **Fourier transform**, an algorithm which allows the frequency and amplitude of each component to be extracted (Figure 2.16).[20]

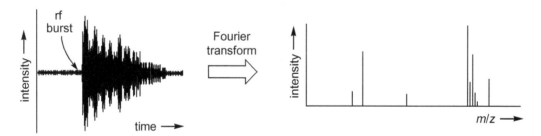

Figure 2.16
An rf burst accelerates the ions, generating a transient ion image current signal (left). The signal is digitised, stored in the computer and a Fourier transform is applied to the data to convert the information into a mass spectrum (right)

Because the ions are measured non-destructively, repeated experiments may be summed to improve the signal-to-noise ratio. The post-excitation orbital radius is collapsed back to near-zero between acquisitions, allowing close to 100 % recovery of ions between 'scans'.[21]

The requirement for extremely low pressure to minimise ion-molecule collisions in the analyser region contributes significantly to the cost of the instrument. Complicated ion optics are required to shepherd the ions from the high pressure of the source (often atmospheric) through multiple differentially pumped (roughing pump – turbomolecular pump – cryogenic pump) regions of the instrument. The superconducting magnet also contributes to the cost but it does have the advantage of providing very stable calibration over a long period of time. As always, however, the best accurate mass measurements require an internal calibrant; these allow FTICR instruments to achieve mass accuracy to well within 1 ppm. Resolution is also astonishing, above 500 000 being almost routine. In a remarkable demonstration, two peptides of nominal mass 904 Da whose composition differed by N_4O vs S_2H_8 or 0.00045 Da were baseline resolved.[22] This mass difference is less than the mass of an electron (0.00055 Da) and represents a mass resolving power of approximately 3,300,000 (Figure 2.17).

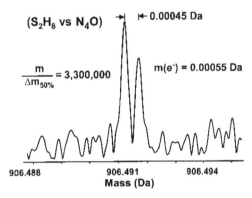

Figure 2.17
Ultra-high-resolution ESI FTICR mass spectrum of monoisotopic $(M + 2H)^+$ of two polypeptide ions. Reproduced by permission of the American Chemical Society

2.6.1 MS/MS

The fact that ions can be detected non-destructively is of great utility for tandem mass spectrometry studies using FTICR. As in the ion trap (Section 2.4), each step is carried out *in time* rather than *in space*. Ions of interest may be selected by expelling all other ions through discharge on the walls of the cell. Calculating the waveform to be applied to the excitation electrodes to achieve this expulsion may be done by means of a 'stored waveform inverse Fourier transform' (SWIFT).[23] Once the trap contains only the ions of interest, they may be excited and fragmented through multiple ion-neutral interactions (collision-induced dissociation, CID). Excitation is usually performed by repeatedly accelerating and decelerating the ions, by a technique called sustained off-resonant excitation (SORI).[24] An alternative is irradiating the ions using a technique called infrared multiphoton dissociation (IRMPD),[25] and this approach has the advantage that no collision gas needs to be introduced (and then pumped away). By whatever method the fragmentation is induced, all the product ions remain trapped in the cell and can be analysed as normal. This process can be repeated again and again, allowing MS/MS/MS... (MS^n) studies. Furthermore, the long period that ions can be stored in the cell allows an almost limitless number of ion-molecule reactivity studies to be carried out.[26] For example, the cluster $[CoRu_3(CO)_{13}]^-$ was stripped of CO ligands using CID to form $[CoRu_3]^-$, then the most intense peak in the isotope envelope was selected for reactivity (by ejecting all others from the cell).[27] Methane was introduced into the cell and allowed

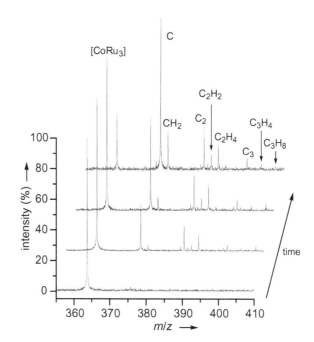

Figure 2.18
(plate 2) Reaction of $[CoRu_3]^-$ with CH_4. Note the build-up of product ions over time

to react with the activated cluster (Figure 2.18, Plate 1). The product ions included species generated by the addition of various C_xH_y fragments to the cluster core.

The newest FTICR instruments are actually often hybrids, with a 'front end' consisting of a quadrupole or ion trap. This facility adds an extra stage (or stages) of MS/MS, but importantly, does the experiment very quickly so the FTICR part of the instrument can be set to work mass analysing the product ions.

2.6.2 Summary

Strengths

- The highest recorded mass resolution of all mass spectrometers – by most criteria, the 'best' mass analyser available;
- Powerful capabilities for ion-molecule chemistry and MS/MS experiments;
- Non-destructive ion detection; ion remeasurement;
- Stable mass calibration in superconducting magnet FTICR systems.

Weaknesses

- Bulky;
- Very expensive;
- Superconducting magnet requires cryogenic cooling;
- Subject to space charge effects and ion-molecule reactions;
- Artifacts such as harmonics and sidebands can be present in the mass spectra.

Mass range and resolution

- Mass range depends on the strength of the magnet, but in practice, data are rarely collected above 3000 m/z except when coupled with a MALDI ion source;
- Resolution routinely in the 100 000s.

Applications

- High-resolution experiments for high-mass analytes;
- Resolving complex mixtures;
- Ion-molecule chemistry.

Popularity

- Highly sought after but access is limited due to the very high cost of the instruments.

References

1. R. G. Cooks, J. H. Benyon, R. M. Caprioli and G. R. Lester, *Metastable Ions*, Elsevier, Amsterdam (1973).
2. K. L. Busch, G. L. Glish and S. A. McLuckey, *Mass Spectrometry/Mass Spectrometry: Techniques and Applications of Tandem Mass Spectrometry*, VCH, New York (1988).
3. P. H. Dawson, Ed. *Quadrupole Mass Spectrometry and its Applications*, Elsevier, Amsterdam (1976). Republished in 1997.
4. R. G. Cooks and R. E. Kaiser, *Acc. Chem. Res.*, 1990, **23**, 213–219; R. E. March, *J. Mass Spectrom.*, 1997, **32**, 351.
5. W. Paul, H. Reinhard and V. Z. Zahn, *Z. Phys.*, 1959, **156**, 1.
6. With Hans Dehmelt, '*for the development of the ion trap technique*'. See http://www.nobel.se/physics/laureates/1989/index.html.
7. P. J. Dyson and J. S. McIndoe, *Inorg. Chim. Acta*, 2003, **354**, 68.
8. K. Cottingham, *Anal. Chem.*, 2003, 315A.
9. E. R. Badman and R. G. Cooks, *J. Mass Spectrom.*, 2000, **35**, 659.
10. M. Guilhaus, V. Mlynski and D. Selby, *Rapid Commun. Mass Spectrom.*, 1997, **11**, 951; K. G. Standing, *Int. J. Mass Spectrom.*, 2000, **200**, 597.
11. W. E. Stephens, *Phys. Rev.*, 1946, **69**, 691.
12. S. M. Colby, T. B. King and J. P. Reilly, *Rapid Commun. Mass Spectrom.*, 1994, **8**, 865; M. L. Vestal, P. Juhasz and S. A. Martin, *Rapid Commun. Mass Spectrom.*, 1995, **9**, 1044.
13. J. H. J. Dawson and M. Guilhaus, *Rapid Commun. Mass Spectrom.*, 1989, **3**, 155.
14. I. V. Chernushevich, W. Ens and K. G. Standing, *Anal. Chem.*, 1999, **71**, 452A.
15. I. V. Chernushevich, A. V. Loboda and B. A. Thomson, *J. Mass Spectrom.* 2001, **36**, 849.
16. A. G. Marshall and C. L. Hendrickson, *Int. J. Mass Spectrom.*, 2002, **215**, 59; A. G. Marshall, C. L. Hendrickson and G. S. Jackson, *Mass Spectrom. Rev.*, 1998, **17**, 1; I. J. Amster, *J. Mass Spectrom.*, 1996, **31**, 1325.
17. M. B. Comisarow and A. G. Marshall, *Chem. Phys. Lett.*, 1974, **25**, 282.
18. A. G. Marshall, *Int. J. Mass Spectrom.* 2000, **200**, 331.
19. S. Guan and A. G. Marshall, *Int. J. Mass Spectrom. Ion Proc.*, 1995, **146–7**, 261.
20. J. W. Cooley and J. W. Tukey, *Math Comput.*, 1965, **19**, 297.
21. J. P. Speir, G. S. Gorman, C. C. Pitsenberger, C. A. Turner, P. P. Wang and I. J. Amster, *Anal. Chem.*, 1993, **65**, 1746.
22. F. He, C. L. Hendrickson and A. G. Marshall, *Anal. Chem.*, 2001, **73**, 647.

23. A. G. Marshall, T.-C. L. Wang and T. L. Ricca, *J. Am. Chem. Soc.*, 1985, **107**, 7893; S. Guan and A. G. Marshall, *Int. J. Mass Spectrom. Ion Proc.*, 1996, **157–8**, 5.

24. J. W. Gaulthier, T. R. Trautmann and D. B. Jacobson, *Anal. Chim. Acta*, 1991, **246**, 211.

25. R. L. Woodin, D. S. Bomse and J. L. Beauchamp, *J. Am. Chem. Soc.*, 1978, **100**, 3248.

26. K. J. Fisher, *Prog. Inorg. Chem.*, 2001, **50**, 343.

27. C. P. G. Butcher, A. Dinca, P. J. Dyson, B. F. G. Johnson, P. R. R. Langridge-Smith and J. S. McIndoe, *Angew. Chem. Int. Ed.*, 2003, **42**, 5752.

3 Ionisation Techniques

Introduction

The principles of operation for the major ionisation techniques that have been developed are described in this chapter. Some, like plasma desorption (PD), are obsolete, but have been included as they provide a historical perspective and serve to illustrate the advantages of more recent developments. Others, such as fast atom bombardment (FAB), are falling in popularity but are still to be found in many mass spectrometry laboratories. Matrix-assisted laser desorption/ionisation (MALDI) is an enormously popular technique in the biosciences, but has been surprisingly little-used by the organometallic and coordination chemist and as such fails to justify its own chapter. In contrast, electrospray ionisation (ESI) is without doubt the dominant ionisation technique currently used by coordination and organometallic chemists. The abundance of applications of ESI MS in inorganic systems are summarised in Chapters 4 to 7. Even so, the discussion is by no means exhaustive and there are many other applications that space considerations unfortunately preclude from inclusion. This clearly indicates the power of the ESI technique.

Electron Ionisation

Perhaps the most well-known ion source uses energetic electrons to ionise gas-phase molecules. **Electron ionisation** (often referred to as electron impact; both abbreviated as EI; the terms are frequently used interchangeably) is extremely common in organic mass spectrometry, as it works most effectively on small, volatile, thermally robust molecules. The sample to be analysed must be in the gas phase, so the technique is ideally suited to be interfaced with a gas chromatograph (GC), providing a sophisticated, structurally informative detector. Involatile liquids and low molecular weight solids are routinely analysed via introduction of the sample directly into the source. A combination of high vacuum and temperature is usually enough to drive most neutral compounds with molecular weights of 1000 Da or less into the gas phase in sufficient concentration (and without significant decomposition) to be successfully analysed. For many years the only ionisation technique available, electron ionisation remains important as the method of choice for non-polar, volatile molecules (Figure 3.1).

An electron ionisation source consists of a heated metal filament (rhenium or tungsten). Electrons boil off the surface of the wire and are accelerated towards an anode. The path

Mass Spectrometry of Inorganic, Coordination and Organometallic Compounds   W. Henderson and J. S. McIndoe
© 2005 John Wiley & Sons, Ltd   ISBNs: 0-470-85015-9 (HB); 0-470-85016-7 (PB)

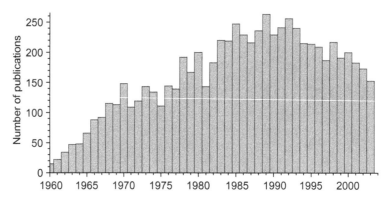

Figure 3.1
The popularity of EI, as measured by papers published.[1] The number of papers is grossly underestimated, but the trend of steady growth, barely affected by the development of newer techniques (until the early 1990s with the introduction of ESI and MALDI) is probably quite accurate

of the electrons is through the volume occupied by vaporised sample molecules and interactions take place between the energetic electrons and the molecules (Figure 3.2).

If an electron transfers enough energy to the molecule, electronic excitation can occur with the expulsion of an electron. In the case of organic molecules, a wide maximum occurs in the efficiency of ion production at an electron energy of around 70 eV (i.e. the electron is accelerated through 70 V). The molecular ion, $[M]^{\bullet+}$ (a radical-cation), frequently is left with energy in excess of that needed for ionisation and the ion fragments to give ions of smaller mass, $[M - X]^+$, and radicals, $X^\bullet$, which are not seen in

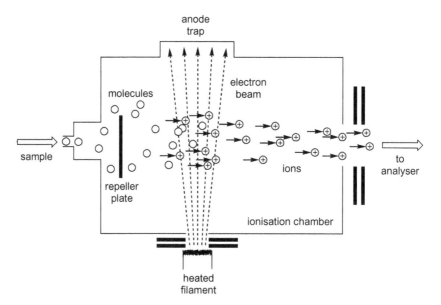

Figure 3.2
Electron ionisation source. The gaseous sample molecules enter the chamber and are ionised by energetic electrons. The resulting ions are drawn out and accelerated through slits into the mass analyser

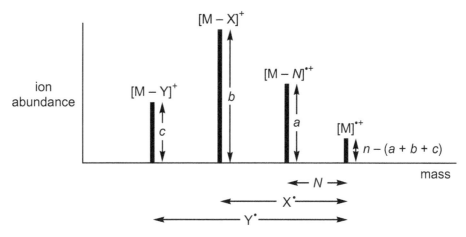

Figure 3.3
An electron ionisation mass spectrum. Note that while neutral fragments and radicals do not appear in the spectrum, their mass may be inferred from the mass difference between precursor and product ion

the mass spectrum as they are not charged. Ions may also fragment via loss of a neutral molecule, N, in which case the fragment ion, $[M - N]^{\bullet+}$, will retain the unpaired electron (Figure 3.3).

$$n.M + e^- \longrightarrow n.[M]^{\bullet+} + 2n.e^-$$
$$then\ n.[M]^{\bullet+} \longrightarrow a.[M - N]^{\bullet+} + N$$
$$b.[M - Y]^+ + Y^\bullet$$
$$c.[M - X]^+ + X^\bullet$$
$$etc.$$

Fragment ions may themselves break down further and/or undergo rearrangement, further complicating the spectrum. While the fragmentation process can compromise detection of the molecular ion, it is generally seen as desirable, as it is highly reproducible, characteristic of the molecule in question and provides useful structural information. In organic chemistry, the spectrum obtained from an analyte (a 'fingerprint') is indexed by recording the intensities of the most abundant peaks and their m/z values, and this information immediately cross-referenced against an electronic library. Known compounds are thus identified rapidly. Compounds that give no hits in the library may be analysed more closely and any structural information inferred from the spectra. Nearly all inorganic and organometallic compounds will fall into this latter category, so understanding of fragmentation processes is important to spectral interpretation.

While it would seem that lowering the electron energy should assist in identification of the molecular ion, in fact the reduction in overall ion current at low energy generally results in a decrease in the absolute intensity of the molecular ion as well, due to the decrease in ionisation efficiency. As a result, while the molecular ion now has greater abundance in comparison to the other ions in the spectrum, it is no easier to detect. The problem of molecular ion identification is illustrated in Figure 3.4, the EI mass spectrum of a crown ether, in which the molecular ion registers an intensity less than 1 % that of the base peak, $[C_2H_5O]^+$.

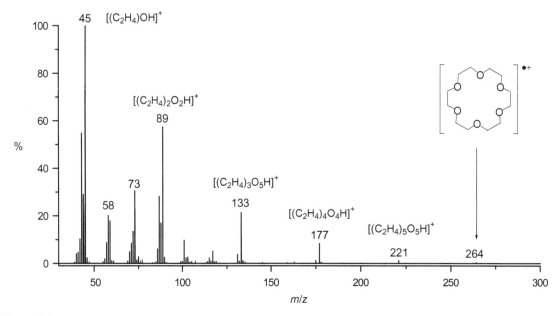

Figure 3.4
The EI mass spectrum of 18-crown-6. Note the vanishingly low intensity of the molecular ion

Generally, the best strategy to convincingly obtain molecular weight information is to switch to a softer ionisation technique, such as ESI. (See Figure 5.10 for ESI mass spectra of metallated crown ethers). The problem of low molecular ion intensity for certain compounds prompted the development of chemical ionisation (CI, see following section).

Because ionisation efficiencies are low in EI, multiple charging is unusual for organic molecules. As a result, spectra generally show only singly charged ions, i.e. all ions appear at their mass values (m/z where $z = 1$). The probability of observing doubly charged ions increases with metal-containing compounds because they tend to be more electropositive than organic compounds.

3.2.1 Fragmentation of Metal-Containing Compounds

The fragmentation of organic compounds has been studied intensively and systematically for many decades.[2] The study of the fragmentation of metal-containing compounds is much less well established, but there are fortunately some basic ground rules that apply in both situations and in many ways the fragmentation patterns are more readily explicable. For organometallic or coordination compounds ML_n where M is a metal and L are atoms or groups of atoms bonded to M, nearly all the ions in the mass spectrum will contain M:

$$\text{i.e.} \quad [ML_n]^{+\bullet} \longrightarrow [ML_{n-1}]^+ + L^\bullet$$
$$\textit{not} \quad [ML_n]^{+\bullet} \longrightarrow ML_{n-1}^\bullet + [L]^+$$

The metal retains the positive charge and the ligand departs as a radical (not detected). This phenomenon occurs essentially because M-containing species are more electropositive and generally have lower ionisation potentials than the ligands. This phenomenon is obvious in cases where the metal is polyisotopic; the isotopic 'signature' of the metal will be repeated in all the fragment ions of the spectrum. The other common fragmentation pathway involves elimination of L as a chemically stable entity:

$$\text{i.e.} \qquad ML_n^{+\bullet} \longrightarrow ML_{n-1}^{+\bullet} + L$$

This process is especially common in organometallic chemistry where the ligands are often neutral molecules, e.g. CO, PR_3, alkene, etc. Neutral ligands that occupy multiple coordination sites such as arenes and diphosphines tend to be considerably more tenacious, and clean cleavage of such ligands is rare. Formally anionic ligands such as cyclopentadienyls and halides are eliminated with difficulty as radicals. Ligands with a formal charge but that can be eliminated as a stable neutral molecule can be removed comparatively easily. For example, the nitrosyl ligand is formally regarded sometimes as NO^+, sometimes as NO^-, but is usually eliminated as neutral nitric oxide, NO. If the ligand ions have the largest peak intensity in the spectrum, thermal decomposition of the sample should always be suspected.

3.2.2 Applications

EI is of somewhat restricted use in organometallic and coordination chemistry, as only volatile, thermally stable compounds may be readily studied. However, some classes of compound do fulfill these requirements, notably neutral metal carbonyl compounds and main group organometallics. Furthermore ligands, often simple organic compounds themselves, are often amenable to characterisation by EI, so this method of ionisation remains critically important. The application of EI to inorganic and organometallic compounds has been well-reviewed in two books published in the early 1970s.[3]

Main group organometallic compounds

Boron hydrides are volatile and boron has two isotopes (^{10}B and ^{11}B, 20 % and 80 % respectively), so B_xH_y compounds seem at first glance to be excellent candidates for analysis by EI MS. However, these compounds are easily decomposed in the hot source and it is likely that in most cases spectra of 'pure' compounds are not observed. Spectra of boranes therefore tend to be spectacularly complicated, exacerbated by the isotopic complexity and facility of decomposition through elimination of hydrogen.

Derivatives of the group 14 elements, such as $SiMe_4$ or $PbEt_4$, are easy to prepare and handle and have convenient volatility, and many compounds of this type have been studied by mass spectrometry. Likewise, Group 15 derivatives, such as phosphines, arsines and stibines, which fulfill the volatility requirement provide good EI mass spectra. Generally, studies have been focused on obtaining molecular weight information. For example, the structure of a compound of empirical formula $GeC_{16}H_{16}$ was originally suggested to be **A**.[4] The mass spectrum showed that the compound was, in fact, $Ge_2C_{32}H_{32}$ **B** with a parent molecular ion in the range 556–568 *m/z*, and had the expected fragments corresponding to the ions $[M - GeMe_2]^{\bullet+}$ and $[Ge_2Me_4]^{\bullet+}$.

Metal carbonyls

Metal carbonyls, $M_n(CO)_m$, are relatively volatile and provide spectra which may be easily interpreted as the dominant fragmentation process is loss of carbonyl ligands as carbon monoxide gas:

$$[M_n(CO)_m]^{\bullet+} \longrightarrow [M_n(CO)_{m-1}]^{\bullet+} \longrightarrow etc \longrightarrow [M_n(CO)]^{\bullet+} \longrightarrow [M_n]^{\bullet+}$$

An example of this process is shown in Figure 3.5.

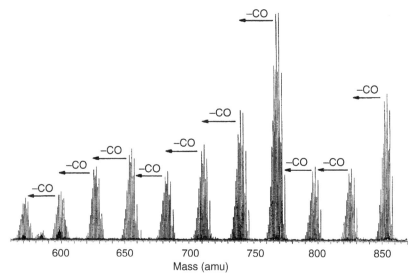

Figure 3.5
EI mass spectrum of $H_2Os_3(CO)_{10}$. Reproduced with permission of the American Chemical Society, from S. L. Mullen, A. G. Marshall, *J. Am. Chem. Soc.* 1988, **110**, 1766

The ability to provide accurate mass information for this type of compound was crucial to the development of polynuclear metal carbonyl chemistry. For example, pyrolysis of $Os_3(CO)_{12}$ produces a range of higher nuclearity clusters e.g. $Os_5C(CO)_{15}$, $Os_6(CO)_{18}$, $Os_7(CO)_{21}$ and $Os_8(CO)_{23}$. In the first instance, EI mass spectrometry was used to establish the correct molecular formula of these compounds and was also used to detect the presence of interstitial atoms (carbides in this case). Other techniques – elemental analysis, IR spectroscopy, NMR spectroscopy – were unable to provide this information.

Organometallic complexes

Other transition-metal complexes to provide good EI mass spectra include cyclopentadienyl complexes such as $Fe(\eta^5\text{-}C_5H_5)_2$ or $Mn(\eta^5\text{-}C_5H_5)(CO)_3$, arene complexes such as $Cr(\eta^6\text{-}C_6H_6)(CO)_3$ and π-allyl complexes such as $[(C_3H_5)_2RhCl]_2$. The main requirements, again, are volatility and thermal stability. Ferrocene has these characteristics in abundance, and the molecular ion strongly dominates the EI mass spectrum of this compound (Figure 3.6).

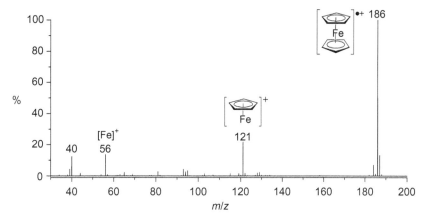

Figure 3.6
EI mass spectrum of ferrocene. Note the especially high abundance of the molecular ion; unusually in EI mass spectra, it is the base peak

Coordination complexes

Coordination complexes meeting the usual requirements can also provide excellent EI spectra, as shown in Figure 3.7 for $Co(acac)_3$. For a useful comparison of ionisation techniques, Figure 5.13 shows ESI mass spectra of the same compound.

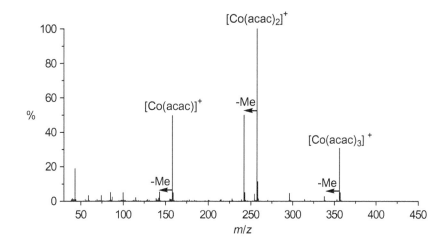

Figure 3.7
EI mass spectrum of $Co(acac)_3$

3.2.3 Summary

Strengths

- Well established and understood;
- Can be applied to virtually all volatile compounds;

- Reproducible mass spectra;
- Fragmentation provides structural information;
- Compounds may be 'fingerprinted' using huge libraries of mass spectra.

Weaknesses

- Sample must be thermally volatile and stable;
- The molecular ion may be weak or absent for many compounds.

Mass range

- *Low* Typically less than 1,000 Da.

Mass analysis

- Any, though most usually encountered in conjunction with GC/MS, which generally use unit mass resolution quadrupole instruments.

Popularity

- High, and steady. Of special importance to GC/MS, and for low molecular weight, non-polar compounds.

Tips

- Excellent for any compound that can be purified by distillation or sublimation.

Chemical Ionisation[5]

The abundance of the molecular ion in electron ionisation mass spectra is often quite low (Figure 3.4). In some cases, fragmentation is so facile that no molecular ions survive the 10^{-5} seconds or so required for mass analysis and only fragments are observed in the spectrum. Because molecular weight information is crucial for structural elucidation, 'softer' ionisation techniques were sought. An early advance, introduced by Munson and Field in 1966,[6] was the development of **chemical ionisation** (CI) mass spectrometry, in which analyte molecules react with ions in the gas phase to generate an analyte ion through transfer of a charged species (usually a proton) between reactants.

In a high vacuum (i.e. low pressure, 10^{-6} mbar) molecules and energetic electrons interact to form ions, as in an electron ionisation source. However, at higher pressure (typically 10^{-3} mbar), the ions formed initially collide with neutral molecules to give different kinds of ions before they can be injected into the mass analyser. This process is exploited in CI, by employing a background pressure of a low molecular weight gas present in considerable excess (10^3–10^4 ×) of the concentration of analyte molecules. For example, at low pressure, methane is ionised to give molecular ions, $CH_4^{\bullet+}$:

$$CH_4 + e^- \longrightarrow CH_4^{\bullet+} + 2e^-$$

but at higher pressures these ions will collide with other methane molecules to give carbonium ions, CH_5^+:

$$CH_4^{\bullet +} + CH_4 \longrightarrow CH_5^+ + CH_3^{\bullet}$$

If there is a substance M present, it may collide with carbonium ions with transfer of a proton (H^+) to give $[M + H]^+$, a so-called **pseudomolecular ion**:

$$M + CH_5^+ \longrightarrow MH^+ + CH_4$$

Because this is a chemical reaction, the methane is called a **reagent gas**. Related reagent gas systems that yield Brønsted acids (BH^+) include ammonia, hydrogen, isobutane and water, and all rely on the analyte having a higher proton affinity than B. CI has the advantage of causing less fragmentation than EI and the quasimolecular ion, MH^+, is often abundant in the spectrum. Note that MH^+ is not a radical cation, so radical-promoted fragmentation pathways are suppressed. Fragmentation primarily involves loss of neutral molecules from the pseudomolecular ion.

Other reagent gas systems have been used that operate by different mechanisms, e.g. benzene as a charge transfer agent (generating $[M]^{\bullet +})$[7] or $SiMe_4$ as a silylating agent (generating $[M + SiMe_3]^+)$,[8] but these are less well developed. **Negative-ion** CI (NCI) spectra may be obtained using reagent gas systems that produce strong Brønsted bases. Most commonly employed is a mixture of nitrous oxide and methane:

$$N_2O + e^- \longrightarrow N_2 + O^{\bullet -}$$
$$O^{\bullet -} + CH_4 \longrightarrow OH^- + CH_3^{\bullet}$$

OH^- will abstract a proton and generate an $[M - H]^-$ ion, provided the analyte has a reasonably acidic proton. A related technique, though not strictly CI, is electron capture ionisation. A buffer gas is used to thermalise electrons and these are captured selectively by molecules with high electron affinity (e.g. polyaromatics or molecules with halogens or nitro groups) to produce $[M]^{\bullet -}$ ions.

CI ion sources are nearly always found in combination with EI sources and such EI/CI sources may be switched from one mode to the other in a matter of seconds. The modifications required to a standard EI source (Figure 3.2) are an inlet for the reagent gas and a vacuum system capable of pumping at higher speed to allow for the differential in pressures between source and mass spectrometer. Also, the ionising electrons are usually accelerated to higher energies ($\sim 200\,eV$) to increase the efficiency of the ionisation process. Because CI involves chemical reactions, the sensitivity of the technique is critically dependent on the conditions of the experiment – reagent gas, pressure, temperature and electron energy are all important. The information provided by CI is complementary to EI and the combination of the two techniques is a powerful tool in structural elucidation.

However, the application of conventional CI to coordination and organometallic chemistry appears to have been rather limited. It relies on the presence of sites on the target molecule that are relatively basic (or acidic), as well as the molecule being of sufficient volatility to be driven into the gas phase. Its relevance is decreasing further because ESI tends to cope well with samples suitable for CI analysis.

Atmospheric pressure chemical ionisation (APCI) is a fairly recently developed variant, which as the name indicates operates at atmospheric pressure. An APCI source is therefore compatible with an electrospray ionisation (ESI) spectrometer (Section 3.9). Indeed, an ESI instrument may have the capability to switch between ESI and APCI modes fairly simply, and dual-purpose probes are sold by many manufacturers. In APCI, the analyte solution is delivered as an electrically neutral spray that is then ionised by means of a corona discharge, which produces ions in an analogous manner to traditional chemical ionisation, for example by reactions with gas-phase protonated solvent molecules or with the ionised nitrogen carrier gas. This contrasts with electrospray ionisation, where the spray of charged droplets is produced by spraying the solution from a metal capillary held at a high potential. Flow rates in APCI (up to $1\,mL\,min^{-1}$) are considerably higher than in ESI (a few $\mu L\,min^{-1}$), due to the appreciably lower sensitivity of the technique. APCI has been little used to date for coordination and organometallic compounds, mostly because its advantages over ESI are relatively few.

The advantages of APCI are most likely to be realised for those studying low molecular weight, non-polar species that are soluble only in non-polar solvents, such as toluene or hexane, where ESI is essentially ineffectual. For example, a series of silsequioxanes (empirical formula $[RSiO_{3/2}]_n$) were characterised in APCI using neat hexane as the solvent, with $[M + H]^+$ ions being observed in each case.[9] Some care should be taken in the analysis of hydrocarbon solvents because the mix of a highly flammable solvent and a corona discharge poses an ignition risk. The risk can be largely obviated by using an inert (nitrogen or argon) desolvation gas and thoroughly purging the source before beginning the experiment.

Rare examples of APCI-MS being used to characterise coordination or organometallic compounds include the analysis of bismuth(III)-thiolate complexes[10] and of platinum(II) complexes with bis(phosphanyl) monosulfides.[11]

3.3.1 Applications

The same restrictions with respect to sample characteristics (neutral, volatile) apply to chemical ionisation as apply to electron ionisation. As such, application of CI has primarily been in instances where fragmentation of the analyte is extensive and the molecular ion is difficult to identify conclusively. Most examples in the literature are for volatile organometallic complexes.

A range of chromium, manganese and vanadium complexes – $CpMn(CO)_3$, $RMn(CO)_3$ (R = acylcyclopentadienyl), $CpMn(CO)_2L$ (L = $NHMe_2$, $OSMe_2$, CNC_6H_{11}, norbornene, norbornadiene),[12] $Cr(\eta^6\text{-aryl})_2$ and $CpVC_7H_6X$ (X = H, OMe, CH_2CO_2Me) – were examined by CI using methane as reagent gas.[13] The ratio of $[M]^{\bullet+}$ (from charge transfer) to $[M + H]^+$ (from protonation) was found to be highly dependent on the type of compound, ligands with heteroatoms tending favouring formation of $[M + H]^+$. $[M]^{\bullet+}$ and $[M + H]^+$ were shown to display different fragmentation patterns, providing complementary structural information. The series of complexes $(\eta^4\text{-diene})Fe(CO)_3$, MCp_2 (M = Fe, Ru, Os, Ni, Co), Cp_2MCl_2 (M = Ti, Zr, Hf) and $M(CO)_6$ (M = Cr, Mo, W) were all successfully characterised using methane CI.[14] The manganese pentacarbonyl complexes $(OC)_3MnM'Ph_3$ (M' = Si, Ge, Sn) were studied using methane, isobutene, ammonia and nitrogen as reagent gases, with methane found to be most reactive.[15] The major fragments in the spectrum from the protonated ion $[M + H]^+$ (which was of low intensity) were $[M - Ph]^+$ and $[M'Ph_3]^+$. Increasing the probe temperature was found to increase the intensity of the $[M - Ph]^+$ ion. A comparison of EI and CI (ammonia reagent gas) in the analysis of Si, Ge, Sn, Pb compounds showed that CI was appreciably better at providing the molecular ion, $[M]^{\bullet+}$, and that both techniques provided useful and complementary information on these compounds, many of which were highly air-sensitive. Transportation to the mass spectrometer was achieved using sealed capillary tubes.[16]

Negative-ion chemical ionisation (NICI) has been applied to $(\eta^6\text{-arene})Cr(CO)_3$, titanocenes and zirconocenes, using methane as the reagent gas.[17] Very little fragmentation was observed. Interestingly, some chlorinated titanocene complexes showed an $[M + Cl]^-$ ion, as well as the radical anion $[M]^{\bullet-}$. Selective ionisation of organometallic compounds using chloride ions has been reported,[18] and NICI of $PtCl_2(PEt_3)_2$ allowed observation of an $[M + Cl]^-$ ion, or $[M + I]^-$ in the presence of MeI.[19] The NICI spectra of metal carbonyl clusters, $M_n(CO)_m$, is exemplified for $(\mu_3\text{-RP})Fe_3(CO)_{10}$ (R = alkyl, p-anisyl, Et_2N), where the highest m/z peak is observed as $[M - CO]^-$, or, if metal-bound hydride ligands are present, $[M - H]^-$.[20]

3.3.2 Summary

Strengths

- May give molecular weight information *via* a pseudomolecular ion $[M + H]^+$, in cases where EI may not give a molecular ion;
- Simple mass spectra, fragmentation reduced compared to EI.

Weaknesses

- Sample must be thermally volatile and stable;
- Fragmentation pattern not informative or reproducible enough for library search;
- Results depend on reagent gas type, pressure or reaction time, and nature of sample.

Mass range

- *Low* Typically less than 1000 Da.

Mass analysis

- Any (see EI).

Popularity

- Never really found favour amongst the inorganic/organometallic community.

Tips

- Try in cases where EI provides a strong ion current but no molecular ion;
- Try APCI for coordination or organometallic compounds in cases where a soft ionisation technique such as ESI (Section 3.9) produces few ions.

Field Ionisation/Field Desorption[21]

Application of a high positive electric potential to an electrode that tapers to a sharp point results in a very high potential gradient (of the order of $10^8 - 10^{10}$ V cm^{-1}) around the regions of high curvature at the tip. A molecule encountering this field has its molecular orbitals distorted and **quantum tunneling** of an electron can occur from the molecule to the positively charged anode. This process is **field ionisation** (FI). The positive ion so formed is repelled by the positive electrode or 'emitter' (an array of very fine needles) and flies off into the mass spectrometer. The practical consequence is that the molecular ion formed in this manner is not excited and very little fragmentation occurs, giving an abundant $[M]^{\bullet +}$ peak. FI was in fact the first of the so-called 'soft' ionisation techniques, in which little energy is imparted to the molecule upon ionisation. Fragmentation of the molecular ion is greatly reduced and $[M]^{\bullet +}$ is often the base peak (most intense) in the spectrum.

Perhaps of more interest to inorganic chemists is the closely related technique of **field desorption** (FD), which is useful for studying non-volatile compounds. The sample is coated directly on the emitter by evaporation of a solution. Introduction to the high vacuum of the source and application of the field can cause desorption of intact molecular ions from the regions of high field strength. The difference between FI and FD is illustrated in Figure 3.8.

For both FI and FD, anode preparation is crucial. Typically, a thin metal wire (~ 10 μm in diameter) is heated to high temperature (>900 °C) in the presence of an intense electric field and a gas such as benzonitrile (chosen for its high dipole moment and low ionisation potential). The gas decomposes to form long, microscopically fine carbon 'whiskers' on the surface of the wire. The sharp points increase the field gradient and make the ionisation process more efficient. Unfortunately, preparation of the emitter requires a separate apparatus, growth of the carbon whiskers may take hours and the resulting

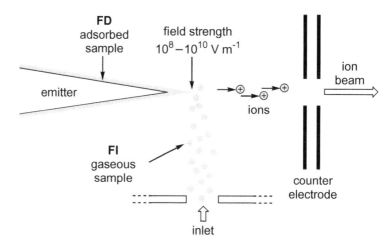

Figure 3.8
The difference between field ionisation (FI) and field desorption (FD). Tunnelling of an electron from a molecule adsorbed on (FD) or adjacent to (FI) the positively charged emitter generates a positive ion which is repelled by the emitter and shoots toward the counter electrode (potential difference $\sim 10\,000$ V) and into the mass spectrometer. Instruments are configured to be able to switch between FI and FD

emitter is rather fragile. Efforts to improve the robustness of the emitter and decrease the time taken to prepare them have included using Group 6 hexacarbonyls in the place of organic gases.[22] The metal carbonyl vapor decomposes on a specially roughened wire to form metallic microneedles. Coating of the sample on to the emitter is a problem for which numerous ingenious techniques were developed, including dipping of the emitter into a solution containing the analyte, coating the emitter using a specially mounted microsyringe, and using an electrospray technique.[23] Only a small proportion of the deposited sample can be adsorbed on the very tip of the emitter whiskers where the field gradient is sufficiently high for electron tunneling to occur, so a degree of analyte mobility is required for a sustained ion current. Generally, heating of the emitter is sufficient to allow ion production for many minutes, but care must be taken to avoid decomposition of the sample.

FD spectra of neutral compounds are generally dominated by the $[M]^{\bullet+}$ ion. Spectra of singly charged salts, usually provide $[cation]^+$ as the base peak in positive ion mode and $[anion]^-$ in the negative ion mode. Low mass salts also produce aggregates of the form $[C_nA_{n-1}]^+$ in positive ion mode and $[C_nA_{n+1}]^-$ in the negative ion mode (C = cation, A = anion). FD is not well suited to the analysis of multiply charged species but this is a problem that most ionisation techniques struggle to overcome, with the exception of electrospray (Section 3.9).

3.4.1 Summary

Strengths

- Very soft ionisation methods, produce abundant molecular ions from wide range of substances;
- Little or no chemical background;
- Works well for neutral organometallics, low molecular weight polymers.

Weaknesses

- Sensitive to sample overloading (10^{-5} g maximum);
- Not suitable for multiply charged compounds;
- Preparation of the (rather fragile) emitter is tedious and requires separate apparatus;
- High fields generally require the use of a sector instrument;
- Relatively slow analysis as the emitter current is increased;
- The sample must be thermally volatile to at least some extent to be desorbed.

Mass range

- Low-moderate, depends on the sample. Typically less than about 2000 to 3000 Da for FD, less than 1000 Da for FI.

Mass analysis

- Usually sector instruments, which are well-suited to the high fields used in FI/FD.

Popularity

- Gain wider acceptance in the 1970s but suffered a catastrophic plummet in popularity with the development of more convenient techniques such as FAB/LSIMS.

Applications/Tips

- Involatile, polar, thermally labile compounds not amenable to EI, *i.e.* the same market now essentially cornered by ESI and MALDI;
- Infrequently encountered today.

Plasma Desorption[24]

The first successful particle bombardment technique was **plasma desorption** mass spectrometry (PDMS) introduced by MacFarlane in 1974.[25] PDMS uses radioactive ^{252}Cf as the particle source, which decomposes into two particles, each about half the nuclear mass and travelling with very high kinetic energies (millions of eV). One fission fragment passes through a thin film of sample, causing the sample to desorb directly as ions (often with the help of a nitrocellulose matrix).[26] The other particle is used to trigger a time-of-flight (TOF) mass analyser (Figure 3.9).

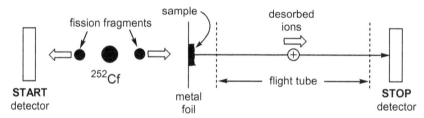

Figure 3.9
Principle of PDMS. A ^{252}Cf nucleus undergoes fission to produce two particles, one of which triggers the timing mechanism of a TOF mass analyser at the same time the other fragment crashes into a thin metal foil, upon which the analyte is mixed with a nitrocellulose matrix. The desorbed ions are accelerated then allowed to drift along the flight tube, separating according to *m/z* ratio, and are detected. Typically 50 000 desorption events are timed over a period of about five minutes to acquire a single full spectrum

Compounds with molecular masses in the tens of thousands have been successfully analysed by PDMS. Data collection is slow, however, as tens of thousands of decay acts must be registered and summed for reasonable signal-to-noise ratios to be obtained. The low ion yield and need for a radioactive source means that PDMS has been to all intents and purposes superseded by more convenient techniques, most notably MALDI (Figure 3.10).

Historically, ^{252}Cf PDMS has an important place, as it was recognised that the key to the desorption/ionisation process was the highly concentrated deposition of energy over a very short period of time. As such, it provided inspiration for the development of the later desorption techniques of SIMS, FAB and MALDI.

3.5.1 Applications

PDMS has somewhat limited effectiveness in providing molecular weight information for thermally fragile inorganic and organometallic compounds. Reasonable results have been obtained for some low molecular weight ionic complexes[28] but, in general, extensive

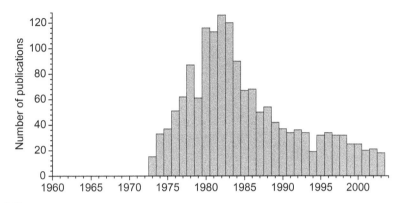

Figure 3.10
The popularity of PDMS, as measured by papers published.[27] Never very popular, the development of FAB/LSIMS in the early 1980s and MALDI in the late 1980s largely killed the technique

fragmentation and/or aggregation occurs of larger compounds. Nonetheless it is the high molecular weight compounds, such as very large metal clusters, that prove most challenging to characterise by other techniques, so these have been the most commonly studied by PDMS.

The series of trinuclear nickel clusters $[Ni_3(\mu_3-L)(\mu_3-I)(\mu_2-dppm)_2]^{n+}$ (L = I$^-$, CO, CNCH$_3$, CN-2,6-Me$_2$C$_6$H$_3$, CN^{i}C$_3$H$_7$, CN^{t}C$_4$H$_9$, CN^{n}C$_4$H$_9$ and NO$^+$, n = 0, 1, 2) were characterised by PD and FAB mass spectrometry.[29] Strong molecular ion peaks were observed for all clusters except L = NO$^+$ for both techniques. Dimers of the trinuclear clusters, $\{[Ni_3(\mu_3-I)(\mu_2-dppm)_3]_2(\mu_3:\mu_3\text{-}\eta^1:\eta^{1'}\text{-CN-R-NC})\}^{2+}$ were prepared by the reaction of two equivalents of L = I$^-$ with one equivalent of the appropriate diisocyanide, CN-R-NC and unambiguously identified by PD and FAB. The clathrochelate complexes $[NBu_4]_2[Fe(X_2Gm)_3(SnCl_3)_2]$ (X = Cl, SPh; Gm = glyoxime) were observed by PDMS in positive ion mode as $[C_3A]^+$ (C = cation, A = dianion) and in negative ion mode as $[A + H]^-$ and $[A - Cl]^-$.[30] Very different results were obtained from a study of high-nuclearity anionic platinum carbonyl clusters with Pt$_{19}$, Pt$_{24}$, Pt$_{26}$ and Pt$_{38}$ close-packed metal cores.[31] An envelope of peaks in the negative-ion spectrum corresponding to successive losses of CO from the intact metal core was observed. More extraordinary was the observation of a series of oligomeric ions formed from cluster aggregation, extending out beyond 100 000 *m/z* (in excess of 500 platinum atoms) in the case of the Pt$_{26}$ cluster (Figure 3.11).[32] A similar but less abundant positive ion spectrum was also observed. The cation associated initially with the cluster was not incorporated into the oligomers.

PDMS has also been used to study the gold phosphine clusters $[Au_6(PPh_3)_6](NO_3)_2$, $[Au_6(PPh_3)_8](NO_3)_2$, $[Au_9(PPh_3)_8](NO_3)_3$, $[Au_{11}(PPh_3)_8Cl_2]Cl$, $Au_{13}(PPh_2Me)_8I_4]I$ and $[CAu_6(PPh_3)_6](BF_4)_2$, which had been previously well-characterised by other methods. Extensive fragmentation of these clusters occurred, to the extent that it was judged impossible to assign a structure or even a formula based on the mass spectral evidence.[33] The same authors[34] examined the large gold cluster first synthesised by Schmid and formulated as $Au_{55}(PPh_3)_{12}Cl_6$,[35] and despite being unable to prove or disprove this assignment (due to detecting only extremely broad 'mass zones' rather than discrete peaks), claimed that the synthesis of the compound was not entirely reproducible and that the product was likely to be inhomogeneous.

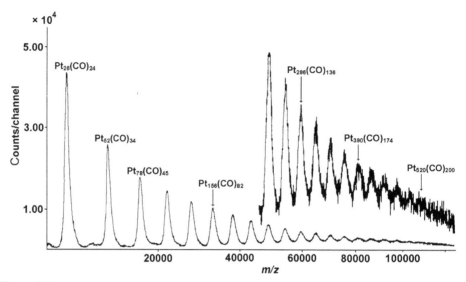

Figure 3.11
The negative-ion plasma desorption mass spectrum of $[Pt_{26}(CO)_{32}]^{2-}$. Reprinted with permission from C. J. McNeal *et al.*, *J. Am. Chem. Soc.*, 1991, **113**, 372. Copyright (1991) American Chemical Society

3.5.2 Summary

Strengths

- Soft ionisation technique, little fragmentation.

Weaknesses

- Obsolete;
- Slow;
- Requires radioactive source.

Mass analysis

- TOF.

Mass range

- *High* Beyond 100 000 Da, but for most organometallic/coordination complexes, considerably less; 50 000 a more realistic upper limit.

Applications/Tips

- Essentially rendered obsolete by the introduction of the cheaper, faster and better performing FAB/LSIMS and MALDI ionisation techniques.

Fast Atom Bombardment/Liquid Secondary Ion Mass Spectrometry[36]

Fast atom bombardment (FAB) was developed in the early 1980s.[37] The technique employs a discharged beam of fast atoms of an inert gas (usually argon or xenon) travelling at 4–10 keV to bombard the sample. If a beam of ions (usually Cs^+, energy of about 30 keV) rather than atoms is used, the technique is called **liquid secondary ion mass spectrometry** (LSIMS). Fast moving atoms are produced by colliding accelerated ions with (effectively) stationary atoms, so LSIMS is somewhat simpler than FAB, as the accelerated ions do not need to be neutralised. Positive or negative ion spectra are produced with similar facility. FAB can be complicated by redox, fragmentation and clustering processes in the study of metal complexes[38] but has become a relatively routine technique for mass spectrometric analysis of organometallic and coordination compounds, whether charged or neutral. Its popularity has suffered in recent years due to the development of ESI and MALDI (Figure 3.12).

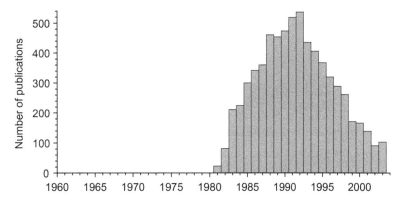

Figure 3.12
The popularity of FAB/LSIMS, as measured by papers published.[39] Popularity of the techniques peaked in the early 1990s, at the point that ESI and MALDI instruments were commercialised

The sample is dissolved in a liquid **matrix** and placed on the metal tip of a probe admitted directly to the ion source. A liquid matrix has the advantage that a fresh surface is maintained, extending the spectral lifetime and enhancing sensitivity. Like PDMS (Section 3.5), the fast-moving particles transfer much of their kinetic energy to the surface, lifting the ions into the gas phase (Figure 3.13). The process is like firing a cannonball at a water surface; most of the energy is dissipated into the liquid from the resulting shock wave but at least some material splashes into the air. As such, heavy atoms (or ions) are preferred, as their momentum, and hence impact on the surface, is greater.

The current of analyte ions can be maintained for prolonged periods, even over several hours, making the technique successful in conjunction with sector instruments scanning over a large mass range. By changing the polarity of the electric field used to extract the ions it is possible to examine positive and negative ions in separate experiments.

3.6.1 Matrices

A FAB matrix must fulfill some basic requirements: (a) low volatility, so they do not evaporate quickly in the vacuum of the mass spectrometer; (b) viscosity must be low to allow diffusion of solutes to the surface; (c) must be chemically inert; (d) must be a good solvent for the sample; (e) must assist ionisation. Some matrices that fulfill these criteria are shown in Figure 3.14.

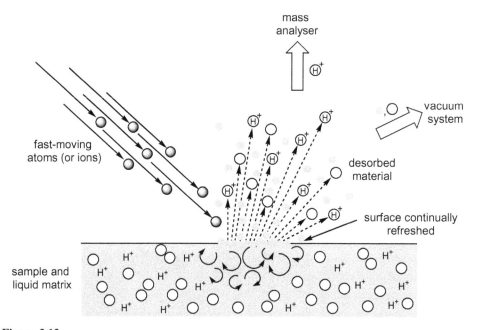

Figure 3.13
A FAB/LSIMS source. The fast moving beam of atoms/ions blasts matrix and analyte into the gas phase. The secondary ions that are mass analysed either are originally charged or acquire a positive charge from protonation (or association with another charged species such as Na$^+$) or a negative charge by deprotonation

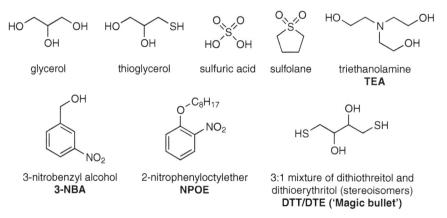

Figure 3.14
Some FAB/LSIMS liquid matrices. 3-NBA and DTT/DTE are popular for organometallic and coordination compounds; NPOE is good for non-polar samples

Matrices are chosen such that interference from ions derived from the matrix itself is minimised, but even so samples with masses of 300 Da or less are best analysed by other means. The continual bombardment of the liquid leads to fragmentation and rearrangement of the matrix itself, and this manifests itself as a background of chemical noise. The relative consistency of intensity across a wide *m/z* range leads to the apt description of these ubiquitous peaks as 'grass'. Its presence does not usually unduly complicate interpretation except in cases where the molecular ion intensity is very low. At low *m/z* values, the spectrum is often noisy, and the presence of peaks attributable to the matrix and clusters thereof can complicate assignment. Switching matrix can obviate inconvenient overlap of sample and matrix peak but in any event it is crucial to have to hand a spectrum of the matrix itself (Figure 3.15 shows an example) to assist assignment.

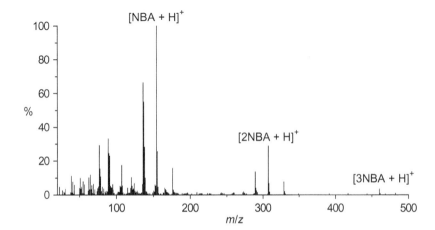

Figure 3.15 The positive-ion LSIMS mass spectrum of 3-NBA

Lots of other matrices have been used with success, seemingly every liquid of low volatility – examples include HMPA, DMSO, crown ethers, liquid paraffin and even pump oils.[40] Neat liquids work well and even slightly impure oils. Ionisation is frequently assisted by the addition of acid (or base), or to promote cationisation (anionisation), alkali metal (halide) salts.

3.6.2 Ions Observed in FAB/LSIMS

Ion formation in FAB or LSIMS arises from charge exchange between the fast-moving incident particles (**A**toms or **I**ons) and the neutral molecules (**M**) in solution, generating $M^{+/-}$ and $A^{+/-}$ or M^+ and **I**. Very frequently, the pseudomolecular ion $[MH]^+$ is observed, arising from deprotonation of the solvent; similarly, adduction with alkali metals ions to provide $[M + Na]^+$, $[M + K]^+$ etc. ions is commonly observed. Ionic compounds are perfectly amenable to the technique and indeed FAB/LSIMS provided one of the first routine means of accessing such compounds mass spectrometrically.

The exact types of (non-fragment) ions observed in FAB/LSIMS depend critically on the nature of the analyte, especially its polarity (Table 3.1). Ionic compounds are transferred into the gas-phase reasonably easily, provided the charge is not more than

Table 3.1 Types of (non-fragment) ions formed in FAB/LSIMS by sample type and ion mode

	Positive ion mode		Negative ion mode	
Sample type	Major ions	Minor ions	Major ions	Minor ions
Ionic, $[C]^+[A]^-$	$[C]^+$, $[nC + (n-1)A]^+$	$[C + A]^{\bullet+}$	$[A]^-$, $[(n-1)C + nA]^-$	$[C + A]^{\bullet-}$
Ionic, $[C]^{2+}[A]^-$	$[C]^+$, $[C + A]^+$	$[C]^{2+}$	$[A]^-$	
Ionic, $[C]^+[A]^{2-}$	$[C]^+$		$[A]^-$, $[C + A]^-$	$[A]^{2-}$
Polar	$[M + H]^+$, $[M + C]^+$	$[M]^{\bullet+}$	$[M - H]^-$	$[M + A]^-$, $[M]^{\bullet-}$
Non-polar	$[M]^{\bullet+}$	$[M + H]^+$	$[M]^{\bullet-}$	$[M - H]^-$

M = molecule, $[M]^{\bullet+}$ = molecular ion, $[M + H]^+$ = protonated molecule, $[M + C]^+$ = pseudomolecular ion generated by association with a cation such as Na^+ or K^+; $[M - H]^-$ = deprotonated molecule, $[M + A]^-$ = pseudomolecular ion generated by association with an anion such as Cl^- or Br^-; C = cation; A = anion.

±2. Doubly charged compounds are often observed reduced or oxidised to the ±1 state, or associate with a singly charged counterion to provide a $[C + A]^{\pm}$ ion (where the charge on one ion is one more than on the other). Neutral compounds acquire charge in the positive ion mode most commonly through association with a proton (protonation) or a cation such as Na^+ or K^+, though the compound must be reasonably polar (contain basic groups on the periphery of the molecule) in order for these to be efficient ionisation routes. Similarly, the negative-ion mode displays $[M - H]^-$ ions, derived from molecules that have acidic hydrogens (especially —OH groups). Anionisation – association of molecule with an anion such as Cl^- or Br^- – is occasionally observed. Cationisation/ anionisation can be promoted by addition of an appropriate salt to the matrix.[41] Non-polar compounds can form radical ions, $[M]^{\bullet\pm}$, but often other ionisation techniques are a better bet for such examples – most FAB/LSIMS instruments are also capable of EI, which is well-suited to non-polar compounds.

Note that all ions may be observed in association with a molecule of matrix, e.g. $[M + H + matrix]^+$, so this possibility should be borne in mind when examining the highest *m/z* peak in the spectrum.

FAB/LSIMS are relatively soft ionisation techniques, so the molecular ion species is quite abundant, with limited fragmentation (though direct comparisons of various ionisation techniques with organometallic compounds show FAB/LSIMS to be appreciably harder than FD or ESI).[42] The presence of even-electron species and limited fragmentation in FAB/LSIMS spectra means that the techniques bears resemblance to chemical ionisation; most fragment ions can be rationalised by elimination of a neutral molecule from the molecular ion. The most common fragment ions are due to loss of a neutral monodentate donor ligand, such as H_2O, NH_3, PR_3 or CO. Complexes with halide ligands (X) often lose HX, concomitant with oxidative addition of a carbon-hydrogen bond of a ligand to the metal center, or alternatively, formation of an $[M - X]^+$ ion is a common ionisation route. Polydentate ligands are much more difficult to remove than their monodentate cousins, and the higher the denticity the more likely it is that fragments of ligand will be removed rather than the intact ligand itself.

3.6.3 Applications[43]

The following gives a selection of examples of coordination and organometallic compounds that have been characterised by the LSIMS/FAB techniques. It is not intended to be all-inclusive, but instead to give a flavour of the range and type of compounds that have been studied, and the information that these techniques provide. In some cases, reference is made to the ESI mass spectra of the same compounds; the comparison is often dramatic.

Ionic complexes

FAB of 18 involatile, cationic organometallic derivatives of bipy complexes of Ru(II) and Os(II) established fragmentation patterns characteristic of the ligand set.[44] Loss of monodentate neutral ligands such as η^2-alkene and CO was the main fragmentation pathway, along with HX elimination. The hydrogen is removed from the bipy ligand to produce a cyclometallated product. Cationic polypyridyl ruthenium complexes behaved in a similar way, with the additional observation of transfer of a fluorine from a PF_6^- counterion to the metal complex.[45] Doubly charged ions are not often observed, but can be for $[RuL_3]X_2$ (L = bipy, phen, and related ligands), and NBA is the preferred matrix (glycerol provided only monocationic species).[46] The LSIMS mass spectrum of $[Ru(bipy)_3]Cl_2$ is shown in Figure 3.16, and fragment ions include prominent reduced species $[Ru(bipy)_x]^+$ (x = 1 – 3) as well as singly charged aggregates of dication and a single monoanion, $[Ru(bipy)_x + Cl]^+$ (x = 1 – 3). For comparison, the ESI mass spectrum of $[Ru(bipy)_3]Cl_2$ is shown in Figure 4.3, which shows a single $[Ru(bipy)_3]^{2+}$ ion with no fragmentation. This clearly demonstrates the gentle nature of ESI compared to LSIMS, and hence explains the increasing popularity of the ESI technique.

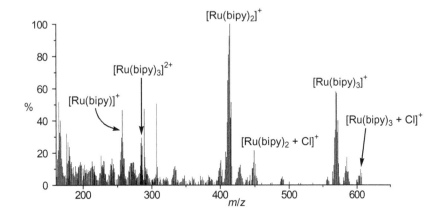

Figure 3.16
Positive ion
LSIMS spectrum
of $[Ru(bipy)_3]Cl_2$
(3-NBA matrix)

Twelve cationic organometallic palladium complexes with terdentate nitrogen ligands were examined using FAB MS.[47] The primary ions formed in the positive ion mode were those corresponding to the intact cation. MS/MS experiments provided evidence for an abundant gas-phase cyclopalladation reaction. A series of cyanometallate complexes $[M^{n+}(CN)_m]^{(m-n)-}$ (M = Fe, Cr, Mn, Ni, Mo or W in various oxidation states) produced several families of ions from each complex in both positive and negative-ion modes, differing from one another by the number of cations associated with the complex ion.[48] Each family consists of members that result from loss of neutral CN (or possibly C_2N_2) groups from the metal complex.

Neutral complexes

Neutral complexes must acquire a charge somehow during the ionisation process, as previously discussed. This is usually via protonation or cationisation. However, in the absence of any polar sites on the molecule, other ionisation routes can be operative the example of a square-planar Pt(II) complex, in which no pseudomolecular ions are observed but where the primary ionisation route is via loss of iodide to give a series of peaks due to $[M - I]^+$ ions and fragments thereof, is shown in Figure 3.17. The methyl

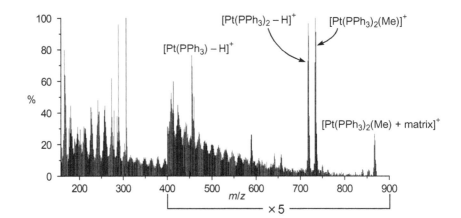

Figure 3.17 Positive ion LSIMS spectrum of Pt(PPh$_3$)$_2$MeI (3-NBA matrix)

group is easily eliminated as CH_4, the hydrogen coming from the phenyl ring of a phosphine and resulting in a cyclometallated complex. Loss of neutral PPh$_3$ is also a major fragmentation pathway. The high noise levels and abundant low-mass peaks due to matrix ions are a typical feature of FAB/LSIMS spectra. Again, the ESI mass spectrum of this complex is much simpler (Figure 4.8), with a high intensity parent-derived $[Pt(PPh_3)_2Me]^+$ ion, though the overall fragmentation pathway, by loss of CH_4, is similar.

Spectra can be quite complex and rearrangement can pose real problems to successful assignment of spectra. A classic example is shown in Figure 3.18, the LSIMS spectrum of Auranofin, a well-known gold-containing drug. The $[M + H]^+$ ion is present, but at very low abundance, and is dwarfed but other ions containing one, two and even three gold atoms.

Formation of $[Au_n]^+$ clusters in the FAB mass spectra of linear Au(I) complexes has been observed and claimed to be indicative of significant intermolecular interactions,[49] but these are not observed in the Auranofin example due to steric crowding about the metal center. Abundant aggregates of the form $[M_a(C{\equiv}CPh)_b(PPh_3)_c]^+$ (M = copper, silver, gold; $4 \geq a \geq b \geq c$) were observed in the FAB mass spectra of the neutral compounds M(C$\equiv$CPh)(PPh$_3$) (M = copper, silver, gold).[50]

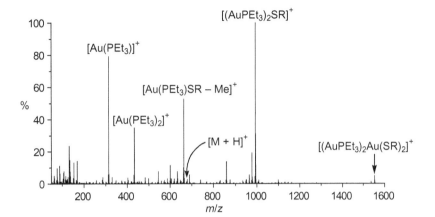

Figure 3.18
Positive-ion
LSIMS spectrum
of Auranofin,
$(Et_3P)Au$
$(SC_{14}H_{19}O_9)$
(3-NBA matrix)

LSIMS provided molecular formula and structural information for diruthenium(II,III) compounds.[51] $Ru_2X(\mu\text{-}O_2CR)_4$ (X = Cl, SCN) show $[M]^+$ or $[M + H]^+$ ions, whereas $Ru_2X(\mu\text{-}O_2CR)_4(OPPh_3)$ (X = Cl, SCN) have no detectable (pseudo)molecular ion but the detected fragment confirmed the proposed stoichiometries. Cluster ions formed from ion/molecule reactions were also detected, of the form $[2M - X]^+$. The neutral rhodium complexes $Rh(PPh_3)_2(CO)Y$ (Y = F, Cl, I, NCO, NCS, OAc, ONO_2, O_2PF_2 or OSO_2CF_3) in a sulfolane matrix provided a detectable $[M]^{\bullet+}$ molecular ion in all cases except for the most weakly coordinating triflate ligand.[52] Fragment ions of high intensity included $[M - CO]^{\bullet+}$, $[M - Y]^+$, $[M - Y - CO]^+$, $[M - Y - PPh_3]^+$ and $[M - Y - PPh_3 - Ph + H]^+$. No ions were observed in the negative-ion spectrum apart from those deriving from the matrix.

A cationic bis(carbene)gold(III) complex, $[Au(CRR')_2I_2]^+$, showed unusual behaviour in that the $[M]^+$ ion was not observed in the mass spectrum when sulfolane was used as the matrix, but rather the odd-electron species $[M - H]^{\bullet+}$.[53]

Clusters

High-nuclearity clusters have been examined successfully using FAB mass spectrometry, such as the decanuclear clusters $[Ru_{10}C_2(CO)_{22}(L)]^{2-}$ and $[Ru_{10}C_2(CO)_{23}(L)]$ (L = diphenylacetylene or norbornadiene),[54] or the mixed-metal clusters $[Re_6Ir\text{-}(\mu_6\text{-}C)(CO)_{20}(AuPPh_3)_{3-n}]^{n-}$ (n = 1, 2, 3).[55] The large transition-metal-gold clusters $[Au_2Pt(PPh_3)_4NO_3]NO_3$, $[Au_6Pt(PPh_3)_7](BPh_4)_2$, $[Au_2Re_2H_6(PPh_3)_6]PF_6$, $[Au_4\text{-}ReH_4\{P(p\text{-}Tol)_3\}_2(PPh_3)_4]PF_6$ and other similar cationic and dicationic clusters of mass 1000–4000 Da were analysed by FAB;[56] these gave well-resolved peaks for either the parent molecular ion $[M]^+$ for the cationic clusters or the ion pair $[M + X]^+$ (where X = counterion) for the dicationic clusters. Fragments were also observed, corresponding to losses of PPh_3, H, CO, Ph and $AuPPh_3$. A positive-ion FAB study of the cluster compounds $[Au_9M_4Cl_4(PPh_2Me)_8][C_2B_9H_{12}]$ (M = gold, silver, copper) showed strong $[M]^+$ molecular ions as well as fragment peaks due to losses of primarily PPh_2Me but also of gold and chlorine.[57] In another study, a range of high-nuclearity anionic ruthenium clusters were examined in negative-ion mode.[58] The anionic clusters gave intact anion peaks $[M]^-$, and all of the dianionic clusters gave protonated monoanions $[M + H]^-$

peaks. Degradation of the cluster metal frameworks of $[HRu_{10}C(CO)_{24}]^-$ and $[Ru_{10}C(CO)_{24}]^{2-}$ into $[HRu_6C(CO)_{16}]^-$ was observed; this degradation process parallels their reactivity in solution. The clusters $Fe_3(CO)_9(\mu_3\text{-}SR)(\mu\text{-}AuPPh_3)$ (R = CHMe_2, CMe_3, C_6H_{11}, Et) fragmented during the FAB process by stepwise loss of CO from the molecular ion $[M]^{\bullet+}$, followed by olefin elimination to give $[Fe_3(SH)(AuPPh_3)]^{\bullet+}$ and finally loss of H$^\bullet$ to afford $[Fe_3(S)(AuPPh_3)]^{+}$.[59]

The positive-ion FAB mass spectrum of $Pt_5(CNC_8H_9)_{10}$ shows a large number of peaks (Figure 3.19).[60] The highest mass peak, A, at $m/z = 2153$, is due to $[M - CNC_8H_9]^+$ and the most intense peak, I, is due to $[M - 7CNC_8H_9]^+$. The remaining peaks, labelled B to S in the figure, are readily assigned to losses of platinum and/or CNC_8H_9 ligands. The spectrum of the related compound $Pt(\mu\text{-}SO_2)_2(CNC_8H_9)_8$ (Figure 3.19b) displays no sign of peaks retaining the SO_2 ligands, and $[M - 2SO_2]^+$ is the highest mass peak (A, $m/z = 2026$) with the remaining peaks due to losses of the CNC_8H_9 ligands $[M - 2SO_2 - xCNC_8H_9]^+$ (x = 1 − 5, B − F).

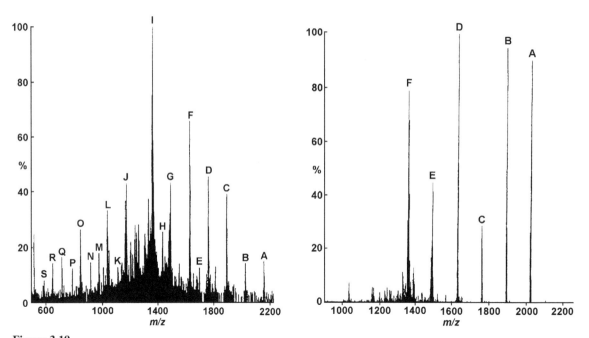

Figure 3.19
The FAB mass spectra of (a) $Pt_5(CNC_8H_9)_{10}$ and (b) $Pt(\mu\text{-}SO_2)_2(CNC_8H_9)_8$. Reprinted with permission by The Royal Society of Chemistry from J. L. Haggitt and D. M. P. Mingos, *J. Chem. Soc., Dalton Trans.*, 1994, 1013

Halide clusters have also been characterised using FAB mass spectrometry. The clusters $[Mo_6Br_8X_6]^{2-}$ (X = CF_3CO_2, SCN, NCO, Cl, Br, I) gave negative-ion spectra in which the highest mass envelopes corresponded to the parent ion paired with one cation, the parent anion, and the parent anion minus one ligand.[61] A very detailed study of iron-sulfur cubane clusters of the type $[ER_4]_2[Fe_4S_4X_4]$ (E = nitrogen or phosphorus, X = Cl, Br, SR) with NBA or NPOE as matrix provided characteristic spectra in the negative-ion mode, dominant high-mass ions corresponding to the intact, oxidised metal core $[Fe_4S_4X_4]^-$ and the dianion associated with a single cation, $[Fe_4S_4X_4 + ER_4]^-$.[62]

Ionic molybdenum-sulfur clusters were analysed with similar facility.[63] Highly charged polyoxometallate clusters such as $[NBu_4]_9[P_2W_{15}Nb_3O_{62}]$ were shown to give the best molecular weight information in the positive-ion mode, with the highest m/z peak corresponding to $[9C + A + H]^+$.[64] The negative-ion mode gave a peak corresponding to $[7C + A + H]^-$, with charge information actually retained (the charge on the anion A of 9^- balances the negative charge on the ion plus seven cations C and one proton), and the spectra provided a richer variety of, and more intense, fragment ions. Importantly, FAB was shown to be a reliable means of correctly formulating the number of oxygen atoms in the heteropolyanion core.

In situ formation of metal complexes

There is a small but notable volume of literature dealing with the observation of new metal complexes forming during the FAB/LSIMS process from precursors added to the matrix. For example, a free base porphyrin (H_2P) and a metal salt (Mg(II), Fe(II), Fe(III), Co(II), Ni(II), Cu(II), Zn(II), Cd(II) and Pb(II)) were subjected to FAB analysis, and porphyrin-metal adduct ions were observed in most cases, excepting Mg(II), Fe(II), Fe(III) and Ni(II).[65] Similarly, the preformed complexes $[MCl(CO)_2(\eta^3\text{-allyl})L_2]$ and $[M(CO)_2(\eta^3\text{-allyl})L_3]X$ (M = molybdenum, tungsten; X = BF_4, PF_6; L = 2-electron donor ligand) exhibit the same fragmentation patterns as spectra produced from a mixture of $[MCl(CO)_2(\eta^3\text{-allyl})(NCMe)_2]$ and the appropriate ligand (L) in glycerol.[66] Without addition of L, the main metal-containing ion observed was $[M(CO)_2(\eta^3\text{-allyl})\text{-(glycerol)}]^+$. Reaction mixtures consisting of an aryl or heterocyclic mercuric acetate, a glycal (enol ether) and $Pd(OAc)_2$ in acetonitrile (MeCN) dissolved in a glycerol/AcOH or triethanolamine matrix were examined by FAB.[67] The mass spectra showed ions assignable to an aryl (heterocyclic) palladium transmetallation product, as well as adducts formed by the addition of aryl- (heterocyclic-) palladium across the enol ether C=C bond, and these species correlate closely with species postulated on the basis of studies of the mechanisms of these reactions.

3.6.4 Summary

Strengths

- Fast and simple;
- Low temperature so thermally sensitive samples can be studied;
- Relatively tolerant of variations in sampling;
- Good for a large variety of compounds;
- Strong ion currents, so good for high-resolution measurements.

Weaknesses

- High chemical background noise;
- Lower m/z range dominated by matrix and matrix/salt cluster ions;
- Analyte must be soluble and stable in the liquid matrix.

Mass analysis

- Typically sector instruments.

Mass range

- *Effective* 200–2000 *m/z*;
- 2000–5000 *m/z*, molecular ions difficult to spot against background;
- Below 200 *m/z*, solvent peaks and chemical noise significant.

Applications

- Polar, involatile compounds of moderate molecular weight;
- Accurate mass information (due to good compatability with sector instruments).

Tips

- The easiest parameter to change in a FAB/LSIMS experiment is the matrix. If interested in a certain class of compound, it is important to screen a 'typical' complex, well-characterised by other means, with range of matrices to find out which works best for that particular type of sample.
- Ensure your sample doesn't react with the matrix. Protic matrices especially may cause degradation and thiol-containing matrices are likely to react with many transition metal complexes.
- It is often difficult to prevent exposure of air-sensitive compounds to the atmosphere in FAB/LSIMS, as the sample *and* matrix need to be placed on a probe before insertion into the mass spectrometer. Generally, it is best to choose the least hygroscopic matrix possible, and store it over a suitable drying agent and under an inert atmosphere. Samples can be made up with matrix easily enough using Schlenk or glovebox handling, and the difficulty is transfer to the mass spectrometer. A strong stream of nitrogen or argon can be employed to bath the sample whilst the transfer is in progress, best done by two people. A nitrogen-flushed plastic glove bag placed over source and sample can also be effective. Samples that are *very* air-sensitive will generally require handling using specially built apparatus.

Matrix Assisted Laser Desorption Ionisation

Laser desorption ionisation (LDI) was first introduced in the 1960s, but it was not until the development of **matrix-assisted** LDI (MALDI) in the mid 1980s and the subsequent commercialisation of instruments based on the new technology that the technique gained real popularity. Tanaka first demonstrated the analysis of high molecular weight biomolecules using a matrix,[68] a discovery which won him a share of the 2002 Nobel Prize for Chemistry.[†] The same year also saw Karas, Hillenkamp and co-workers publish the first[69] of many papers on the use of organic matrices in MALDI,[70] a development that has propelled the technique to enormous success (Figure 3.20).

[†]'For their development of soft desorption ionisation methods for mass spectrometric analyses of biological macromolecules', shared with John Fenn (who developed electrospray as an ionisation technique). See *http://www.nobel.se/chemistry/laureates/2002/*. At the time of the award, Tanaka was working in relative obscurity having made essentially a one-off contribution, whereas Karas and Hillenkamp were widely acknowledged as the leaders in the field, their organic matrices having proved more lastingly useful. As a result, the decision was controversial in the mass spectrometry community, but the story is a classic example of how the Nobel prize is awarded for 'changing the way people think', rather than for a body of work, no matter how important or accomplished.

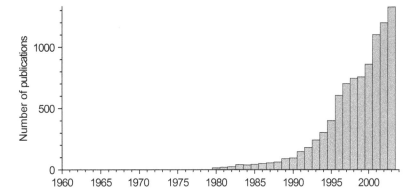

Figure 3.20
The explosive
growth in
popularity of
MALDI[71]

The explosive growth in popularity of MALDI is second only to the parallel phenomenon of electrospray ionisation (Section 3.9). The widespread adoption of MALDI instruments has been driven largely by the success of the technique in studying biological macromolecules and, to a somewhat lesser extent, in polymer chemistry and in routine characterisation of polar organic molecules. MALDI is yet to make the same sort of impact in coordination and organometallic chemistry but applications are beginning to emerge and the popularity of the technique is likely to increase amongst this audience, especially if new matrices suitable for the analysis of compounds sensitive to moisture are developed.

Laser desorption methods ablate material from a surface using a pulsed laser, typically nitrogen lasers (337 nm wavelength) or frequency-tripled Nd:YAG lasers (355 nm), with pulse widths of 5 ns or less. Pulsed lasers can deliver a large density of energy into a small space in a very short period of time, of the order of 10^6–10^8 W cm^{-2}. The energy density is much higher than can be achieved with a continuous laser and is sufficient to desorb and ionise molecules from a solid surface. However, it does require a mass analyser compatible with pulsed ionisation methods, most commonly a time-of-flight (TOF) instrument (Section 2.5). Direct laser desorption/ionisation (LDI) relies on very rapid heating of the sample or sample substrate to vaporise molecules so quickly that they do not have time to decompose. LDI is good for low- to medium-molecular weight compounds and surface analysis.

The LDI technique really took off with advances in both technology and methodology. Technological advances have included more sophisticated lasers and a raft of improvements in TOF design (reflectron, delayed extraction, etc.; Section 2.5.1) as well as better timing electronics. The big improvement in methodology came with the introduction of matrix-assisted laser desorption/ionisation (MALDI) mass spectrometry. MALDI involves the use of a solid matrix that is co-crystallised along with the sample. A dense gas cloud is formed upon irradiation, which expands supersonically (at about Mach 3) into the vacuum. It is believed that analyte ionisation occurs in the expanding plume as the direct result of interactions between analyte neutrals, excited matrix radicals/ions and protons and cations such as sodium and silver. A variety of chemical and physical pathways have been suggested for MALDI ion formation, and interested readers are directed to an informative review.[72] The inherent 'soft' ionisation of MALDI

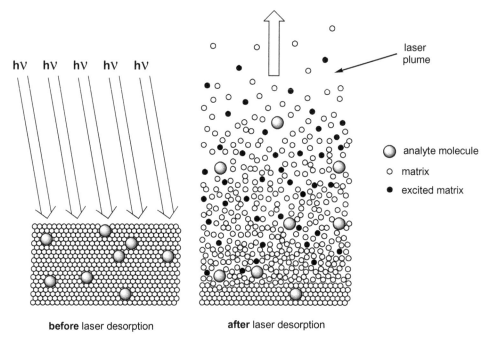

Figure 3.21
Analyte molecules are co-crystallised with the matrix (left). Irradiation of the solid matrix/analyte with a laser beam causes absorption of light by the matrix and an explosive phase transition occurs in which a jet of matrix neutrals blasts into the gas phase, entraining the (largely undisturbed) analyte molecules (right). The soft ionisation is a result of the large excess of matrix (relative to analyte) having a high molar absorptivity coefficient and therefore absorbing most of the energy imparted by the laser

has proved to be tremendously effective for determining the molecular weight of high-mass molecules such as polymers or biomolecules such as proteins, as such molecules are lifted intact into the gas phase along with the matrix (Figure 3.21).

The major advantage of (MA)LDI is that the ionisation process produces primarily singly charged ions, so even quite complex mixtures can be examined. This is true even for compounds that have a high nominal charge and in the analysis of coordination and organometallic compounds it is crucial to realise that some charge information is lost during the MALDI process. The sign of the charge is retained – positively charged complexes will always produce monocations, and negatively charged complexes mono-anions – but the magnitude is always ±1, regardless of whether the original species was singly or multiply charged. This phenomenon is due to the high probability of neutralisation for multiply charged ions by the surfeit of electrons present in the plume from matrix photoionisation.[73] The plume carries a net positive charge excess due to rapid loss of electrons, so complete neutralisation is not a problem. This also explains why negative ion spectra are usually less intense than seen in positive-ion mode.

Atmospheric pressure (AP) MALDI is a relatively recent innovation in which the ion source is operated outside of the vacuum and the ions are transported by quadrupoles or hexapoles through differentially pumped regions to the mass analyser (usually an ion trap or orthogonal TOF).[74] Development has been driven by advantages such as interchange-ability with ESI sources and the ability to carry out MS^n studies (with ion traps).

However, AP MALDI is restricted in mass range to the upper mass limit of the analyser itself.

3.7.1 Matrices

MALDI matrices must fulfill a certain set of requirements; they must: (i) absorb light at the laser wavelength; (ii) promote ionisation of the analyte; (iii) be soluble in a solvent common to the analyte; and (iv) co-crystallise with the sample. Most matrices are organic acids that have a strong UV chromophore, typically an aromatic ring (Figure 3.22).

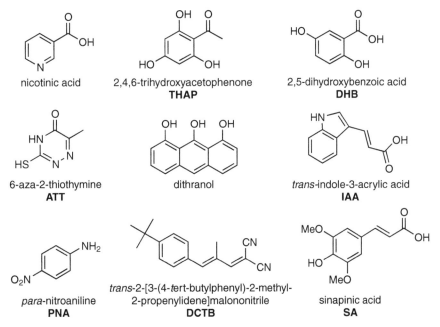

nicotinic acid

2,4,6-trihydroxyacetophenone
THAP

2,5-dihydroxybenzoic acid
DHB

6-aza-2-thiothymine
ATT

dithranol

trans-indole-3-acrylic acid
IAA

para-nitroaniline
PNA

trans-2-[3-(4-*tert*-butylphenyl)-2-methyl-
2-propenylidene]malononitrile
DCTB

sinapinic acid
SA

Figure 3.22
Some of the organic molecules used as MALDI matrices. The neutral (ATT, dithranol, DCTB) or basic (PNA) matrices are usually favoured in the analysis of organometallic/coordination compounds

Because the co-crystallised matrix/sample mixture is not homogeneous by nature, there is an inherent low shot-to-shot reproducibility in the technique. To overcome this, instruments have the facility to fire anywhere on the sample and some are fitted with cameras to aid in the search for the 'sweet spot' on the slide. Sample preparation is crucial to the successful MALDI experiment and a great variety of methods have been developed. A fast and simple method involves placing a drop of sample solution on the slide and a drop of matrix solution (usually at least 100 × more concentrated) is added on top. The solutions are mixed using the dropper tip and dried quickly under a stream of nitrogen. Other methods of sample preparation are more elaborate and include steps such as vacuum drying, allowing the drop of analyte solution to evaporate on top of a dried droplet of matrix, various layering procedures, spin-coating of premixed sample solutions, or even electrospraying solutions on to the target slide.[75] All strive for the creation of as homogeneous and intimate a layer of analyte and matrix microcrystals as possible. However, the advantages for the coordination/organometallic chemist of the more elaborate techniques are likely to be minimal – generally, there is no shortage of sample, nor are the compounds of interest often of very high molecular weight, and the presence of buffers is rarely an problem. An exception to this rule is possibly a solventless method used for the characterisation of insoluble polymers – a 1:1 mixture of matrix and analyte are ground into a fine powder and hydraulically pressed into a thin flat disc (just as in making a KBr disc for infrared spectroscopic analysis).[76] The disc is attached to a MALDI slide using double-sided tape and analysed as normal. Straight analysis of the powder also appears to work well.[77] This approach appears well-suited to coordination/ organometallic compounds that have limited solubility and thus limited options for characterisation.

The acidic functionality present on most of the matrices provides plenty of protons to assist in formation of quasimolecular ions, $[M + H]^+$, and the strong chromophore means the matrix absorbs most of the energy of the laser, protecting the analyte molecules from radiation damage and hence minimising fragmentation. However, the acidity can be destructive to compounds sensitive to protons, so many of these matrices (designed primarily for the analysis of proteins/peptides, oligonucleotides, polymers etc.) are unsuitable. If the target compound has no basic sites, is already charged or decomposes in acidic environments, a neutral matrix such as dithranol, DCTB or ATT is probably be the best place to start. For observation of non-covalent complexes, basic matrices such as PNA are best to avoid denaturing. Development of specialist matrices for the coordination/organometallic chemist has been unforthcoming, mostly due to a lack of awareness and hence demand amongst this community, but it seems likely that this situation will be improved upon in the coming years. The recent introduction of solid supports that play the part of the matrix is a promising development, desorption-ionisation on porous silicon (DIOS) being particularly notable.[78] The carefully fabricated[79] surface contains tiny pores that trap analyte molecules placed on it. Hydrosilation stabilises the porous silicon surface, enabling it to be used repeatedly with little degradation, and also makes it possible to tailor the surface for particular types of samples. DIOS may prove to be a popular approach for low (<3000 Da) molecular weight samples, though it has yet to be applied to coordination/organometallic compounds.

3.7.2 MALDI of Air-Sensitive Samples

A fundamental difficulty with conducting MALDI experiments of air-sensitive compounds is that the vast majority of systems have no facility for protecting the sample from the air while introducing it to the instrument. Because such a tiny quantity of material is generally used, any exposure to air is often fatal to the sample. The matrix must be kept rigorously dry and free of water. A basic precaution is to prepare MALDI slides in a glovebox and coat the sample spot with a final layer of matrix solution for protection. The sample may be taken to the instrument under an inert atmosphere and transferred quickly under a stream of inert gas, but this procedure is awkward and still leaves the possibility of decomposition. Use of a particularly large blob of sample plus matrix may help, as repeated laser shots at the same point on the target may remove the decomposed surface material and reveal fresh sample beneath. A crystal of the compound to be studied can be affixed directly to the slide with a piece of double-sided sticky tape – crystals are usually much more robust than powdered samples but the advantage of using a matrix is then lost (and many inorganic chemists would rightly argue that if there are crystals, there are plenty of other options for analysis!). The most elegant solution to the problem is to

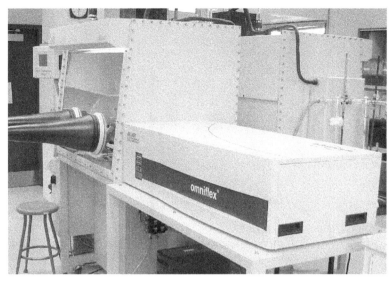

Figure 3.23
MALDI-TOF mass spectrometer coupled directly to an inert-atmosphere glovebox for the analysis of air- and moisture-sensitive compounds. Photo taken in the laboratory of Prof. Deryn Fogg, University of Ottawa

connect a MALDI instrument directly to a glovebox, an approach introduced by Fogg (Figure 3.23).[80] The interface takes up a relatively small proportion of the box and this approach entirely eliminates the danger of decomposition of the sample prior to analysis.

3.7.3 Applications

Complexes of biomolecules

The great success of MALDI in the characterisation of peptides, proteins and (oligo)nucleotides has led to a natural extension to compounds in which these biomolecules are coordinated to metal centres. Generally, the same matrices and conditions are used for the metal complexes as were successfully employed for the 'ligands'. MALDI analysis of complexes of transition metals with peptides and proteins is complicated by the difficulty in distinguishing between specific and non-specific adducts. The spectrum is usually dominated by the metal-free, protonated adduct unless a large excess of the transition metal ion is present, in which case non-specific adducts are also observed. Enhancement of the signal from the specific adduct can be achieved in one of several ways: (i) use of a non-acidic matrix such as *para*-nitroaniline, which reduces the intensity of the protonated adduct; (ii) removal of excess metal ions by precipitation of the hydroxide at high pH; (iii) addition of a competing complexing agent, such as a cyclodextrin (Figure 3.24).[81]

The complexation of a variety of different metals to peptides have been studied with some success. For example, platinum adducts from the reaction of $[Pt(en)]^{2+}$, $[Pt(dien)]^{2+}$ or cisplatin with polypeptides were characterised using positive-ion MALDI.[82] Similarly, dipeptide complexes of Cu(II), Ni(II) and Zn(II) were investigated and MALDI used to assist in the identification of a trinuclear copper complex.[83]

Lanthanide complexes decorated with ligands that carry an activated ester group to enable protein functionalisation at lysine residues were shown to react with the model

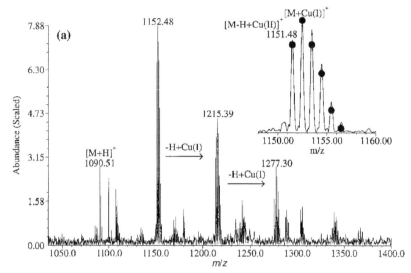

Figure 3.24

Positive ion MALDI-FTICR mass spectrum of a peptide-copper complex recorded using *p*-nitroaniline as matrix. The calculated isotope pattern is for a 40 % contribution from $[M + Cu(I)]^+$, and 60 % from $[M - H + Cu(II)]^+$. Increasing the pH or adding γ-cyclodextrin suppressed the ions with multiple Cu(I) adducts. Reproduced by permission of John Wiley & Sons

protein bovine serum albumin to produce a shift in mass corresponding to addition of an average of eight lanthanide complexes (Figure 3.25) per protein molecule.[84]

The negative-ion mode was shown to be useful in monitoring the transformation of the modified amino acid in *trans*-$[Pt(CF_3)\{Z\text{-}Ala(CN)OCH_3\}(PPh_3)_2][BF_4]$ to an oxazoline ligand.[85] The appearance of a negative ion in this case is particularly surprising because the original complex is cationic. However, this observation is not unique – the products of photoreactive Ru^{2+} complexes with oligodeoxynucleotides (DNA fragments) have been characterised using the negative ion mode.[86]

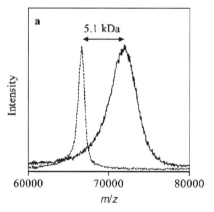

Figure 3.25

Positive ion MALDI-TOF mass spectrum of a bovine serum albumin (dotted line) and labelled with ca. 8 Tb^{3+} complexes (solid line). Reprinted with permission from N. Weibel et al, *J. Am. Chem. Soc.*, 2004, **126**, 4888. Copyright (2004) American Chemical Society

Carboranes

Carboranes are cage-shaped molecules in which (primarily) boron and carbon form the polyhedral skeleton, and each may be substituted on the outside of the cage by a proton, halogen, alkyl group, etc. Frequently, these are anionic and negative-ion MALDI has been used with success in obtaining molecular weight information on such compounds. For example, the carboranes $[1\text{-}H\text{-}CB_{11}X_{11}]^-$, $[1\text{-}Me\text{-}CB_{11}X_{11}]^-$, $[1\text{-}H\text{-}CB_{11}X_{11}]^-$, $[1\text{-}H\text{-}CB_9X_9]^-$, $[1\text{-}NH_2\text{-}CB_9X_9]^-$, $[1\text{-}H\text{-}CB_9Me_9]^-$ and $[1\text{-}H\text{-}CB_{11}Y_5X_6]^-$ (X, Y = Cl, Br, I) all provide $[M]^-$ ions and the complex isotope envelope for each greatly aids confident assignment. MALDI was used to monitor the extent of halogenation of these carboranes, and to analyse the resulting mixtures of polyhalogenocarboranes.[87] $[Co(1,2\text{-}C_2B_9H_{11})_2]^-$ and thioether derivatives thereof have been similarly characterised.[88]

Supramolecular chemistry

Chemists have made use of MALDI to characterise supramolecular assemblies, a task made inherently difficult by the relative weakness of the bonding involved compared to covalently bonded molecules. Particularly useful for such characterisation is the addition of Ag^+, which interacts with aromatic rings and unsaturated bonds (both common in supramolecular chemistry) to provide $[M + Ag]^+$ ions.[89] If the supramolecular assembly includes cationic metal centres, even this precaution may not be necessary, as positive ions can be generated by loss of an anion, $[M - A]^+$. Such an example[90] is shown in Figure 3.26.

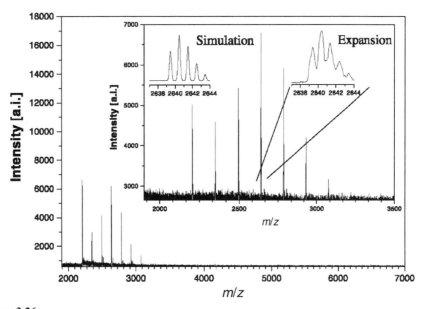

Figure 3.26
Positive ion MALDI-TOF mass spectrum of a $[2 \times 2]$ Co(II) metal coordination array, dithranol matrix. The peaks in the spectrum correspond to $[M - nPF_6]^+$ ($n = 8 - 3$). Reproduced by permission of Kluwer Academic Publishers

All ions are singly charged, even when eight anionic counterions are removed. Loss of charge information upon ionisation is a feature of MALDI spectra, presumably due to the strong flux of free electrons present in the energetic laser-generated plume. Self-assembled tetranuclear copper and nickel complexes[91] and β-cyclodextrin based supra-molecular copper complexes[92] have also been characterised by MALDI.

Porphyrin complexes constitute a special case in analysis of supramolecular complexes, because they possess a strong UV chromophore themselves. As such, there is often no need for a matrix to obtain good spectra and this is an important property as porphyrins can often be easily demetallated by weak acids (such as typical MALDI matrices).[93] Essentially, this type of analysis can be classed simply as LDI, as no matrix is used to assist in the desorption process. Excellent spectra of high-mass covalently bound arrays have been obtained (Figure 3.27); the example shown consists of a porphyrin pentamer.[94]

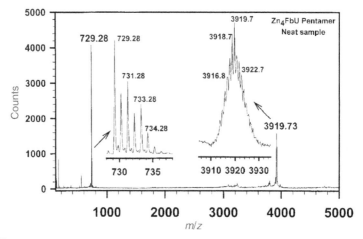

Figure 3.27
Positive ion MALDI-TOF mass spectrum of a central porphyrin ring covalently functionalised with four Zn(II) porphyrins. Reproduced by permission of Wiley Interscience

Numerous examples of the employment of MALDI for the characterisation of porphyrin complexes are available in the literature. A 5935 *m/z* dicationic hexaporphyrin complex of ruthenium was characterised using MALDI with a dithranol matrix; the base peak was the monocationic $[M]^{+}$.[95] MALDI was used to establish the absence of structural defects in a series of dendritic iron porphyrins,[96] and to characterise oligomers of linear ethynylporphyrin-bridged platinum(II) complexes.[97] Non-covalent arrays of ruthenium porphyrin complexes using phosphines were investigated using MALDI,[98] and the same authors have published a useful survey of MALDI-TOF as applied to supramolecular metalloporphyrin assemblies.[99] It reiterates that the use of a matrix is rarely necessary for successful characterisation (and that in fact protic matrices may induce ligand exchange reactions) highlights the use of porphyrins as useful matrices[100] in their own right, and demonstrates the importance of tuning the laser power to optimise signal intensity.

Fullerenes

Fullerenes are polyhedral carbon allotropes – the archetypal compounds being the soccerball shaped C_{60} and the elongated rugbyball-shaped C_{70} – that were originally detected using mass spectrometry techniques.[101] Since their discovery in 1985, fullerenes have attracted immense interest for both their fundamental properties and applications.[102] This was an important discovery of a new molecular allotrope of an element and not surprisingly a very wide range of derivatives have been synthesised, including metal complexes. The development of mass spectrometric methods for their characterisation has been welcomed, because these molecules are typically non–polar, and many, like C_{60}, have relatively low solubilities in many common solvents. While ESI MS is successful for the analysis of fullerenes (Section 4.6) and especially labile derivatives in solution, the parent fullerenes are probably best analysed using LDI techniques.

Fullerenes can easily be analysed by MALDI MS, giving the intact $[M]^{\bullet+}$ or $[M]^{\bullet-}$ ions in positive- and negative-ion modes respectively, Figure 3.28. As for other strongly UV-absorbing molecules such as porphyrins, a matrix is typically not necessary. Derivatives such as fullerene oxides[103] and metallofullerenes (containing an encapsulated metal or metals)[104] can also be easily characterised. Fullerenes such as C_{60} have themselves been used as a MALDI matrix, for example in the analysis of diuretic drugs;[105] the advantage of C_{60} over 'conventional' organic matrices (such as those in Figure 3.22) is that it provides low background noise in the low mass region ($<m/z$ 720).

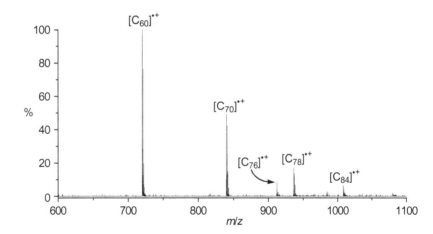

Figure 3.28
Positive-ion MALDI-TOF mass spectrum of a fullerite mixture

Dendrimers, oligomers, polymers

Polymer characterisation by MALDI provides very detailed information on composition and molecular weight distribution. Many examples exist in the literature and judging by the rate of publication, it is a technique that is growing quickly in popularity. A simple example of the kind of information provided is depicted in Figure 3.29 for a relatively low molecular weight polydimethylsiloxane, $R(Me_2SiO)_nR$.[106]

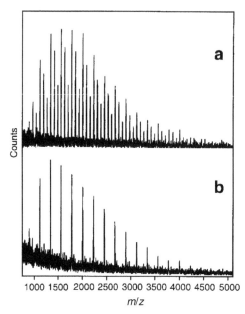

Figure 3.29
Positive ion MALDI-TOF mass spectra of (a) *sec*-butyllithium and (b) *n*-butyllithium initiated
polymerizations of hexamethylcyclotrisiloxane. Reproduced by permission of Elsevier

Cationisation often occurs via association of a metal, most usually Na^+ (or K^+) for
polymers containing heteroatoms, providing monocationic ions of the type $[M + Na]^+$.[107]
Ag^+ is similarly effective in the characterisation of polymers containing double bonds,
e.g. polystyrenes or polydienes.[108] Organometallic complexes have even been used as
cationisation agents, $[Ru(C_5H_5)(NCMe)_3][PF_6]$ reacting with polystyrene to provide
$[M + Ru(C_5H_5)]^+$ ions in the mass spectrum (Figure 3.30).[109]

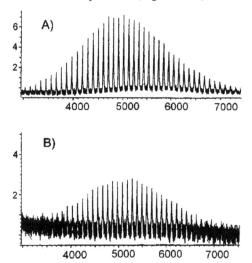

Figure 3.30
Positive ion MALDI mass spectrum of polystyrene with $M_n = 5050$ (top) with one equivalent of
$[Ru(C_5H_5)(NCMe)_3]^+$ and (bottom) with one equivalent of Ag^+. Reproduced by permission of
Elsevier

The success of the MALDI technique in these systems has resulted in a natural extension to polymeric materials that contain covalently-bound metals. These systems are often tough to characterise by other methods. Dendritic systems based on cationic Pd(II) complexes with molecular weights as high as 30 000 m/z were characterised using MALDI, with the highest mass peak in nearly all cases corresponding to $[M - BF_4]^+$ (BF_4^- was the counterion in this system).[110] All signals were monocationic, despite the presence of numerous cationic metal centres. A similar study on dendrimers with diphosphine functionality on the periphery was successful (Figure 3.31) but when derivatised with cationic rhodium complexes, the metallated dendrimers proved to be inaccessible to MALDI analysis.[111]

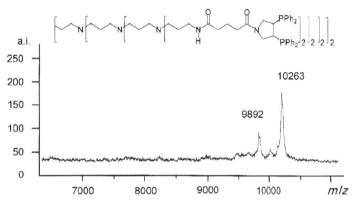

Figure 3.31
Positive ion MALDI-TOF mass spectrum of a third-generation phosphine functionalised dendrimer. Reproduced by permission of Wiley-VCH

Similar results were obtained for a series of phenylacetylide dendrimers – characterisation by MALDI worked well for the organic dendrimer (typically as the $[M + Na]^+$ sodiated adduct), but when metallated with $Co_2(CO)_6$ groups, no signals could be obtained.[112] Dendrimer analysis by MALDI-TOF becomes progressively more difficult as the size of the dendrimer increases – for example, platinum(II) acetylide dendrimers were synthesised out to the sixth generation (number of platinum atoms = 3, 9, 21, 45, 93, 189 in generations 1 to 6) but only the first generation at 2869 m/z proved amenable to MALDI.[113] Small metal-containing dendrimers can be relatively easily characterised using MALDI; for example, hydrosilation of the hexasubstituted benzene $C_6(CH_2CH_2HC=CH_2)_6$ with $FcSiMe_2H$ (Fc = ferrocenyl) generated a first generation dendrimer of molecular weight 1868 m/z (Figure 3.32).[114]

Oligomers containing metallocenes appear to be quite amenable to MALDI analysis. A series of metallocene-containing rings and chains were characterised primarily by MALDI, providing key information on the product mixture – molecular weight, polydispersion and mean ring size (Figure 3.33).[115] The ions formed in this case are radical cations, $[M]^{\bullet+}$ and this observation is in keeping with previous studies which showed that the MALDI spectra of non-polymer polymers containing ferrocene, ferrocenylnaphthalene and ruthenocenylnaphthalene groups in their repeating units contained radical cations.[116] This unusual behaviour – most organic samples and biomolecules undergo cationisation via addition of a proton or alkali metal – is probably due to the ease with which metallocenes may be oxidised.

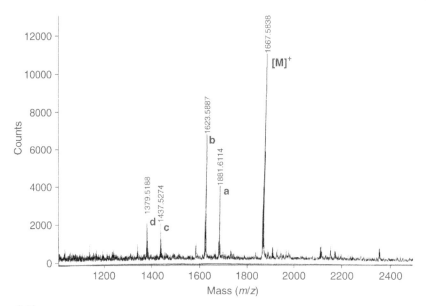

Figure 3.32
MALDI TOF mass spectrum of $[C_6\{(CH_2)_4SiMe_2Fc\}_6]$. M: molecular peak; (a) $[M–FcH]^+$; (b) $[M–FcSiMe_2H]^+$; (c) $[M–Fc–FcSiMe_2H]^+$; (d) $[M–2FcSiMe_2H]^+$. Note that the peaks **a** and **c** correspond to silylene terminated branches. Reproduced by permission of Elsevier

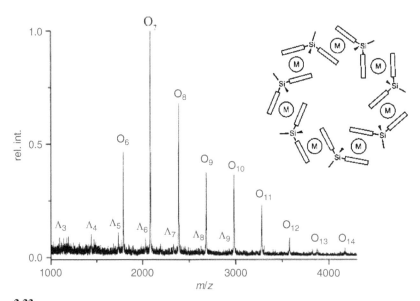

Figure 3.33
MALDI-TOF mass spectrum of the products obtained from the reaction of **LLi**$_2$ with $[FeCl_2(thf)_{1.5}]$ at 25 °C under high dilution (**L** = ligand with two doubly silyl-bridged cyclopentadienyl anions). $\bigcirc_i$ and $\wedge_j$ are rings and chains of the general formula $(LFe)_i$, and $HL(LFe)_jH$, respectively. The inset figure is $\bigcirc_7$. Reproduced by permission of Wiley-VCH

Clusters

MALDI analysis of neutral transition metal carbonyl clusters generates intense spectra in both positive and negative ion modes in which the molecular ion is just another peak in the midst of copious signals produced from fragmentation and/or aggregation processes.[117] The metal-carbonyl bond absorbs strongly in the UV region and a matrix is not required for successful analysis. One of the first clusters examined was the hexaruthenium cluster $Ru_6C(CO)_{17}$.[118] An extensive series of high mass signals were observed in the negative-ion mode that could be attributed to $[\{Ru_6C(CO)_n\}_m]^-$ where m ranged from one to at least 25. Studies have been carried out on $M_3(CO)_{12}$ (M = iron, ruthenium, osmium), $Ru_3(CO)_{12-n}(PPh_3)_n$ (n = 1, 2, 3), $H_3M_3(CO)_{12}$ (M = manganese, rhenium), $Rh_6(CO)_{16}$, $Os_6(CO)_{18}$ and numerous other clusters,[119] and a number of conclusions may be drawn from these investigations.[120]

1. No useful molecular weight information is generated for any neutral clusters and molecular ions are absent in the negative-ion spectra in all cases.
2. Clusters which include metal atoms with two or less metal-metal bonds are prone to fragment and the nuclearity of the observed species often bear no relationship with the nuclearity of the target cluster.
3. Clusters that contain metal atoms with three or more metal-metal bonds tend to aggregate in multiples of the original nuclearity.
4. Open clusters may not aggregate at all, as they can compensate for decarbonylation by polyhedral rearrangement to a more compact metal core.
5. Partial decarbonylation is a characteristic feature of all spectra and the extent of decarbonylation is more pronounced for the higher aggregates.
6. Substantial M-M bond formation and metal core rearrangement likely occurs upon aggregation.

Some of these features are exemplified in Figure 3.34, the negative-ion LDI mass spectrum of $Ir_4(CO)_{12}$. Note the absence of any signal due to molecular ion species; the spectrum consists entirely of fragments and gas-phase aggregation products.

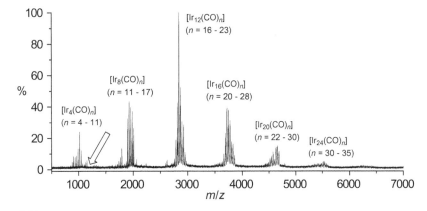

Figure 3.34
The negative-ion LDI-TOF mass spectrum of $Ir_4(CO)_{12}$ in the range 500–7000 *m/z*. The arrow indicates where the molecular ion $[Ir_4(CO)_{12}]^{\bullet-}$ would be expected. Reproduced by permission of The American Chemical Society

Anionic clusters provide exceptions to the rules above, as their spectra are absent any aggregation products and only a monoanionic molecular ion is observed along with fragments corresponding to loss of CO.[121] Note that again, the magnitude of the charge is lost – monoanions are also observed even from dianionic or trianionic precursors.

The phosphorus-containing clusters $Fe_4(\eta^5-C_5H_4Me)_4(CO)_6P_8$ and $Fe_6(\eta^5-C_5H_4Me)_4(CO)_{13}P_8$ have been successfully characterised by LDI-FTICR mass spectrometry.[122] A comparative analysis of LDI-FTICR and EI-FTICR mass spectra of four organometallic complexes of varying volatilities, including the dimers $Fe_2(\eta-C_5Me_5)_2(\mu-P_2)_2$ and $Co_2(\eta-C_5Me_5)_2(\mu-P_2)_2$, concluded that LDI gave either similar or much superior information to that provided (if at all) by EI, especially when the target complex was thermally unstable.[123] The halide clusters $[(Nb_6X_{12})X_2(H_2O)_4].4H_2O$ (X = Cl, Br) were studied using LDI-FTICR mass spectrometry.[124]

The ion $[(Nb_6X_{12})X_2]^-$ and fragment ions involving losses of one to three X were the most abundant in the spectrum, indicating the stability of the core cluster, Nb_6X_{12}. The first example of the application of MALDI to metal halide cluster chemistry was in a study of the clusters $Cs_3Re_3X_{12}$ (X = Cl, Br) and $(NBu_4)_2[Re_2Cl_8]$.[125] MALDI produced the better spectra than LDI (i.e. without a matrix) and characteristic fragments due to loss of X gave peaks that were used to identify the number of rhenium atoms and the identity of the ligands, by the isotope patterns and m/z values. The same authors have since successfully extended the utility of the technique to the cluster anions $[Re_3(\mu_3-E)(\mu-Cl)_3Cl_6]^{2-}$ (E = S, Se, Te).[126]

Coordination complexes

MALDI has been applied in a growing number of cases to simple coordination complexes. For example, a series of large palladium-containing heterocycles were structurally elucidated using MALDI, the presence of numerous basic groups on the periphery of the complexes facilitating the association of H^+, Na^+ or K^+.[127] The octakis(tetracarbonylcobaltio)octasilsesquioxane $[(CO)_4Co]_8Si_8O_{12}$ similarly associated an alkali metal to give a spectrum of $[M + K]^+$, along with a single fragment ion $[M - Co(CO)_4 + K]^+$.[128] A dinuclear oxo-bridged iron complex provided an $[M - H]^-$ ion in the negative-ion MALDI spectrum, despite the absence of any acidic hydrogens on the complex;[129] this result shows the worth of examining both ion modes as a matter of routine, no matter the expected result. Triangular organometallic complexes (molecular weights above 2500 Da) with cis-$Pt(PEt_3)_2$ metallocorners and bis-alkynyl bridging ligands provided $[M]^{\bullet+}$ or $[M + Li]^+$ ions.[130] Silver thiolates and silver phenylacetylide were characterised by MALDI using a dithranol matrix and oligomeric ions of the form $[(AgX)_n + Ag]^+$ (X = SR, C_2Ph) were observed.[131]

A useful paper describes in detail the MALDI spectra of ruthenium-osmium bipyridyl complexes, comparing the effects of different matrices and providing detailed assignments.[132] LDI spectra were shown to be complicated by the presence of ions due to the loss of hydrogen, fragmentation through loss of bipyridyl ligands, addition of fluorine (from BF_4^- or PF_6^- counterions) (Figure 3.35). Considerable improvement in data quality and simplicity was obtained by the use of a matrix, sinapinic acid proving much superior to DHB. Also improved was the calibration and isotopic distribution matching, and none of the fragments seen in LDI were observed. An equally detailed investigation of one monometallic and three trimetallic cationic coordination complexes with chelating pyridine/pyrazine ligands also compares LDI and MALDI as characterisation methods.[133]

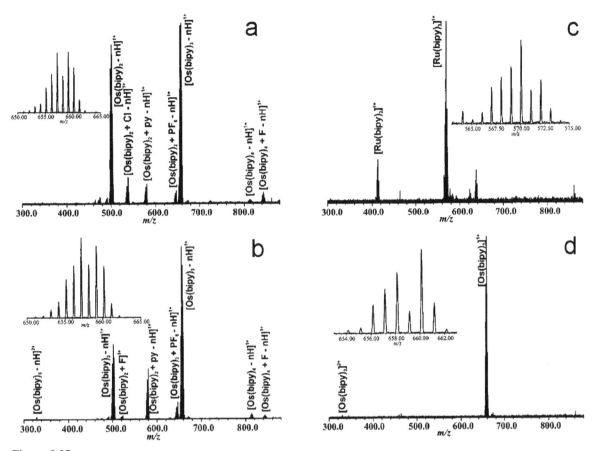

Figure 3.35
The positive-ion LDI-FTICR mass spectrum of (a) [Os(bipy)$_3$]Cl$_2$ and (b) [Os(bipy)$_3$](PF$_6$)$_2$; MALDI spectra of (c) [Ru(bipy)$_3$]Cl$_2$ and (d) [Os(bipy)$_3$]Cl$_2$ with sinapinic acid as matrix. Insets are expanded views of M$^+$ isotopic clusters. Reproduced by permission of Elsevier

For the monometallic complex, there was little to choose between the techniques, but spectra of the trimetallic complexes were vastly improved by the use of a matrix. Molecular ions of the form [M − nPF$_6$]$^+$ were observed, along with intermediate-mass fragments. Usefully, a wide range of matrices were tried, with the best results obtained from sinapinic acid and ATT, followed by HCCA (α-cyano-4-hydroxycinnamic acid) and DHB. Matrix-to-analyte ratio was varied from 200:1 to 5000:1, with the best results obtained at about 500:1 to 2000:1.

3.7.4 Summary

Strengths

- Rapid and convenient molecular weight determination;
- Soft ionisation technique, little fragmentation;

Weaknesses

- MS/MS difficult, so structural information often limited;
- Requires a mass analyser that is compatible with pulsed ionisation techniques;
- Not compatible with LC/MS;
- Fragmentation and aggregation processes can complicate spectra;
- Singly charged ions appear exclusively regardless of the original charge state of the sample;
- Spectra often very sensitive to choice of matrix.

Mass range

- *Very high* Beyond 500 000 Da, but for most organometallic/coordination complexes, considerably less; 50 000 is a more realistic upper limit.

Mass analysis

- Nearly always TOF, as this is the only method that allows access to the high mass monocharged ions produced by MALDI. However, MALDI sources have been developed for FTICR and ion trap instruments for the study of lower mass ions.

Applications

- Complexes with acidic or basic groups on the periphery;
- Porphyrin complexes;
- Fullerenes;
- Dendrimers, oligomers, polymers;
- Singly charged complexes;
- Multiply charged complexes, but expect complicated spectra.

Tips

- Select a matrix as similar to the target analyte as possible: try neutral matrices first;
- If compound has a good UV chromophore, try to collect data without a matrix (LDI);
- Always check both positive- and negative-ion modes.

Inductively Coupled Plasma Mass Spectrometry

Inductively coupled plasma mass spectrometry (ICP-MS) is essentially a form of elemental analysis, as it involves the conversion of all analytes to singly charged gas phase ions and so provides no chemical information regarding the molecular form or oxidation state of the sample. ICP-MS is a highly sensitive technique used to measure trace elements in a variety of solid and liquid materials. Samples are dispersed into a stream of argon gas and carried to a **plasma torch**. A rapidly oscillating electromagnetic field imparts extremely high kinetic energies to ions and electrons and the resulting plasma reaches effective temperatures of about 10 000 °C. Such fierce conditions promptly render the molecules of a sample down to ions of their constituent elements through repeated collisions with the extraordinarily energetic plasma.

Liquids can be introduced to the torch by aspirating the sample into the flow of argon gas using a nebuliser. For solids, the samples are held in a cell, flushed with argon and

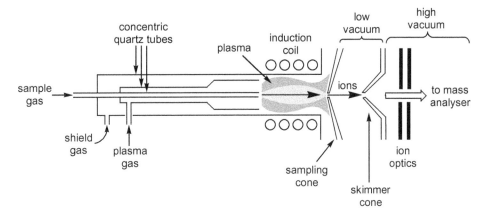

Figure 3.36
Inductively coupled plasma source. Argon gas ($\sim$1 L min^{-1}) sweeps the sample through the central quartz tube into the plasma 'flame', which is heated by a water-cooled induction coil powered by an rf generator. The plasma gas, also argon, flows through ($1 - 2$ L min^{-1}) the second of the tubes. To prevent the plasma melting the outermost quartz tube, a cooling shield gas flows at a high rate ($\sim$10 L min^{-1}) through the outermost tube. The plasma is drawn through the sampling orifice (also water-cooled) into a region of lower pressure where the neutrals are pumped away, the electrons discharged and the positive ions accelerated through the skimmer orifice into the high vacuum of the mass analyser

ablated from the surface (typically using a laser) or directly inserted into the flame. After the sample has been torn apart at high temperature, the resulting ions are drawn through a series of orifices and ion optics into the high vacuum, room temperature mass analyser and measured for abundance and *m/z* (Figure 3.36).

ICP-MS is used mainly for the determination of trace element concentrations in inorganic solids and liquids. Elements from lithium to uranium can be determined. The instruments are highly sensitive and have a detection limit of a few parts per quadrillion in solution. For solids analysis by laser ablation, a determination of the concentration of 40 trace elements is achieved just a few minutes. Applications include fluid inclusion analysis, forensic and fingerprinting applications and the analysis of trace elements in human teeth and bones. High resolution instruments can be used for isotope ratio determinations. Applications include the analysis of iron and nickel isotopes in meteorites, boron isotope ratio determinations as well as single crystal dating techniques.

3.8.1 Applications

For organometallic and coordination chemists, ICP-MS has somewhat limited utility, due to the lack of structural information provided. Nonetheless, it can provide useful information on the ratios of various elements relative to others and so the technique finds some use in studies of heteronuclear cluster compounds. For example, ICP-MS analysis showed that the mixed-metal cluster [Pt(PPh$_3$)(AuPPh$_3$)$_6$(Cu$_4$Cl$_3$PPh$_3$)](NO$_3$) had a Pt:Au:Cu ratio of 1:5.9:4.0.[134] Similarly, ICP-MS provided a Pt:Au:Cu ratio of 1:7.9:2.0 for the cluster [Pt(H)(AuPPh$_3$)$_8$(CuCl)$_2$](NO$_3$) and a Pt:Au:P ratio of 1:8.9:8.1 for [Pt(H)(AuPy)(AuPPh$_3$)$_8$](NO$_3$)$_2$.[135] ICP analysis has also been applied to the heterometallic cluster [Mo$_6$BiS$_8$(H$_2$O)$_{18}$]$^{8+}$, confirming the Mo:Bi:S ratio as 6:1:8.[136] The clusters [Mo$_6$SnSe$_4$(H$_2$O)$_{12}$]$^{6+}$ and [Mo$_6$SnSe$_8$(H$_2$O)$_{18}$]$^{8+}$ were characterised in similar fashion.[137] The technique is a promising one for application to heteronuclear clusters for which single crystals can not be obtained.

3.8.2 Summary

Strengths

- Trace element analysis;
- Any sort of sample may be analysed;
- Depth profiles of surfaces can be generated.

Weaknesses

- No structural information.

Mass range

- *N/A* Creates singly charged elemental ions; no need for mass range to extend beyond 255 m/z, $[UOH]^+$ is the highest mass species observed.

Mass analysis

- Usually quadrupole, but TOF instruments are a recent development (these have the advantage of a high duty cycle and hence can provide good time resolution on the change of speciation).

Applications

- Analytical chemistry, geochemistry, trace element analysis;
- Elemental analysis.

Popularity

- High and improving, though yet to make a significant impact in the study of organometallic or coordination compounds. Very important in the distinctly separate field of analytical inorganic chemistry (trace analysis).

Tips

- Best applied where establishment of the proportions of two or more heavy elements is important.

Electrospray Ionisation[138]

The electrospray process involves the creation of a fine aerosol of highly charged micro droplets in a strong electric field. It was first developed over 80 years ago and found applications in electrostatic painting, fuel atomisation, drug delivery and rocket propulsion. Electrospray as an ionisation technique for mass spectrometry was developed by Dole and co-workers in the late 1960s[139] and considerably improved upon by Yamashita and Fenn who in 1984 coupled an electrospray source to a quadrupole mass analyser.[140] The ability of electrospray ionisation (ESI) to provide mass spectrometric access to large,

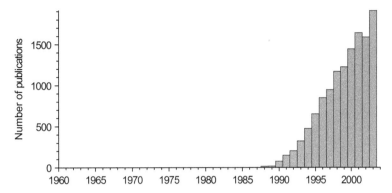

Figure 3.37
The rapid growth in popularity of electrospray as an ionisation technique in mass spectrometry[142]

thermally fragile polar molecules – especially proteins[141] – led to an explosion of interest in the technique which has essentially continued unabated (Figure 3.37). Fenn was recognised for his contributions by the award of the 2002 Nobel Prize for Chemistry.[†]

The tremendous growth in popularity of the electrospray technique is representative of its powerful abilities to investigate species in solution even in vanishingly low concentrations and in complex mixtures. Organometallic and coordination chemists have recognised the power of ESI MS for solving analytical problems and the technique is becoming ever more established. The contents of this book are representative of the special usefulness of ESI MS to this community – half of the pages are devoted to applications of this ionisation technique.

A continuous flow of solution containing the analyte from a highly charged (2–5 kV) capillary generates an electrospray. The solution elutes from the capillary into a chamber at atmospheric pressure, producing a fine spray of highly charged droplets due to the presence of the electric field (Figure 3.38), a process called **nebulisation**. A combination of thermal and pneumatic means is used to desolvate the ions as they enter the ion source. Solution flow rates can range from less than a microliter per minute to several milliliters per minute and so ESI MS is ideally suited to coupling with high-performance liquid chromatography (HPLC).[‡]

The solvent contained in the droplets is evaporated by a warm counter-flow of nitrogen gas until the charge density increases to a point at which the repulsion becomes of the same order as the surface tension. The droplet then may fragment in what is termed a '**Coulomb explosion**', producing many daughter droplets that undergo the same process, ultimately resulting in bare analyte ions. An alternative picture is one in which the ions 'evaporate' from the surface of the droplet.[143] Whatever the exact mechanism, ESI is a very 'soft' means of ionisation that causes little or no fragmentation of the sample.

[†]For the development of soft desorption ionisation methods for mass spectrometric analyses of biological macromolecules', shared with Koichi Tanaka (who introduced MALDI as an ionisation technique). See *http://www.nobel.se/chemistry/laureates/2002/*, or J. B. Fenn, '*Electrospray wings for molecular elephants (Nobel lecture)' Angew. Chem., Int. Ed.*, 2003, **42**, 3871–3894.

[‡]So much so that ESI mass spectrometers are usually sold along with an HPLC as a package deal, hence the common term 'LCMS'.

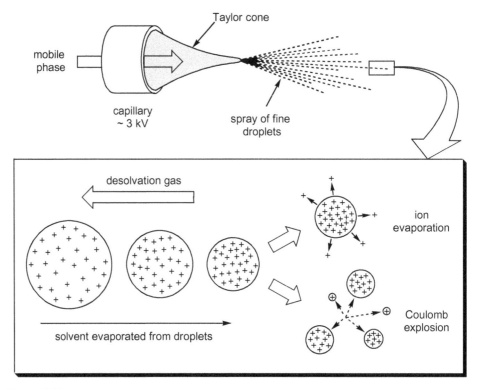

Figure 3.38
The desolvation process. A solution containing the analyte is pumped through a charged (and often heated) capillary. The liquid emerging from the capillary elongates into a 'Taylor cone', from the end of which a very fine spray of charged droplets develops (top). The droplets so formed are rapidly dried by a flow of warm bath gas (usually nitrogen) and the loss of solvent causes the charge density to rise to the point that ion evaporation and/or Coulomb explosion occurs to produce desolvated, gas-phase analyte ions

The electrospray ion source is at very high pressure (atmospheric) with respect to the very low pressure that is required for ion separation by a mass analyser, so the interface between the two involve a series of skimmer cones (acting as small orifices) between the various differentially pumped regions (Figure 3.39). Early designs had the capillary exit pointing directly into the mass analyser but to limit contamination practically all modern designs have an orthogonal (or at least off-axis) spray direction.

The ions are drawn into the spectrometer proper through the skimmer cones. A voltage can be applied (the **cone voltage**), which will accelerate the ions relative to the neutral gas molecules. This leads to energetic ion-neutral collisions and fragmentation due to what is termed **collision induced dissociation** (CID) (Section 2.2). The remaining bath gas is pumped away in stages (in order to attain the high vacuum necessary for separation of the ions) and the ions are focused through a lensing system into the mass analyser.

3.9.1 Electrochemistry in the ESI Process

The production of charged droplets in electrospray is fundamentally dependent on electrochemical processes occurring at the capillary (which acts as one electrode).[144]

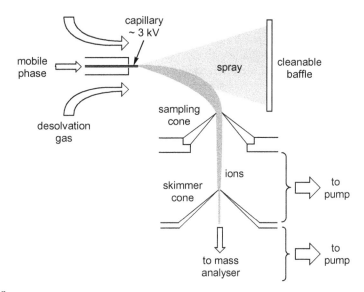

Figure 3.39
An electrospray source. The spray of droplets is dried by a flow of desolvation gas (usually nitrogen) and a portion of the spray is pneumatically drawn into the vacuum of mass spectrometer through a narrow orifice (the sampling cone). The pressure at this stage is approximately 1 mbar, as the remaining solvent and desolvation gas is pumped away (in some instruments, this mixture is dried further by passage through a heated capillary). The ions then pass through a skimmer (extraction) cone and are guided electrostatically into the high vacuum of the mass analyser. The number of stages of differential pumping are dependent on the vacuum requirements of the mass analyser (ion trap < quadrupole < TOF $\sim$ sector < FTICR)

As many ions must be oxidised at the capillary as positive ions are extracted to the mass analyser (vice versa for negative ion mode). The circuit is completed when the ions are discharged in the mass analyser or detector (the counterelectrode). The stainless steel capillary itself may also be oxidised. The electrochemical reactions occurring at the capillary necessarily alter the composition of the solution entering the emitter and, as a result, the ions observed in ESI MS are not necessarily those present in the solution that was used. This has important consequences for the organometallic and coordination chemist, as metal complexes frequently have a rich electrochemistry and the oxidation state of the metal species may be affected. Oxidation (positive ion mode) and reduction (negative ion mode) reactions alter the valence state of the metal under study:

$$M^{n+} + me^- \longrightarrow M^{(n-m)+}$$
$$M^{n-} - me^- \longrightarrow M^{(n-m)-}$$

For easily reduced ions such as Ag^+, deposition of the metal can occur at the capillary, resulting in loss of up to half the Ag^+ present in solution. Electrochemical processes are also responsible for Cu^{2+} and Hg^{2+} appearing as $[Cu]^+$ and $[Hg]^+$ ions in the mass spectrum.[145] Neutral species can also be oxidised or reduced at the probe tip to produce $[M]^{\bullet+}$ or $[M]^{\bullet-}$ and examples of these processes will be given in the subsequent sections for example see the example of ferrocene oxidation in Section 7.4.1. Electrochemical reactions can allow the characterisation of complexes that otherwise have no simple means of acquiring a charge; in this instance, electrospray acts as a true ionisation method rather than just a means of transporting preformed ions into the gas phase. However, the downside is that the charge on complexes observed in the ESI mass spectrum can in some cases bear little relationship to the species actually present in solution.

An early demonstration of this effect was performed by Fenselau and co-workers,[146] who showed that multiply charged negative ions, $[M - nH]^{n-}$, of proteins appeared in the spectrum even under conditions of low pH. In fact, the signal-to-noise ratio *improved* in the negative ion mode with decreasing pH, suggesting that the increased conductivity of the solution assisted in the formation of negative ions. This observation leads to the peculiar advice that in some cases it is beneficial to lower the pH to assist in the observation of deprotonated species, and conversely raising the pH can assist in a stronger signal for protonated species, because both H^+ and OH^- are highly effective at increasing the solution conductivity.

3.9.2 Multiply Charged Species

The appearance of multiply charged species enables ESI to characterise compounds whose molecular weight would otherwise be far in excess of that accessible to most mass analysers (with the notable exception of TOF). Biological macromolecules tend to accumulate one unit of charge for every 1 – 2000 Da, so nearly all proteins, for example, produce signals in the region of 1 – 2000 *m/z*, regardless of their actual molecular weight. This effect is best illustrated with an example. Horse heart myoglobin has a molecular weight of 16 951.49 Da and is commonly used as a calibrant.[147] Under typical ESI MS conditions, it is protonated between 11 and 20 times to provide individual peaks with different charge states between 1542.086 *m/z* and 848.557 *m/z* (Figure 3.40).

The provision of multiple independent values for the molecular weight considerably improves the reliability of the data. Deconvolution programs integrated into the mass spectrometer data system can transform the multiple peaks into a single peak at the correct molecular weight automatically.

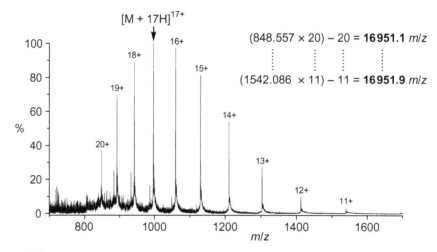

Figure 3.40
Positive-ion ESI mass spectrum of horse-heart myoglobin, a protein with a molecular weight of 16 951.49 Da. Mobile phase: water/methanol/formic acid

All biological macromolecules share the property of a surfeit of basic sites which can associate a proton, or equally, many acidic sites that can lose one. However, most organometallic and coordination compounds have few such sites and, equally, it is also rare that they attain such high molecular weights. It is, in fact, highly unusual for such complexes to acquire multiple charges and it can generally be assumed that all species derived from a neutral molecule carry just a single charge.

Compounds that are *inherently* multiply charged, on the other hand, are quite common in inorganic and organometallic chemistry, but the charge information is only retained reliably if the charge density is reasonably low. Multiply charged anions in particular have a limited existence in the gas phase due to Coulomb repulsion between the excess charges[148] and ions such as $[CO_3]^{2-}$ or $[PO_4]^{3-}$ do not exist at all in the gas phase. Examples of other oxyanions can be found in Sections 4.4 and 5.10.1. Loss of an electron from multiply charged anions often occurs spontaneously and electron autodetachment can sometimes be observed in the fragmentation pattern.[149] Loss of an electron is very obvious in the spectrum, because the z in the m/z value changes by a whole unit, so a dianion losing an electron will appear to double in mass. Be suspicious of any assignment of multiply charged anions below $\sim$500 m/z – compelling evidence in the form of a well-resolved isotope pattern showing spacing between peaks of $1/z$ Da is required. Similar evidence is required for polycationic complexes. Charge diminution in this case often occurs via the complex behaving as an acid, e.g. $[Fe(H_2O)_6]^{3+} + H_2O$ produces $[Fe(H_2O)_5(OH)]^{2+} + [H_3O]^+$; this is discussed further in Section 4.2.2.

3.9.3 Nanospray

Nanospray is a miniaturised electrospray source that operates at flow rates of the order of nanolitres per minute. Such low flow rates are achieved using special microcapillaries and the solution is drawn by capillary action through an orifice of micron diameter. Extraordinary senstitivities may be achieved using this technique and it is of great importance to protein chemists, for example, whose samples are often extremely precious and consumption even on the scale of conventional ESI is significant. This is far less likely to be an issue for the inorganic or organometallic chemist, who can usually sacrifice the tenth of a milligram or so of sample required for a normal ESI experiment without concern. As such, the high cost of the one-use capillaries used in nanospray makes the technique rather less attractive.

However, there are some rather special advantages of nanospray that may find specialist application. First, the very fact that each capillary is single use eliminates the possibility of cross contamination, so nanospray represents a convenient way of getting very clean spectra unaffected by previous users. This is a boon for those infrequently using an open-access instrument. Second, very high concentrations may be used without fear of contaminating the source with analyte – nanospray has been used spray neat ionic liquids successfully.[150] This ability could prove highly advantageous in the examination of systems where speciation is thought to change at high concentrations. Third, the extremely small opening to the capillary means that the rate at which the atmosphere diffuses into the capillary when exposed to air is negligible. This means that capillaries may be filled with even extremely air-sensitive material in an inert-atmosphere glovebox, then transported without precautions to the mass spectrometer for analysis.[151] This represents an enormous advantage for many inorganic and organometallic chemists, because while it is not difficult to exclude air and moisture from the sample introduction process in ESI MS by the simple expedient of using gas-tight syringes and a syringe pump, excluding moisture from the entire mass spectrometer requires the connivance of all other users for a substantial period (days). This expedient may not be in the least bit practical for an open access instrument, where many users will be running samples using all sorts of protic and hydroscopic solvents (alcohols and of course water itself). Nanospray offers a simple, if somewhat expensive, way of obtaining mass spectrometric data on highly air-sensitive samples even on a machine used for other purposes.

3.9.4   ESI MS: Practical Considerations

When the ESI mass spectrum of an inorganic (or for that matter *any*) substance is to be recorded, there are often a number of experimental variables that must be considered. The appropriate choice of these is essential if the best quality ESI MS data are to be obtained. Some of these variables are discussed in this section.

Solvents

If a *charged* analyte is to be studied by ESI MS, then practically any moderately polar solvent can theoretically be used (dichloromethane being the least polar solvent usually employed). The solvent provides a convenient method for producing a dilute solution of the analyte. It is preferable that the solvent has reasonable volatility, such that the drying gas can remove solvent from the charged droplets produced in the ESI process (Section 3.9) without recourse to excessively high temperatures that might cause sample decomposition. Solvents such as methanol, ethanol, acetonitrile and water are all commonly used, because of their moderate boiling points ($<100\,^{\circ}C$) and polarities. In many cases, mixtures of water plus an organic solvent (often acetonitrile) are used. Such polar, protic solvents and mixtures thereof are typically able to dissolve the miniscule quantities of most compounds – even sparingly soluble ones – so that MS analysis can be carried out. However, even relatively involatile solvents such as ethylene glycol ($HOCH_2CH_2OH$) have been used for ESI-MS analyses.[152] Of course, if a sparingly soluble compound is analysed, caution may need to be exercised in analysing the resulting spectra, because ions formed from soluble impurities might dominate the spectra.

For the analysis of moisture-sensitive compounds, protic solvents have to be avoided, and solvents for air-sensitive compounds need to be rigorously deoxygenated. Useful alternatives are acetonitrile, dichloromethane, 1,2-dichloroethane or tetrahydrofuran. If a neutral compound is to be analysed in a polar aprotic solvent, because of the lack of protons or other cations it may be necessary to add an ionisation agent. Potassium iodide has good solubility in many polar aprotic solvents, and can be used to generate $[M + K]^+$ ions.[153]

Sample preparation

Sample preparation for ESI-MS analysis simply involves preparing a dilute solution of the analyte sample in a suitable solvent. The concentrations required are typically low. In inorganic chemistry research, sufficient sample is usually available, such that a solution of concentration maximum 1 mg mL^{-1} (and preferably much less than this) provides good quality spectra. Generally, sample concentration is not critical, but high concentrations of dissolved solids will result in more frequent source cleaning being required, and may exacerbate the effects of species that linger in capillary systems, as described later. As a useful rule of thumb, because many inorganic samples are coloured, anything more than an extremely pale-coloured solution is probably far too concentrated.

One technique that may be used to provide useful concentrations of compounds that might be sparingly soluble in a desired solvent (such as methanol) is to dissolve the analyte in a few drops of a polar aprotic solvent such as dichloromethane, pyridine, $HC(O)NMe_2$ (DMF) or Me_2SO (DMSO) and then dilute with the bulk solvent.

Centrifugation of samples is strongly recommended prior to analysis since small particulate matter can block the capillary systems of the instrument. Plastic Eppendorf vials and a mini-centrifuge provide a cheap, effective method.

Ionisation agents

Depending on the type of compound that is to be analysed, it may be necessary to add an ionisation agent to the analyte solution, in order to promote the formation of ions, and, in some cases, to simplify spectra if ionisation is proceeding through several pathways.

Ionisation of a neutral analyte often occurs by protonation, or alternatively cationisation, with an adventitious cation present in the solvent used, such as Na^+, K^+ or NH_4^+. Such cations are present in solvents, especially polar, protic solvents (water especially), which have been stored in glass bottles. Acetonitrile often contains significant traces of NH_3 and NH_4^+, as well as CN^-. In some cases, adduct ions with several cations can occur, giving $[M + H]^+$, $[M + NH_4]^+$, $[M + Na]^+$ and $[M + K]^+$ ions, that are respectively $+1$, $+18$, $+23$ and $+39$ *m/z* heavier (for a monocation) than the neutral analyte M. Verification, and simplification of the spectrum, is easily achieved by spiking the analyte solution with a small quantity of one of the cations, for example by adding a drop of dilute (aqueous) sodium chloride solution whereupon the intensity of the $[M + Na]^+$ ion should increase.

Many different ionisation reagents are potentially useful, and examples are given in the subsequent chapters. Some of the main ionisation agents used involving inorganic or organometallic systems are summarised in Table 3.2.

Table 3.2 Ionisation agents for inorganic compounds

Ionisation agent	Application
Acid (e.g. HCO_2H, CH_3CO_2H)	Protonation of neutral analytes
Alkali metal or ammonium salts	Cationisation of neutral analytes
Silver salts¶	Cationisation of neutral analytes with polarisable electron density such as phosphines, arsines, thioethers, arenes, alkenes, alkynes *etc.* (Sections 6.3.1 and 7.7)
Alkoxide ions	Neutral metal carbonyls (Section 7.2.3)
Ferrocene derivatives	Chemically- or electrochemically-oxidisable derivatives can be formed, for e.g. steroid analysis (Section 7.4.2)
Pyridine	Analysis of neutral metal halide complexes via anion displacement (Section 5.3)
Chloride	Formation of $[M + Cl]^-$ ions with H-bonding analytes e.g. aniline and *cis*-$[PtCl_2(NH_3)_2]$ (Section 5.13.1)

¶ The $PhHg^+$ cation behaves similarly.

Cross contamination

Perhaps one of the major disadvantages of ESI compared to MALDI is the requirement for an analyte *solution*, which passes through a capillary system of the instrument, with the resulting risk of 'contamination', whereby some species have a tendency to linger in the capillary and/or injector systems of the spectrometer. This effect is particularly pronounced for species with high ionisation efficiency, such as large organic cations (e.g. Ph_4P^+, PPN^+ etc). In our experience, this can cause major problems, either by the observation of unwanted ions, suppressing signals due to the desired species, or at worst completely obscuring them if their m/z values are the same. Table 3.3 lists some of the common species (with their m/z values in some cases) that, in our experience, have been found to linger.

Table 3.3 Species found to linger in the spectrometer

Contaminant	Ions (m/z)
Quaternary phosphonium, arsonium and ammonium cations such as $[Ph_4P]^+$, $[Ph_4As]^+$, $[octylNMe_3]^+$, $[Ph_3PNPPh_3]^+$ (PPN^+) etc.	$[M]^+$; $[PPN]^+$ (538), $[Ph_4P]^+$ (339) etc.
$CpRuX(PPh_3)_2$ (X = halide)[#]	$[CpRu(PPh_3)_2]^+$ (691), $[CpRu(PPh_3)_2(MeCN)]^+$ (732)
Silver(I) salts[#]	$[Ag]^+$ (107, 109), $[Ag(MeCN)]^+$ (148, 150), $[Ag(MeCN)_2]^+$ (189, 191)
Alkali metal ions	$[Na]^+$ (+23), $[K]^+$ (+39) (not a problem themselves, but cationise neutrals very readily and are hard to eliminate from the instrument)
Gold-trialkyl(aryl)phosphine complexes	e.g. $[Au(PPh_3)_2]^+$ (721)
Iodide	$[I]^-$ (127)
Tetrafluoroborate, hexafluorophosphate	$[BF_4]^-$ (87), $[PF_6]^-$ (145)
Polyethers	A series of ions differing by m/z 44 (ethylene oxide) or m/z 58 (propylene oxide) repeat units

[#]Data are given for acetonitrile solutions, where the solvent shows relatively good coordinating properties to the metal centre.

Silver(I) ions have a strong tendency to linger in the instrument and if compounds with a strong affinity for silver (especially phosphines, but also arsines, thioethers, alkenes, arenes *etc.*) are subsequently injected, these can form (unexpected) adducts with the silver ions. Reduction of silver ions in the ESI capillary system can also occur, providing a possible mechanism for this observation. The presence of silver-containing species can fortunately be readily discerned by the distinctive isotopic composition of silver [^{107}Ag (51 %) and ^{109}Ag (49 %)]. Removal of the silver can be facilitated by flushing the capillary system with a solvent that has a strong affinity for silver, such as pyridine.

Some of the materials used as mass calibrants in electrospray systems can also cause difficulties. Polyethylene glycols, of the general composition $HO(CH_2CH_2O)_nH$, can be used for positive-ion calibration, and can cause difficulties when used in high concentrations. Because they often cover a wide m/z range, they can cause particular problems. Alkali metal (e.g. sodium, caesium) iodides are also used as calibrants in both positive- and negative-ion modes, however we have found that the iodide ion can linger, and cause difficulties in exchanging with chloride ligands of platinum group metal complexes.

The best way to avoid cross-contamination is to minimise the concentration of the (problematical) analyte solution; it is better to start with an initial solution that is far more dilute than might be required and then increase the concentration as necessary to get good signal-to-noise. Flushing the capillary systems with a range of solvents in a sequence of polarities (e.g. aqueous formic acid – water – methanol – acetonitrile – dichloromethane and, finally, the next mobile phase solvent) will often enhance the cleanup process. Most modern ESI MS spectrometers use a syringe pump input and, as a recommendation, all users should be encouraged to use their own syringes, capillary tubing and adaptors, to minimise cross-contamination between workgroups. In this case, it is only the stainless steel capillary which is common to all users, so contamination is minimised. If one type of capillary tubing may is susceptible to lingering effects, switching to another material may prove effective; several types e.g. poly(etherether ketone) (PEEK), fused silica are commercially available. In cases of extreme contamination the only practical solution may be to flush the entire capillary system for several hours. This obviously has implications for high throughput use, so the timing of analysis of samples containing lingering ions needs to be carefully considered.

Cone voltage

Of any experimentally-variable parameter, the *cone voltage* (Sections 2.2 and 3.9) is probably the most important in determining exactly what ions are observed in the ESI mass spectrum. Many of the examples given in Chapters 4 to 7 of this book illustrate the dramatic effect of cone voltage. The actual value of this parameter required is highly dependent on the exact system under study, and users are encouraged to investigate a range of cone voltages for each sample. While a higher cone voltage may result in fragmentation, it also typically results in increased signal intensity, so a compromise is usually necessary.

3.9.5 Applications

Chapters 4 to 7 cover the many applications of electrospray ionisation mass spectrometry in inorganic and organometallic chemistry. It is a technique that has rapidly found use in a wide range of areas in inorganic and organometallic chemistry.

3.9.6 Summary

Strengths

- Good for charged, polar or basic compounds;
- Multiple charging allows the detection of high mass compounds at accessible *m/z* ratios;
- Low chemical background and excellent detection limits;
- Can control presence or absence of fragmentation by controlling the cone voltage;
- Compatible with MS/MS methods.

Weaknesses

- No good for uncharged, non-basic, non-polar compounds;
- Sensitive to contaminants such as alkali metals or basic compounds;
- Relatively complex hardware compared to other ion sources due to the requirement of differential pumping.

Mass range

- *High* Multiple charging allows access to molecules of extraordinarily high molecular weight. Most samples are analysed below 2000 *m/z*.

Mass analysis

- Any, though infrequently found in conjunction with sector instruments (there are experimental difficulties inherent in coupling an atmospheric pressure source with the high voltages used in sector mass analysis). Quadrupoles a popular choice.

Popularity

- Very high, and applications in inorganic and organometallic chemistry are growing fast. Especially important in solution chemistry.

Tips

- Any compound soluble and stable in a moderately polar solvent is fair game for electrospray analysis – if ionic, analysis should be trivial. If neutral, try one of the derivatising agents. Check both ionisation modes.
- Air-sensitive samples may be run without difficulty, but recall that you are at the mercy of the previous user. If your samples are extremely water-sensitive, the instrument will need to be run free of protic solvents for a long period before use (a day, preferably more).

References

1. SciFinder Scholar, search for 'electron ionisation'. Note that a search of this nature will underestimate the number of papers especially as a technique becomes more established, because articles employing the technique in a routine way will not be flagged.
2. F. W. McLafferty and J. Choi, *Interpretation of Mass Spectra*, 4th Ed., University Science Books, Mill Valley (1996).
3. M. R. Litzow and T. R. Spalding, '*Mass Spectrometry of Inorganic and Organometallic Compounds*', Elsevier, Amsterdam (1973); J. Charalambous, Ed., '*Mass Spectrometry of Metal Compounds*', Butterworths, London (1975).
4. F. Johnson, R. S. Golke and W. A. Nasutavicus, *J. Organomet. Chem.*, 1965, **3**, 233.
5. A. G. Harrison, '*Chemical Ionisation Mass Spectrometry*', CRC Press, Boca Raton (1992).
6. M. S. B. Munson and F. H. Field, *J. Am. Chem. Soc.*, 1966, **88**, 2621.
7. S. C. Subba Rao and C. Fenselau, *Anal. Chem.*, 1978, **50**, 511.
8. T. J. Odiorne, D. J. Harvey and P. Vouros, *J. Phys. Chem.*, 1972, **76**, 3217.
9. R. Bakhtiar, *Rapid Commun. Mass Spectrom.*, 1999, **13**, 87.
10. L. Agocs, N. Burford, J. M. Curtis, G. B. Yhard, K. Robertson, T. S. Cameron, and J. F. Richardson, *J. Am. Chem. Soc.*, 1996, **118**, 3225.
11. P. Mastrorilli, C. F. Nobile, G. P. Suranna, F. P. Fanizzi, G. Ciccarella, U. Englert and Q. Li, *Eur. J. Inorg. Chem.*, 2004, 1234.
12. J. Mueller and C. Haensch, *J. Organomet. Chem.*, 1984, **262**, 323.
13. J. Mueller, E. Baumgartner and C. Haensch, *Int. J. Mass Spectrom. Ion Proc.*, 1983, **47**, 523.
14. D. F. Hunt, J. W. Russell and R. L. Torian, *J. Organomet. Chem.*, 1972, **43**, 175.
15. D. Perugini, G. Innorta, S. Torroni and A. Foffani, *J. Organomet. Chem.*, 1986, **308**, 167.

16. S. P. Constantine, D. J. Cardin and B. G. Bollen, *Rapid Commun. Mass Spectrom.*, 2000, **14**, 329.

17. J. Tirouflet, J. Besancon, B. Gautheron, F. Gomez and D. Fraisse, *J. Organomet. Chem.*, 1982, **234**, 143

18. R. P. Morgan, C. A. Gilchrist, K. R. Jennings and I. K. Gregor, *Int. J. Mass Spectrom. Ion Proc.*, 1983, **46**, 309.

19. A. Turco, A. Morvillo, U. Vettori and P. Traldi, *Inorg. Chem.*, 1985, **24**, 1123.

20. K. Kroll and G. Huttner, *J. Organomet. Chem.*, 1987, **329**, 369.

21. L. Prókai, '*Field Desorption Mass Spectrometry*', Marcel Dekker, New York (1990).

22. H. B. Linden, E. Hilt and H. D. Beckey, *J. Phys. E*, 1978, **11**, 1033; H. B. Linden, H. D. Beckey and F. Okuyama, *Appl. Phys.*, 1980, **22**, 83.

23. C. J. McNeal, R. D. MacFarlane and E. L. Thurston, *Anal. Chem.*, 1979, **51**, 2036.

24. R. D. MacFarlane, *Acc. Chem. Res.*, 1982, **15**, 268; B. Sundqvist and R. D. MacFarlane, *Mass Spectrom. Rev.*, 1985, **4**, 421.

25. D. F. Torgerson, R. P. Skowronski and R. D. MacFarlane, *Biochem. Biophys. Res. Commun.*, 1974, **60**, 616; R. D. MacFarlane and D. F. Torgerson, *Science*, 1976, **191**, 920.

26. P. Roepstorff, *Acc. Chem. Res.*, 1989, **22**, 421.

27. SciFinder Scholar, search for 'plasma desorption' and 'mass spectrometry'. Note that a search of this nature will underestimate the number of papers especially as a technique becomes more established, because articles employing the technique in a routine way will often not be flagged.

28. P. A. Van Veelen, U. R. Tjaden, J. Van der Greef and R. Hage, *Org. Mass Spectrom.*, 1991, **26**, 74.

29. D. A. Morgenstern, C. C. Bonham, A. P. Rothwell, K. V. Wood and C. P. Kubiak, *Polyhedron*, 1995, **14**, 1129.

30. Y. Z. Voloshin, O. A. Varzatskii, A. V. Palchik, N. G. Strizhakova, I. I. Vorontsov, M. Y. Antipin, D. I. Kochubey and B. N. Novgorodov, *New J. Chem.*, 2003, **27**, 1148.

31. J. M. Hughes, Y. Huang, R. D. MacFarlane, C. J. McNeal, G. J. Lewis and L. F. Dahl, *Int. J. Mass Spec. Ion Proc.*, 1993, **126**, 197.

32. C. J. McNeal, J. M. Hughes, G. J. Lewis and L. F. Dahl, *J. Am. Chem. Soc.*, 1991, **113**, 372.

33. C. J. McNeal, R. E. P. Winpenny, R. D. MacFarlane, L. H. Pignolet, L. T. J. Nelson, T. G. Gardner, L. H. Irgens, G. Vigh and J. P. Fackler, *Inorg. Chem.*, 1993, **32**, 5582.

34. J. P. Fackler, C. J. McNeal, L. H. Pignolet and R. E. P. Winpenny, *J. Am. Chem. Soc.*, 1989, **111**, 6434.

35. G. Schmid, *Chem. Rev.*, 1992, **92**, 1709 and references therein.

36. C. Fenselau and R. J. Cotter, *Chem. Rev.* 1987, **87**, 501.

37. M. Barber, R. S. Bordoli, R. D. Sedgwick and A. N. Tyler, *J. Chem. Soc., Chem. Commun.*, 1981, 325; M. Barber, R. S. Bordoli, G. J. Elliot, R. D. Sedgwick and A. N. Tyler, *Anal. Chem.*, 1982, **54**, 645A.

38. T. J. Kamp, *Coord. Chem. Rev.*, 1993, **125**, 333.

39. SciFinder Scholar, search for 'FAB' or 'LSIMS'. Note that a search of this nature will underestimate the number of papers especially as a technique becomes more established, because articles employing the technique in a routine way will often not be flagged.

40. A. K. Abdul-Sada, A. M. Greenway and K. R. Seddon, *Eur. Mass Spectrom.*, 1996, **2**, 77; J. M. Miller, *Adv. Inorg. Chem. Radiochem.*, 1984, **28**, 1–27.

41. G. Siuzdak, S. V. Wendeborn and K. C. Nicolaou, *Int. J. Mass Spectrom. Ion Proc.*, 1992, **112**, 79.

42. L. A. P. Kane-Maguire, R. Kanitz and M. M. Sheil, *J. Organomet. Chem.*, 1995, **486**, 243. F. Bitsch, G. Hegy, C. Dietrich-Buchecker, E. Leize, J.-P. Sauvage and A. van Dorsselaer, *New J. Chem.*, 1994, **18**, 801.

43. J. M. Miller, *Mass Spectrom Rev.*, 1989, **9**, 319; M. I. Bruce and M. J. Liddell, *Appl. Organomet. Chem.*, 1987, **1**, 191.

44. R. L. Cerny, B. P. Sullivan, M. M. Bursey and T. J. Meyer, *Inorg. Chem.*, 1985, **24**, 397.

45. X. Liang, S. Suwanrumpha and R. B. Freas, *Inorg. Chem.*, 1991, **30**, 652.

46. J. M. Miller, K. Balasanmugam, J. Nye, G. B. Deacon and N. C. Thomas, *Inorg. Chem.*, 1987, **26**, 560.

47. K. A. N. Verkerk, C. G. de Koster, B. A. Markies, J. Boersma, G. Van Koten, W. Heerma and J. Haverkamp, *Organometallics*, 1995, **14**, 2081.

48. K. R. Jennings, T. J. Kemp and B. Sieklucka, *J. Chem. Soc., Dalton Trans.*, 1988, 2905.

49. A. L. Bandini, G. Banditelli, D. Favretto and P. Traldi, *Rapid Commun. Mass Spectrom.*, 2000, **14**, 1499.

50. M. I. Bruce and M. J. Liddell, *J. Organomet. Chem.*, 1992, **427**, 263.

51. M. C. Barral, R. Jimenez-Aparicio, J. L. Priego, E. C. Royer, E. C. and F. A. Urbanos, *Inorg. Chim. Acta*, 1998, **277**, 76.

52. D. M. Branan, N. W. Hoffman, E. A. McElroy, D. L. Ramage, M. J. Robbins, J. R. Eyler, C. H. Watson, P. DeFur and J. A. Leary, *Inorg. Chem.*, 1990, **29**, 1915.

53. A. L. Bandini, G. Banditelli, G. Minghetti, B. Pelli and P. Traldi, *Organometallics*, 1989, **8**, 590.

54. K. Lee and J. R. Shapley, *Organometallics*, 1998, **17**, 4368.

55. L. Q. Ma, U. Brand and J. R. Shapley, *Inorg. Chem.*, 1998, **37**, 3060.

56. P. D. Boyle, B. J. Johnson, B. D. Alexander, J. A. Casalnuovo, P. R. Gannon, S. M. Johnson, E. A. Larka, A. M. Mueting and L. H. Pignolet, *Inorg. Chem.*, 1987, **26**, 1346.

57. R. C. B. Copley and D. M. P. Mingos, *J. Chem. Soc., Dalton Trans.*, 1996, 491.

58. T. Chihara, L. P. Yang, Y. Esumi and Y. Wakatsuki, *J. Mass Spectrom.*, 1995, **30**, 684.

59. E. Delgado, E. Hernandez, O. Rossell, M. Seco and X. Solans, *J. Chem. Soc., Dalton Trans.*, 1993, 2191.

60. J. L. Haggitt and D. M. P. Mingos, *J. Chem. Soc., Dalton Trans.*, 1994, 1013.

61. S. M. Malinak, L. K. Madden, H. A. Bullen, J. J. McLeod and D. C. Gaswick, *Inorg. Chim. Acta*, 1998, **278**, 241.

62. W. L. Lee, D. A. Gage, Z. H. Huang, C. K. Chang, M. G. Kanatzidis and J. Allison, *J. Am. Chem. Soc.*, 1992, **114**, 7132.

63. K. Hegetschweiler, T. Keller, W. Amrein and W. Schneider, *Inorg. Chem.*, 1991, **30**, 873.

64. A. Trovarelli and R. G. Finke, *Inorg. Chem.*, 1993, **32**, 6034.

65. M. R. M. Domingues, M. G. Santana-Marques and A. J. Ferrer-Correia, *Int. J. Mass Spectrom. Ion Proc.*, 1997, **165/166**, 551.

66. B. J. Brisdon and A. J. Floyd, *J. Organomet. Chem.*, 1985, **288**, 305.

67. H. T. Kalinoski, U. Hacksell, D. F. Barofsky, E. Barofsky and G. D. Daves, Jr., *J. Am. Chem. Soc.*, 1985, **107**, 6476.

68. K. Tanaka, H. Waki, Y. Ido, S. Akita, Y. Yoshida and T. Yoshida, *Rapid Commun. Mass Spectrom.*, 1988, **2**, 151.

69. M. Karas, D. Bachmann, U. Bahr and F. Hillenkamp, *Int. J. Mass Spectrom. Ion Proc.*, 1987, **78**, 53.

70. F. Hillenkamp and M. Karas, *Int. J. Mass Spectrom.*, 2000, **200**, 71.

71. SciFinder Scholar, search for 'laser desorption' and 'mass spectrometry'. Note that a search of this nature will underestimate the number of papers especially as a technique becomes more established, because articles employing the technique in a routine way will often not be flagged.

72. R. Zenobi and R. Knochenmuss, *Mass Spectrom. Rev.*, 1998, **17**, 337.

73. M. Karas, M. Gluckmann and J. Schafer, *J. Mass Spectrom.*, 2000, **35**, 1.

74. S. C. Moyer and R. J. Cotter, *Anal. Chem.*, 2002, 469A.

75. MALDI Mass Spectrometry, *Analytix 6* (2001). Sigma-Aldrich analytical newsletter, available online at http://www.sigmaaldrich.com/img/assets/4242/fl_analytix6_2001_new.pdf.

76. R. Skelton, F. Dubois and R. Zenobi, *Anal. Chem.*, 2000, **72**, 1707.

77. M. J. Felton and C. M. Harris, *Anal. Chem.*, 2003, 54A.

78. J. Wei, J. M. Buriak and G. Siuzdak, *Nature*, 1999, **399**, 243; Z. Shen, J. J. Thomas, C. Averbuj, K. M. Broo, M. Englehard, J. E. Crowell, M. G. Finn and G. Siuzdak, *Anal. Chem.*, 2001, **73**, 612.

79. S. Tuomikoshi, K. Huikko, K. Grigoras, P. Östman, R. Kostiainen, M. Baumann, J. Abian, T. Kotiaho and S. Franssila, *Lab Chip*, 2002, **2**, 247.

80. D. Fogg, University of Ottawa, personal communication.

81. B. Salih, C. Masselon and R. Zenobi, *J. Mass. Spectrom.*, 1998, **33**, 994.

82. V. Marchan, V. Moreno, E. Pedroso and A. Grandas, *Chem. Eur. J.*, 2001, **7**, 808.

83. K. Osz, K. Varnagy, H. Sueli-Vargha, D. Sanna, G. Micera and I. Sovago, *J. Chem. Soc., Dalton Trans.*, 2003, 2009.

84. N. Weibel, L. J. Charbonniere, M. Guardigli, A. Roda and R. Ziessel, *J. Am. Chem. Soc.*, 2004, **126**, 4888.

85. L. Campardo, M. Gobbo, R. Rocchi, R. Berani, M. Mozzon and R. A. Michelin, *Inorg. Chim. Acta*, 1996, **245**, 269.

86. D. Ossipov, S. Gohil and J. Chattopadhyaya, *J. Am. Chem. Soc.*, 2002, **124**, 13416.

87. Z. Xie, C.-W. Tsang, E. T.-P. Sze, Q. Yang, D. T. W. Chan and T. C. W. Mak, *Inorg. Chem.*, 1998, **37**, 6444; C.-W. Tsang, Q. Yang, E. T.-P. Sze, D. T. W. Chan, T. C. W. Mak and Z. Xie, *Inorg. Chem.*, 2000, **39**, 3582; C.-W. Tsang, Q. Yang, E. T.-P. Sze, D. T. W. Chan, T. C. W. Mak and Z. Xie, *Inorg. Chem.*, 2000, **39**, 5851.

88. F. Teixidor, J. Pedrajas, I. Rojo, C. Vinas, R. Kivekaes, R. Sillanpaeae, I. Sivaev, V. Bregadze and S. Sjoeberg, *Organometallics*, 2003, **22**, 3414.

89. K. A. Jolliffe, M. C. Calama, R. Fokkens, N. M. M. Nibbering, P. Timmermann and D. N. Reinhoudt, *Angew. Chem. Int. Ed.*, 1998, **37**, 1247.

90. U. S. Schubert and C. Eschbaumer, *J. Incl. Phenom. Macrocyclic Chem.*, 1999, **35**, 101.

91. R. Kraemer, I. O. Fritsky, H. Pritzkow and L. A. Kovbasyuk, *J. Chem. Soc., Dalton Trans.*, 2002, 1307.

92. W. B. Jeon, K. H. Bae and S. M. Byun, *J. Inorg. Biochem.*, 1998, **71**, 163.

93. D. Fenyo, B. T. Chait, T. E. Johnson and J. S. Lindsay, *J. Porphyrins Phthalocyanines*, 1997, **1**, 93.

94. N. Srinivasan, C. A. Haney, J. S. Lindsay, W. Zhang and B. T. Chait, *J. Porphyrins Phthalocyanines*, 1999, **3**, 283.

95. H. Kitagishi, A. Satake and Y. Kobuke, *Inorg. Chem.*, 2004, **43**, 3394.

96. P. Weyermann, F. Diederich, J.-P. Gisselbrecht, C. Boudon and M. Gross, *Helv. Chim. Acta*, 2002, **85**, 571; P. Weyermann and F. Diederich, *Helv. Chim. Acta*, 2002, **85**, 599.

97. A. Ferri, G. Polzonetti, S. Licoccia, R. Paolesse, D. Favretto, P. Traldi and M. V. Russo, *J. Chem. Soc., Dalton Trans.*, 1998, 4063.

98. S. L. Darling, E. Stulz, N. Feeder, N. Bampos and J. K. M. Sanders, *New J. Chem.*, 2000, **24**, 262.

99. E. Stulz, C. C. Mak and J. K. M. Sanders, *J. Chem. Soc., Dalton Trans.*, 2001, 604.

100. F. O. Ayorinde, K. Garvin and K. Saaed, *Rapid Commun. Mass Spectrom.*, 2000, **14**, 608; F. O. Ayorinde, K. Garvin and K. Saaed, *Rapid Commun. Mass Spectrom.*, 1999, **13**, 2472.

101. H. W. Kroto, J. R. Heath, S. C. O'Brien, R. F. Curl, and R. E. Smalley, *Nature*, 1985, **318**, 162.

102. K. M. Kadish and R. S. Ruoff Eds., *Fullerenes: chemistry, physics and technology*, Wiley-Interscience, 2000. See also the thematic issue of *Acc. Chem. Res.*, 1992, **25**, issue number 3, on fullerene chemistry.

103. S. G. Penn, D. A. Costa, A. L. Balch and C. B. Lebrilla, *Int. J. Mass Spectrom. Ion Proc.*, 1997, **169/170**, 371; S. Lebedkin, S. Ballenweg, J. Gross, R. Taylor and W. Krätschmer, *Tetrahedron Lett.*, 1995, **36**, 4971.

104. J. Xiao, M. R. Savina, G. B. Martin, A. H. Francis, and M. E. Meyerhoff, *J. Am. Chem. Soc.*, 1994, **116**, 9341.

105. J. P. Huang, C. H. Yuan, J. Shiea and Y. C. Chen, *J. Analyt. Toxicol.*, 1999, **23**, 337.

106. A. M. Hawkridge and J. A. Gardella, Jr., *J. Am. Soc. Mass Spectrom.*, 2003, **14**, 95.

107. U. Bahr, A. Deppe, M. Karas, F. Hillenkamp and U. Giessmann, *Anal Chem.*, 1992, **64**, 2866.

108. P. O. Danis, D. E. Karr, Y. Xiong and K. G. Owens, *Rapid Commun. Mass Spectrom.*, 1996, **10**, 862.

109. E. Royo and H.-H. Brintzinger, *J. Organomet. Chem.*, 2002, **663**, 213.

110. H.-J. van Manen, R. H. Fokkens, N. M. M. Nibbering, F. C. J. M. van Veggel and D. N. Reinhoudt, *J. Org. Chem.*, 2001, **66**, 4643.

111. G. D. Engel and L. H. Gade, *Chem. Eur. J.*, 2002, **8**, 4319.

112. C. Kim, I. Jung, *Inorg. Chem. Commun.*, 1998, **1**, 427; C. Kim, I. Jung, *J. Organomet. Chem.*, 1999, **588**, 9.

113. K. Onitsuka, A. Shimizu and S. Takahashi, *Chem. Commun.*, 2003, 280.

114. J. Ruiz, E. Alonso, J.-C. Blais and D. Astruc, *J. Organomet. Chem.*, 1999, **582**, 139.

115. F. H. Köhler, A. Schell and B. Weber, *Chem. Eur. J.*, 2002, **8**, 5219; F. H. Köhler and A. Schell, *Rapid Commun. Mass Spectrom.*, 1999, **13**, 1088.

116. P. Juhasz and C. E. Costello, *Rapid Commun. Mass Spectrom.*, 1993, **7**, 343.

117. J. S. McIndoe, *Trans. Met. Chem.*, 2003, **28**, 122.

118. M. J. Dale, P. J. Dyson, B. F. G. Johnson, P. R. R. Langridge-Smith and H. T. Yates, *J. Chem. Soc., Chem. Commun.*, 1995, 1689; M. J. Dale, P. J. Dyson, B. F. G. Johnson, C. M. Martin, P. R. R. Langridge-Smith and R. Zenobi, *J. Chem. Soc., Dalton Trans.*, 1996, 771.

119. G. Critchley, P. J. Dyson, B. F. G. Johnson, J. S. McIndoe, R. K. O'Reilly and P. R. R. Langridge-Smith, *Organometallics*, 1999, **18**, 4090; P. J. Dyson, A. K. Hearley, B. F. G. Johnson, J. S. McIndoe and P. R. R. Langridge-Smith, *Inorg. Chem. Commun.*, 1999, **2**, 591; P. J. Dyson, J. E. McGrady, M. Reinhold, B. F. G. Johnson, J. S. McIndoe and P. R. R. Langridge-Smith, *J. Clust. Sci.*, 2000, **11**, 391; P. J. Dyson, A. K. Hearley, B. F. G. Johnson, J. S. McIndoe and P. R. R. Langridge-Smith, *J. Chem. Soc., Dalton Trans.*, 2000, 2521.

120. P. J. Dyson, A. K. Hearley, B. F. G. Johnson, P. R. R. Langridge-Smith and J. S. McIndoe, *Inorg. Chem.*, 2004, **43**, 4962.

121. P. J. Dyson, B. F. G. Johnson, J. S. McIndoe and P. R. R. Langridge-Smith, *Inorg. Chem.*, 2000, **39**, 2430.

122. M. E. Barr, B. R. Adams, R. R. Weller and L. F. Dahl, *J. Am. Chem. Soc.*, 1991, **113**, 3052.

123. A. Bjarnason, R. E. Desenfants, M. E. Barr and L. F. Dahl, *Organometallics*, 1990, **9**, 657.

124. S. Martinovic, L. P. Tolic, D. Srzic, N. Kezele, D. Plavsic, L. Klasinc, *Rapid Commun. Mass Spectrom.*, 1996, **10**, 51.

125. N. C. Dopke, P. M. Treichel and M. M. Vestling, *Inorg. Chem.*, 1998, **37**, 1272.

126. R. W. McGaff, N. C. Dopke, R. K. Hayashi, D. R. Powell and P. M. Treichel, *Polyhedron*, 2000, **19**, 1245.

127. S. Cerezo, J. Cortes, J.-M. Lopez-Romero, M. Moreno-Mana, T. Parella, R. Pleixats and A. Roglens, *Tetrahedron*, 1998, **54**, 14885.

128. M. Rattay, D. Fenske and P. Jutzi, *Organometallics*, 1998, **17**, 2930.

129. T. Glaser, R. H. Pawelke and M. Heidemeier, *Z. Anorg. Allg. Chem.*, 2003, **629**, 2274.

130. S. J. Lee, A. Hu and W. Lin, *J. Am. Chem. Soc.*, 2002, **124**, 12948.

131. Y. H. Xu, X. R. He, C. F. Hu, B. K. Teo and H. Y. Chen, *Rapid Commun. Mass Spectrom.*, 2000, **14**, 298.

132. J. E. Ham, B. Durham and J. R. Scott, *J. Am. Soc. Mass Spectrom.*, 2003, **14**, 393.

133. S. W. Hunsucker, R. C. Watson and B. M. Tissue, *Rapid Commun. Mass Spectrom.*, 2001, **15**, 1334.

134. T. G. M. M. Kappen, P. P. J. Schlebos, J. J. Bour, W. P. Bosman, J. M. M Smits, P. T. Beurskens and J. J. Steggerda, *J. Am. Chem. Soc.*, 1995, **117**, 8327.

135. T. G. M. M. Kappen, P. P. J. Schlebos, J. J. Bour, W. P. Bosman, J. M. M Smits, P. T. Beurskens and J. J. Steggerda, *Inorg. Chem.*, 1995, **34**, 2133.

136. D. M. Saysell and A. G. Sykes, *Inorg. Chem.*, 1996, **35**, 5536.

137. R. Hernandez-Molina, D. N. Dybtsev, V. P. Fedin, M. R. J. Elsegood, W. Clegg and A. G. Sykes, *Inorg. Chem.*, 1998, **37**, 2995.

138. N. B. Cech and C. J. Enke, *Mass Spectrom. Rev.*, 2001, **20**, 362; S. J. Gaskell, *J. Mass Spectrom.*, 1997, **32**, 677; S. A. Hofstadler, R. Bakhtiar and R. D. Smith, *J. Chem. Ed.*, 1996, **73**, A82, A84; R. Bakhtiar , S. A. Hofstadler and R. D. Smith, *J. Chem. Ed.*, 1996, **73**, A118; C. E. C. A. Hop and R. Bakhtiar, *J. Chem. Ed.*, 1996, **73**, A162, A164.

139. M. Dole, L. L. Mack, R. L. Hines, R. C. Mobley, L. D. Ferguson and M. B. Alice, *J. Chem. Phys.*, 1968, **49**, 2240.

140. M. Yamashita and J. B. Fenn, *J. Phys. Chem.*, 1984, **88**, 4451; M. Yamashita and J. B. Fenn, *J. Phys. Chem.*, 1984, **88**, 4671; C. M. Whitehouse, R. N. Dreyer, M. Yamashita and J. B. Fenn, *Anal. Chem.*, 1985, **57**, 675.

141. J. B. Fenn, M. Mann and C. K. Meng, *Science*, 1989, **246**, 64.

142. SciFinder Scholar, search for 'electrospray' and 'mass spectrometry'. Note that a search of this nature will underestimate the number of papers especially as a technique becomes more established, because articles employing the technique in a routine way will often not be flagged.

143. J. V. Iribarne and B. A. Thomson, *J. Chem. Phys.*, 1976, **64**, 2287.

144. (a) J. F. de la Mora, G. J. Van Berkel, C. G. Enke, R. B. Cole, M. Martinez-Sanchez and J. B. Fenn, *J. Mass Spectrom.*, 2000, **35**, 939; (b) P. Kebarle, *J. Mass Spectrom.*, 2000, **35**, 804.

145. G. J. Van Berkel, *J. Mass Spectrom.*, 2000, **35**, 773.

146. M. A. Kelly, M. M. Vestling, C. C. Fenselau and P. B. Smith, *Org. Mass Spectrom.*, 1992, **27**, 1143.

147. J. Zaia, R. S. Annan and K. Biemann, *Rapid Commun. Mass Spectrom.*, 1992, **6**, 32.

148. X.-B. Wang and L.-S. Wang, *Phys. Rev. Lett.*, 1999, **83**, 3402.

149. C. P. G. Butcher, B. F. G. Johnson, J. S. McIndoe, X. Yang, X. B. Wang and L. S. Wang, *J. Chem. Phys.*, 2002, **116**, 6560.

150. P. Dyson, I. Khalaila, S. Luettgen, J. S. McIndoe and D. Zhao, *Chem. Commun.*, 2004, 2204.

151. Prof. Peter Chen, personal communication.

152. R. Guevremont, J. C. Y. Le Blanc and K. W. M. Siu, *Org. Mass Spectrom.*, 1993, **28**, 1345.

153. R. Saf, C. Mirtl, and K. Hummel, *Tetrahedron Letters*, 1994, **35**, 6653.

4   The ESI MS Behaviour of Simple Inorganic Compounds

Introduction

The electrospray mass spectrometric behaviour of various inorganic compounds are surveyed in this and subsequent chapters. Of course, inorganic compounds cover a diverse range of materials, from simple metal salts, through to coordination and organometallic compounds, as well as 'newer' types of materials such as metal carbonyl clusters, fullerenes, dendrimers, supramolecular assemblies and bioinorganic materials. ESI is the most widely applied MS technique used for inorganic samples today. This dominance is reflected in the heavy emphasis placed on the technique in this and subsequent chapters. By contrast, relatively little work has concerned the MALDI analysis of inorganic materials; this is discussed separately in Section 3.7.

As a starting point, this chapter will focus on relatively simple compounds such as metal salts, main group oxoanions, and fullerenes, in order to focus on some of the basic patterns of ESI MS behaviour shown by inorganic materials. Throughout the text, examples selected from readily accessible materials are used to illustrate the behaviour patterns of the various sample types.

Simple Metal Salts

Since simple metal salts are among the most fundamental inorganic substances known, it is not surprising that many studies have investigated metal salt solutions by ESI MS, using both positive- and negative-ion modes.[1] The behaviour of a metal salt solution is highly dependent on the solvent, the charge on the cation, and on other factors such as cation size, the counteranion and electrochemical properties (as will be illustrated later for the special case of copper(II) and other reducible ions).

4.2.1   Salts of Singly-Charged Ions, M^+X^-

A solution of a simple alkali metal halide salt such as caesium bromide in a coordinating solvent such as acetonitrile (MeCN) gives a series of positive ions in the ESI mass spectrum (Figure 4.1) dependent on the cone voltage. At a low cone voltage (5 V, Figure 4.1a) the solvated caesium ion is observed, $[Cs(MeCN)_x]^+$, where x = 1 or 2. The solvent

Mass Spectrometry of Inorganic, Coordination and Organometallic Compounds   W. Henderson and J. S. McIndoe
© 2005 John Wiley & Sons, Ltd   ISBNs: 0-470-85015-9 (HB); 0-470-85016-7 (PB)

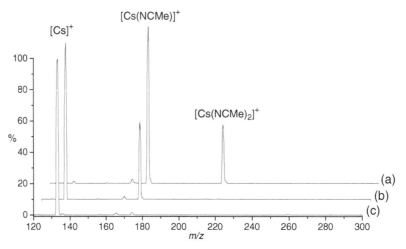

Figure 4.1
Positive-ion ESI mass spectra for an acetonitrile solution of CsBr at cone voltages of (a) 5 V, (b) 15 V and (c) 25 V, showing the solvation of the Cs^+ ion at lower cone voltages, and the bare Cs^+ ion at the higher cone voltage

molecules coordinate to the caesium ion and help to stabilise the positive charge. However, at higher cone voltages (15 V, Figure 4.1b; 25 V, Figure 4.1c) the solvent molecules are stripped away, leaving the bare Cs^+ cation. Hence, the actual species observed is highly dependent on the experimental conditions used, specifically the cone voltage (Sections 2.2 and 3.9.3).

The ESI response of a cation can vary considerably with its identity. This is nicely illustrated by Figure 4.2, which shows the positive-ion ESI mass spectrum of an

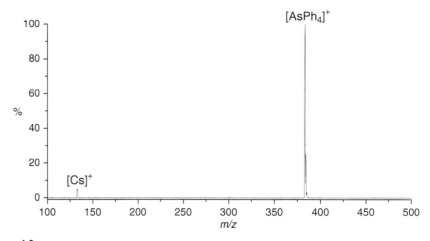

Figure 4.2
The significantly different ESI responses ('ionisation efficiencies') of the Cs^+ and $[Ph_4As]^+$ cations is clearly demonstrated in the positive-ion ESI mass spectrum of an equimolar mixture of CsCl and $[Ph_4As]Cl$ in methanol-water (1:1), at a cone voltage of 50 V. The high ionisation efficiency of the $[Ph_4As]^+$ ion is due to its high surface activity

equimolar solution of caesium chloride and an organic cation, tetraphenylarsonium chloride, [Ph$_4$As]Cl, in a 1:1 methanol-water solution. The spectrum is clearly dominated by the [Ph$_4$As]$^+$ cation, which has a significantly greater ESI response than the Cs$^+$ ion. Several theories have been proposed to account for this behaviour, which are dependent on large organic cations having a greater surface activity than a simple metal cation.[2] According to the ion evaporation model (Figure 3.38), ions that are less strongly solvated will have higher evaporation rates from the evaporating droplet, and will thus give a higher ESI response. In an alternative model,[2] the fixed amount of charge in a liquid droplet resides on the surface of the droplet; the inside of the droplet will be electrically neutral, most likely from ion-paired analytes that will not be observed in the mass spectrum. Hence, analytes with a higher surface activity (and a reduced tendency for ion-pair formation) will tend to show a greater ESI response.

The relative solvating power of different solvents also influences the ions observed, and can be easily investigated by the use of mixed solvent systems. In the case of silver nitrate in a mixed solvent system of acetonitrile and methanol, species containing exclusively MeCN ligands are seen, consistent with the superior donor ligand properties of MeCN towards Ag$^+$ ions. The spectrum at a low cone voltage (5 V; Figure 4.3a) also gives some information on the coordination properties of silver, which tends to prefer relatively low coordination numbers. Thus, the [Ag(MeCN)$_2$]$^+$ ion (*m/z* 189, 191) is essentially the only species in the spectrum. The tris(acetonitrile) complex is not observed. When the cone voltage is raised (20 V, Figure 4.3b; 50 V, Figure 4.3c) the acetonitrile ligands are stripped away, leaving the bare Ag$^+$ ion; it is noteworthy that the cone voltages required to remove the coordinated solvent molecules are higher than in the Cs$^+$ case, consistent with stronger metal-ligand complexes for silver. All of the silver species display the

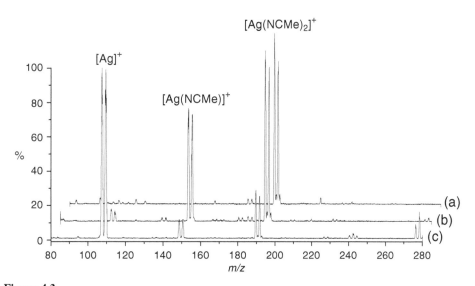

Figure 4.3
Positive-ion ESI mass spectra of AgNO$_3$ in a mixed methanol-acetonitrile solution, at cone voltages of (a) 5 V, (b) 20 V and (c) 50 V, showing the formation of [Ag]$^+$, [Ag(MeCN)]$^+$ and [Ag(MeCN)$_2$]$^+$ in various ratios at the different cone voltages

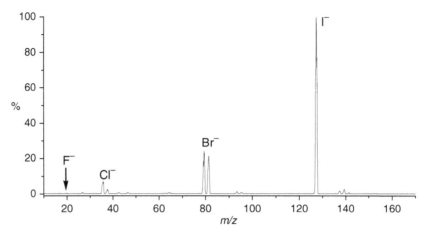

Figure 4.4
Negative-ion ESI mass spectrum of an equimolar mixture of NaF, NaCl, NaBr and NaI, in methanol at a cone voltage of 100 V, showing the decreasing ion intensities in the order $I^- > Br^- > Cl^- > F^-$

characteristic isotope distribution pattern of silver (^{107}Ag 100 %, ^{109}Ag 92.9 % relative abundances).

The use of negative-ion mode allows anions to be studied in exactly the same way. Thus, halide salts show the simple unsolvated halide ion at higher cone voltages; the negative-ion ESI mass spectrum of an equimolar mixture of NaF, NaCl, NaBr and NaI is shown in Figure 4.4. The spectrum is dominated by an intense signal due to the monoisotopic iodide ion (m/z 127), with successively weaker signals due to bromide (m/z 79, 81), chloride (m/z 35, 37) and almost undetectable fluoride (m/z 19). The Cl^- and Br^- ions display the characteristic isotopic signatures of these elements (see Appendix 1). The iodide ion has a high ionisation efficiency in ESI compared with the other halides; compare the high ionisation efficiency of $[Ph_4As]^+$ compared with Cs^+, Figure 4.2. Using low energy ionisation conditions, solvation of anions can be observed. Thus, ESI MS analysis (in the m/z range 0–170) of F^-, Cl^-, Br^- and I^- gives the ions $[F(H_2O)_n]^-$ (n = 0–5), $[Cl(H_2O)_n]^-$ (n = 0–4), $[Br(H_2O)_n]^-$ (n = 0–3) and $[I(H_2O)_n]^-$ (n = 0–2). The greater degree of solvation of the smaller, higher charge-density fluoride than the larger, low charge-density iodide is noteworthy, and correlates well with the decreasing tendency to participate in hydrogen bonding in the sequence $F^- > Cl^- > Br^- > I^-$.[3]

For many metal salts, aggregate (or 'salt cluster') ions are also observed when spectra are recorded over wider m/z windows. For a salt M^+X^- the simplest ions are of the type $[(MX)_n + M]^+$ (in positive-ion mode) and $[(MX)_n + X]^-$ in negative ion mode, though higher-nuclearity ions with higher charges can also be observed. For example, alkali metal chloride salts, and many sodium salts NaX ($X^- = HCO_2^-$, $CH_3CO_2^-$, ClO_3^-, NO_2^- and NO_3^-) behave in this way, and the cone voltage is the principal influence that affects the distribution of the various ions.[4] Many other salts would be expected to show the same behaviour.

A typical ESI spectrum of such a metal salt, sodium iodide (which also contains a trace of caesium to provide an additional low m/z Cs^+ ion) is shown in Figure 4.5. The upper

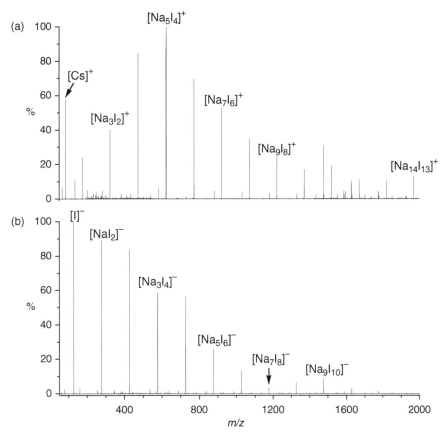

Figure 4.5
ESI mass spectra of a 'calibration standard' solution of sodium iodide in methanol, containing a trace of added Cs^+ (to provide an additional low m/z ion: (top) positive ion mode; (bottom) negative-ion mode. The spectra show the formation of positive $[(NaI)_x + Na]^+$ and negative $[(NaI)_y + I]^-$ ions respectively

spectrum (a) shows the ions observed in positive-ion mode and the lower spectrum (b) shows the negative-ion spectrum. This aggregate ion formation finds practical application for mass calibration in both positive- and negative-ion modes.[5] Caesium iodide and sodium iodide form aggregate ions over a wide mass range, and because of the monoisotopic nature of sodium (^{23}Na), caesium (^{133}Cs) and iodine (^{127}I) the resulting ions occur as single, sharp peaks that provide accurate mass standards.

Although the various ions can be easily assigned in this case, analysis of samples that contain significant concentrations of salts (for example, buffers, widely used on protein studies) may cause problems due to overlapping of peaks. Furthermore, suppression of the signal from the desired analyte can occur; with the bulk of the charged species being metal salt-derived ions, the analyte will bear a relatively small fraction of the charge in the ES ionisation process, resulting in low signal intensity.

The ESI MS Analysis of Ionic Liquids and Dissolved Catalysts

Ionic liquids are cation/anion combinations – i.e. salts – that resist crystallisation so remain as liquids at ambient temperature. There is an exponentially increasing interest in ionic liquids as solvents because of their excellent properties, especially their zero volatility, good solvating power, low toxicity, non-flammability and electrochemical inertness.[6] The majority so far reported are organic cations such as alkylimidazolium with anions such as $[AlCl_4]^-$, $[BF_4]^-$, or $[PF_6]^-$. Because ionic liquids are salts of the type A^+X^-, the analysis of these materials by ESI MS gives a series of aggregate ions, for example [bmim]PF_6 (bmim = 1-butyl-3-methylimidazolium) gives positive ions $[bmim]^+$, $[(bmim)_2(PF_6)]^+$, $[(bmim)_3(PF_6)_2]^+$, $[(bmim)_4(PF_6)_3]^+$, $[(bmim)_5(PF_6)_4]^+$, and $[(bmim)_6(PF_6)_5]^+$ in the m/z range 100 – 1600.[7]

ESI MS can also be used to characterise small amounts of dissolved transition metal catalyst in the presence of abundant ions from the ionic liquid. For example, a solution of the complex $[Ru(\eta^6\text{-}p\text{-cymene})(\eta^2\text{-triphos})Cl]^+$ **4.1** (as its PF_6^- salt) in [bmim]PF_6 solution can be observed as the parent ion (m/z 895) in positive-ion ESI MS, after dilution with methanol. ESI MS has also been used to identify traces of chloride impurity in [bmim]BF_4, by the characteristic chloride isotope pattern at m/z 35 and 37.[8] Traces of chloride in ionic liquids can affect both the physical properties such as viscosity, and also the performance of transition metal catalysts. ESI MS has also been applied to the analysis of **undiluted** ionic liquids, which could be useful for the analysis of solutes that are reactive towards diluents such as methanol.[9] However, this could lead to significant contamination of the instrument, and it has been suggested that *solutions* of ionic liquids better facilitate the analysis of these substances and that the ESI MS analysis of neat ionic liquids is in fact assisted by evaporation of residual traces of volatile solvents.[10]

Overall, ESI MS provides a convenient method for the analysis of ionic liquids and dissolved species such as catalysts and impurities.

4.1

4.2.2 Salts of Multiply-Charged Ions, e.g. $M^{2+}(X^-)_2$, $M^{3+}(X^-)_3$ etc.

The pattern of observed ions can become considerably more complex when the ion(s) have higher charges. For a solution containing a dication, such as Mg^{2+}, Ca^{2+} or Zn^{2+}, solvated ions are observed in the gas phase, but are typically more strongly solvated than monocations.[11,12] Salt cluster ions are also observed for MX_2 salts, for example calcium chloride ($CaCl_2$) gives ions of the type $[(CaCl_2)_n + CaCl]^+$, and $[(CaCl_2)_n + Cl]^-$, as well as more highly-charged salt cluster ions.[13]

A trivalent ion is generally not seen in its simple hydrated form, $[M(H_2O)_n]^{3+}$, in the gas phase, and the same is true of formally tetravalent ions such as Th^{4+}. The key factor here is not the absolute charge on the ion but the **charge density**. Highly charged ions will typically undergo a **charge reduction** process, which is often deprotonation of a coordinated solvent molecule (such as water, methanol etc) that contains a labile, 'acidic' proton. This process has been discussed extensively in the literature.[12,14,15,16] The initial process for a solvated trivalent metal ion is:

$$[M(H_2O)_n]^{3+} + H_2O \longrightarrow [M(OH)(H_2O)_{n-1}]^{2+} + H_3O^+$$

The resulting dication is more stable than the trication towards further charge reduction, though depending on the conditions, such as increased cone voltage, further charge reduction may occur. Alternatively, if the counteranion is itself a good ligand, then coordination to a multiply-charged metal ion often occurs so that the resulting ions are singly charged. Ultimately, at very high cone voltages, ligands are stripped away from the metal centre, which is reduced to a low positive oxidation state (typically M^+ for many metals, but sometimes oxo ligands are retained due to the strong oxophilicity of the metal). This use of an electrospray instrument is termed '*bare metal mode*', sometimes referred to as *elemental mode*.

An example of the behaviour shown by a formally highly-charged cation is uranyl nitrate, $UO_2(NO_3)_2$. Uranium is formally in the hexavalent state in this compound, but the UO_2^{2+} (uranyl) group has a high stability and complexes are typically considered as those of the UO_2^{2+} dication. Nevertheless, at high cone voltages the UO_2^{2+} dication can be fragmented. Positive-ion ESI mass spectra for $UO_2(NO_3)_2$ are given in Figure 4.6. The spectra illustrate the key features of ion formation by coordination of a nitrate anion to the uranyl dication, solvation of the resulting cation, and (at higher cone voltages) charge reduction to lower oxidation state uranium species. At a low cone voltage (5 V,

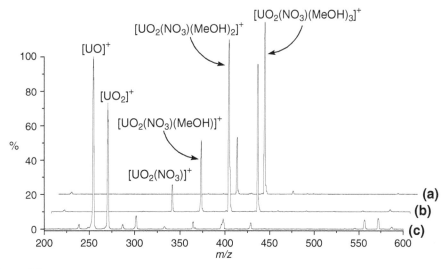

Figure 4.6
Positive-ion ESI mass spectra of uranyl nitrate, $UO_2(NO_3)_2$ in methanol at cone voltages of (a) 5 V, (b) 20 V and (c) 200 V

Figure 4.6a), the only major species observed are $[UO_2(NO_3)(MeOH)_2]^+$ (*m/z* 398) and $[UO_2(NO_3)(MeOH)_3]^+$ (*m/z* 429). When the 5 V spectrum is recorded using ethanol as the solvent, the analogous species $[UO_2(NO_3)(EtOH)_3]^+$ is observed at *m/z* 470. This allows methoxy species such as $[UO_2(OMe)(MeOH)_4]^+$, which also has *m/z* 429, to be ruled out. At 20 V (Figure 4.6b), the methanol solvent molecules are stripped away, leading to the bare $[UO_2(NO_3)]^+$ cation (*m/z* 334), while in the less volatile ethanol solvent, $[UO_2(NO_3)(EtOH)_2]^+$ (*m/z* 424) is the base peak. At a very high cone voltage (200 V, Figure 4.6c) reduction of the uranium from the hexavalent state occurs giving pentavalent $[UO_2]^+$ (*m/z* 271) and trivalent $[UO]^+$ (*m/z* 255).

The latter spectrum is an example of the use of an ESI instrument in 'elemental mode', as discussed earlier in this section. Other examples are thorium(IV) nitrate and indium(III) chloride (discussed below) and the conversion of the $[Me_3Pb]^+$ cation to $[MePb]^+$ and then bare $[Pb]^+$, discussed in Chapter 6.

The related actinide nitrate $Th(NO_3)_4$ also illustrates the general features of this type of compound, though the spectra are more complex. The nitrate anion acts as a bidentate chelating ligand towards the formal Th^{4+} ion (which has no existence in an unligated state due to its high charge). In addition the methanol solvent is able to be deprotonated giving the OMe^- (methoxide) ligand. The observed ions in the positive-ion ESI mass spectrum are all monocations containing coordinated anions; spectra of $Th(NO_3)_4$ in methanol solution at various cone voltages are given in Figure 4.7. Various solvated ions are formed at low cone voltages (e.g. 20 V), for example the ion at *m/z* 514 is assigned as $[Th(NO_3)_3(MeOH)_3]^+$. The assignment of this ion is confirmed by running the spectrum of $Th(NO_3)_4$ in ethanol solution, which gives a corresponding $[Th(NO_3)_3(EtOH)_3]^+$ ion at

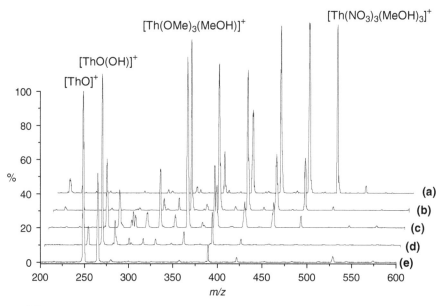

Figure 4.7
Positive-ion ESI mass spectra of $Th(NO_3)_4$ in methanol at a range of cone voltages: (a) 20 V, (b) 40 V, (c) 60 V, (d) 100 V and (e) 160 V

m/z 556. At lower m/z values, and increasingly at higher cone voltages, ions are formed by replacement of NO_3^- by OMe^-. For example, the ion at m/z 420 is assigned to $[Th(OMe)_3(MeOH)_3]^+$ (m/z 420); $[Th(NO_3)_3]^+$ would have m/z 418. At an intermediate cone voltage (60 V, Figure 4.7c) the $[Th(OMe)_3(MeOH)]^+$ ion at m/z 357 dominates the spectrum. At very high cone voltages, for example 160 V (Figure 4.7e) 'bare' ions (in which all the original nitrate and alcohol ligands have been removed) are observed– $[ThO]^+$ at m/z 248 and $[ThO(OH)]^+$ at m/z 265.

Main group metal salts behave in a similar way; positive-ion ESI mass spectra for indium (III) chloride ($InCl_3$) at a range of cone voltages are shown in Figure 4.8. At low cone voltages (5 and 20 V, Figure 4.8a and b respectively), the spectra are dominated by the singly charged solvated ions $[InCl_2(MeOH)_x]^+$ ($x = 3, 4$). These ions have distinct isotope patterns due to the presence of two chlorine atoms. On increasing the cone voltage, the methanol molecules are stripped away, giving the bare $[InCl_2]^+$ ion, Figure 4.8c. At a high cone voltage (150 V, Figure 4.8d), reduction occurs giving the In^+ cation. The ESI MS experiment is thus providing information on the stability of lower oxidation states of a main group metal. In(I) is a well-known oxidation state for this element, stabilised by the inert pair effect, and it can easily be generated using mass spectrometry. For some very easily-reduced metal ions such as copper(II), reduced species are observed even at very low cone voltages; this is discussed in greater detail in Section 4.2.4.

Although trivalent M^{3+} ions have never been seen in the gas phase solvated by water or methanol molecules alone, by changing to a different electrospray solvent, such as acetonitrile (MeCN), the solvated $[La(MeCN)_x]^{3+}$ ion can be seen, even with as few as

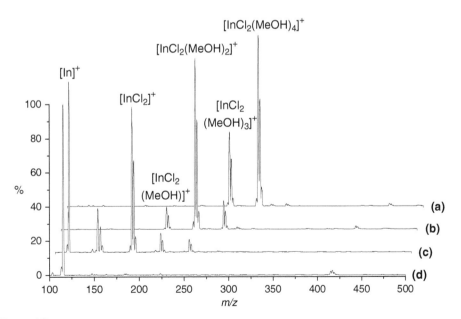

Figure 4.8
Positive-ion ESI mass spectra of $InCl_3$ in methanol at cone voltages of (a) 5 V, (b) 20 V, (c) 80 V and (d) 150 V, showing the formation of the solvated cations $[InCl_2(MeOH)_x]^+$ ($x = 0-4$) at low to intermediate cone voltages, and reduction to In^+ at high cone voltages

three or four MeCN solvate molecules. Dimethyl sulfoxide ($SOMe_2$), is a similar aprotic solvent that has been used to generate gas-phase solvated trivalent cations $[M(SOMe_2)_n]^{3+}$ of both main group elements (aluminium, gallium, indium, bismuth), transition metals (vanadium, iron, chromium), and selected lanthanides (lanthanum, ytterbium) and scandium.[17] Acetonitrile and dimethyl sulfoxide are better donor ligands than water or methanol, thus stabilising the charge on the metal ion. The same behaviour was also seen in Figure 4.2 in the preference of silver ions for acetonitrile over methanol. Because acetonitrile and dimethyl sulfoxide lack labile, 'acidic' protons, the charge reduction process for multiply-charged cations (analogous to equation 4.1) is suppressed.[16,18]

4.2.3   Negative-Ion ESI Mass Spectra of Metal Salts

The spectra of metal salts can be considerably simplified when negative-ion spectra are recorded. Metal salts themselves will typically generate negatively charged species in solution that can be detected by ESI MS; of course, the extent to which this occurs is dependent on the coordinating ability of the anion. Negative-ion formation can be enhanced by adding an excess of an anion, for example as HCl or HNO_3 (which are volatile, and so will not cause instrument blockages). In this way, metal *cations* can be converted into (singly-charged) *anions* of the type $[MX_n]^-$. As an example, while the positive-ion ESI mass spectra of $Th(NO_3)_4$ in methanol (Figure 4.7) are relatively complex, the negative-ion spectrum (Figure 4.9) is in contrast very simple, comprising only a single ion due to the $[Th(NO_3)_5]^-$ ion at *m/z* 542. Many other metal salts behave in the same way, for example $Bi(NO_3)_3$ gives $[Bi(NO_3)_4]^-$, $Pr(NO_3)_3$ gives $[Pr(NO_3)_4]^-$ and $UO_2(NO_3)_2$ gives $[UO_2(NO_3)_3]^-$. Al^{3+} and Y^{3+} ions also give $[M(NO_3)_4]^-$ ions with excess added HNO_3.[19]

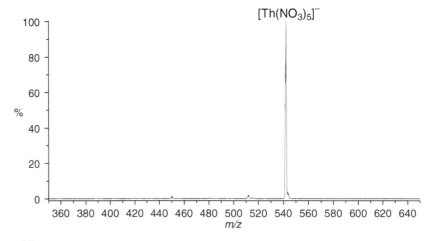

Figure 4.9
Negative-ion ESI mass spectrum of $Th(NO_3)_4$ in methanol at a cone voltage of 20 V, showing the exclusive formation of the $[Th(NO_3)_5]^-$ ion

The formation of anions can help to provide information on metal ion speciation. One example is the conversion of $[Fe(H_2O)_6]^{3+}$ with excess Cl^- into the very stable $[FeCl_4]^-$ ion, which is detected by ESI MS. Such negative ions can be more stable towards the moderate CID conditions that are needed to remove associated solvent molecules, and so give easily characterised ions that help to preserve the metal in its original oxidation state, though reduction can still occur. By removing coordinated water molecules and the high positive charge on a $[M(H_2O)_n]^{x+}$ cation, the main driving force for the charge reduction process is thus removed. In this way, a wide range of metal ions in combination with anions such as iodide, chloride, and fluoride have been found to give simple metal-halide anions in ESI MS.[20]

4.2.4 ESI MS Behaviour of Easily-Reduced Metal Ions: Copper(II), Iron(III) and Mercury(II)

The reduction processes (operating at elevated cone voltages, predominantly in positive ion mode) described for various metal ions in the preceding section are amplified for easily reducible metal ions such as copper(II) and iron(III). Copper(II) is a metal ion that shows mild oxidising characteristics, being reduced in the process to copper(I). The ESI MS behaviour shown by an aqueous solution of a copper(II) salt such as $CuCl_2$ under gentle ionisation conditions shows differences to that of a non-reducible metal salt such as calcium chloride. While the expected salt cluster ions, $[(CuCl_2)_n + CuCl]^+$ are observed, there are also cluster ions which contain additional CuCl moieties, containing copper in the $+1$ oxidation state.[13]

Several explanations appear to have been put forward in order to explain the gas phase reduction of copper(II) species. One explanation is that an electric discharge between the ESI capillary and the sampling cone (which is visibly observed as a blue discharge) produces gaseous electrons that can reduce metal ions (with appropriate reduction potentials) on the surface of a droplet, or in the gas phase:[13,21]

$$Cu^{2+} + e^- \longrightarrow Cu^+$$

Alternatively, charge transfer between the metal complex and gas phase solvent molecules has been proposed to account for the reduction of various copper(II) complexes. A series of solvents were examined, and the solvent ionisation energy was correlated with the amount of reduced copper(I) observed in the ESI mass spectrum. Complexes that are more readily reduced in solution also tend to be more easily reduced in the gas phase, and (solution) redox potentials were also found to be important.[22]

The overall effect of the above mass spectrometric processes are that copper(I)-containing ions are widely observed in solutions of copper(II) salts, the relative intensity of which is dependent on various factors such as the solvent, the stability of the metal-ligand combination, and the cone voltage. As the cone voltage is increased, the relative intensity of copper(I) species increases at the expense of copper(II). The solvent can also have a very dramatic effect, since the ESI MS analysis of Cu^{2+} ions in acetonitrile solution yields only copper(I) ions, due to the well-known stabilisation of copper(I) by this ligand.[23,24,25]

This behaviour is illustrated by the ESI MS spectra for $CuCl_2$, which yields the bare Cu^+ ion, in addition to solvated analogues such as $[Cu(H_2O)]^+$ and $[Cu(MeOH)]^+$.[26] For example, positive-ion ESI mass spectra for a methanol solution of $CuCl_2$ at cone voltages

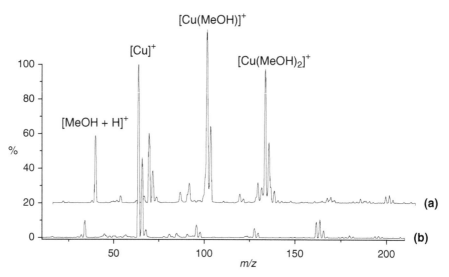

Figure 4.10
The reduction of a copper(II) solution [CuCl$_2$] during ESI MS analysis is clearly shown by the formation of various solvated Cu$^+$ species at a cone voltage of 40 V (a). At a higher cone voltage (60 V, b) the methanol molecules are stripped away, leaving the bare Cu$^+$ ion

of 40 and 60 V are shown in Figure 4.10. At both cone voltages only copper(I) species are observed; at the lower cone voltage the Cu$^+$ ion is solvated by methanol molecules, which are stripped away at the higher cone voltage, leaving the bare Cu$^+$ ion. However, addition of a ligand known to form stable complexes with copper(II) can alter the species observed; thus when EDTA is added to a copper(II) solution, this forms a very stable complex with Cu^{2+} ions, and in the ESI mass spectrum, only copper(II)-EDTA species are seen.[26]

Reduction of copper(II) during ESI MS analysis has been observed in the presence of various ligands such as: 15-crown-5, 2,2'-bipyridine,[27] urea,[28] glycine[29] and other amino acids,[30] and the tripeptide GlyHisLys.[21] Other easily-reduced metal ions show similar ESI MS behaviour to copper(II), also giving reduced ions under low cone voltage conditions. Iron(III) is an example, due to the similar stabilities of the II and III oxidation states.

Metal ions such as iron(III) and copper(II) also easily give reduced species in *negative*-ion mode; ESI MS analysis of Fe dissolved in dilute nitric acid yields a mixture of iron(III) and iron(II) ions–[Fe(NO$_3$)$_4$]$^-$ and [Fe(NO$_3$)$_3$]$^-$ respectively, and copper(II) behaves similarly.[19]

In this case, electrolysis of the *solution* at the metal capillary is proposed to be the main source of reduced species through the simple redox reactions:

$$Cu(II) + e^- \longrightarrow Cu(I)$$
$$Fe(III) + e^- \longrightarrow Fe(II)$$

Contrast the reduction process in *positive*-ion mode: Under the same conditions, yttrium and aluminium produce only [Y(NO$_3$)$_4$]$^-$ and [Al(NO$_3$)$_4$]$^-$, because the trivalent state of these metals is highly stable and cannot be reduced in aqueous solution.[19] It is even possible to fully reduce metal ions such as Cu^{2+}, Ag^+ and Hg^{2+} to the metals on the electrospray emitter when operated in negative-ion mode. On switching to positive-ion mode, the deposited metals are then re-oxidised and released back into solution.[31] This could partly account for the tendency for silver ions to linger in ESI instruments (Section 3.9.3).

Polyanions Formed by Main Group Elements

Negative-ion ESI MS has been used to probe speciation in aqueous solutions of main group polyanions, such as polyselenides[32] and polyiodides,[33] and the technique should find application to the study of other polyatomic main group species in due course.

Polyselenides are air-sensitive species but can be analysed successfully using inert atmosphere conditions. A solution of Na_2Se_4 yielded many polyselenide anions, including Se_n^{2-}, protonated HSe_n^- and sodiated $NaSe_n^-$ ($n = 2-5$). All observed ions were hydrated, and ions with fewer waters of hydration typically existed as the protonated or sodiated ions. The tendency for more highly charged anions to be more strongly hydrated is also seen in other anion systems such as SO_4^{2-} (Section 4.4).[34,35]

Polyiodides have been known for many years, and many different structural types have been characterised in the solid state. However, relatively little is known about their solution speciation, so ESI MS is a potentially very useful technique for investigating polyhalogen anions, including interhalogen species formed by more two or more different halogens. Solutions of I_2 plus X^- ($X = Cl$, Br, I, NCS) in water reveal members of the series X_3^-, X_2I^-, XI_2^- and I_3^-, due to rapid redistribution reactions. No adducts were observed between I_2 and OH^-, NCO^- (cyanate), N_3^- (azide) or NO_2^- (nitrite). In the case of I_2 plus iodide, there was no evidence for the I_5^- anion which has been proposed previously, but instead, the equilibrium $I_3^- + I^- = I_4^{2-}$ is compatible with the ESI MS results and earlier literature. Aqueous solutions of I_2 plus Cl^- contain the ICl_4^- ion, which undergoes decomposition in the gas phase to give the previously unknown ICl_3^- radical anion.

When a mixture of iodine and pyridine is dissolved in methanol, a brown solution is formed, and positive-ion ESI MS shows the ion $[py_2H]^+$ as the base peak at *m/z* 160, together with $[py_2I]^+$ (*m/z* 285) and $[pyI]^+$ (*m/z* 206), depending on the cone voltage.[36] Iodine is known to form solvated cationic species when dissolved in polar coordinating solvents and ESI MS can be used to provide evidence on their existence.

Oxoanions Formed by Main Group Elements

Non-metal oxyanions have been widely studied by ESI MS and related techniques. In general, at low cone voltages, species with a low charge density give the parent ions with little or no fragmentation or aggregation to reduce the charge. For more highly-charged species, charge-reduction processes may occur. Ions having a higher charge density may undergo *protonation* or *metallation*, or at higher cone voltages, *fragmentation*, in order to reduce the charge density. Under gentle ionisation conditions, solvated ions can also often be observed.

As an illustrative example, negative-ion ESI MS spectra for $NaIO_3$ are shown in Figure 4.11. At a low cone voltage (25 V, Figure 4.11a) the $[IO_3]^-$ ion (*m/z* 175) is essentially

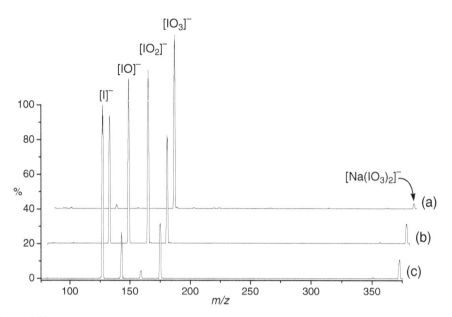

Figure 4.11
Negative-ion ESI mass spectra for $NaIO_3$ under different cone voltages, showing the parent $[IO_3]^-$ ion at low cone voltages (a, 25 V), and lower oxidation state iodine species at higher cone voltages (b, 100 V; c, 180 V)

the only species observed, except for a trace of I^- (*m/z* 127). As the cone voltage is increased, to 100 V (Figure 4.11b) and then 180 V (Figure 4.11c), fragmentation occurs, giving the species $[IO_2]^-$ (*m/z* 159), $[IO]^-$ (*m/z* 143) and finally I^-; all of these species are known, stable iodine species in their own right. The stabilisation of the $[IO_3]^-$ anion by coordination to sodium cations, forming the $[Na(IO_3)_2]^-$ adduct ion, can be clearly seen in the higher cone voltage spectra. The reduction of gas-phase ions to lower oxidation state species at elevated cone voltages is a common feature for these oxyanions and also for many other inorganic species, such as transition metal oxoanions (such as ReO_4^-, Section 5.10.1), as well as metal cations (Section 4.2.2).

Quantitative analysis of oxyhalide anions such as $[ClO_3]^-$, $[ClO_4]^-$ and $[IO_3]^-$ (and various halide ions) by ESI MS has been carried out.[3] Electrospray has also been coupled to ion chromatography for the sub-ppb level analysis of a number of oxyhalide anions such as $[BrO_3]^-$, $[ClO_3]^-$, $[ClO_2]^-$ and $[IO_3]^-$. The oxyanions are initially extracted using an ion exchange column, and then eluted online into the mass spectrometer.[37]

A comprehensive ESI MS study of the behaviour of sulfur oxyanions has been carried out by Stewart and co-workers.[35] For the ions $[S_2O_8]^{2-}$ and $[SO_4]^{2-}$, spectra under low fragmentation conditions tend to be dominated by hydrated ions. The bare $[S_2O_8]^{2-}$ ion is observed, being large enough to stabilise the two negative charges. For sulfate, four water molecules were found to be required to maintain the charge, after which further loss of water resulted in formation of the $[HSO_4]^-$ ion:

$$[SO_4(H_2O)_4]^{2-} \longrightarrow [HSO_4]^- + [OH(H_2O)_n]^- + (3\text{-}n)H_2O$$

For thiosulfate, $[S_2O_3]^{2-}$, which is less stable than sulfate, low energy fragmentation conditions give a series of hydrated thiosulfate ions $[S_2O_3(H_2O)_x]^{2-}$ ($x = 3 - 9$). Upon CID, the principal ion produced is $[S_2O_3]^-$ (*m/z* 112) and solvated analogues containing one and two water molecules. Sulfite, $[SO_3]^{2-}$, was found to react with the methanol solvent used, giving solely the $[MeOSO_2]^-$ ion (*m/z* 95). Oxidation of thiosulfate by iodine resulted in formation of the $[S_4O_6]^{2-}$ ion and solvated analogues thereof. The $[S_4O_6]^{2-}$ ion could be desolvated by CID, but then underwent fragmentation to species such as $[S_2O_3]^-$, $[S_3O_3]^-$ and $[SO_3]^-$. Reaction of $[S_2O_3]^{2-}$ with MnO_2 was also studied, and furnished the $[S_2O_6]^{2-}$ ion, in addition to $[HSO_4]^-$, $[CH_3OSO_3]^-$, $[HS_2O_6]^-$ and some other ions. Silver(I) complexes of thiosulfate ions were also detected in the reaction mixture of AgCl and $[S_2O_3]^{2-}$. A separate study of the stability and hydration of gas phase sulfur oxyanions generated by ESI has been carried out;[34] the ions $[SO_4]^{2-}$, $[S_2O_6]^{2-}$ and $[S_2O_8]^{2-}$ were found to exist as strongly hydrated species in the gas phase. CID of $[SO_4(H_2O)_n]^{2-}$ led to loss of H_2O down to n approximately three, at which stage charge separation occurred, giving $[HSO_4]^-$ and OH^-; naked $[SO_4]^{2-}$ could not be detected, in accord with the study by Stewart *et al*. In contrast, $[S_2O_6]^{2-}$ and $[S_2O_8]^{2-}$ showed greater stability in the gas phase because of the more extensive charge delocalisation.

Given their important practical applications in numerous consumer and industrial products, an ESI MS study has been carried out on a range of phosphate and polyphosphate anions, specifically orthophosphate **4.2**, pyrophosphate **4.3**, tripolyphosphate **4.4**, trimetaphosphate **4.5** and tetrapolyphosphate **4.6**.[38] The expected behaviour was seen, with the smaller ions orthophosphate and pyrophosphate appearing as singly-charged (protonated) ions $[H_2PO_4]^-$ and $[H_3P_2O_5]^-$ respectively, and the larger polyphosphates giving primarily doubly–charged species, e.g. $[HP_3O_9]^{2-}$ for trimetaphosphate. Negative-ion ESI MS was also successfully applied to the analysis of polyphosphate mixtures and to the quantitative analysis of solutions containing polyphosphate ions, including natural systems such as stream water.

Silicate oligomers have been characterised in aqueous solutions using ESI MS.[39] Solutions of silica dissolved in $(R_4N)OH$ (R = Me, Et) were analysed by negative-ion ESI MS and results compared with data from ^{29}Si NMR spectroscopy. The complex speciation of the solutions was confirmed by the study, which revealed a range of linear, cyclic and polyhedral silicate oligomers.

Borane Anions

Several studies have concerned the analysis of boranes by ESI MS. A wide variety of anionic borane salts e.g. $[B_{11}H_{13}-C_6H_{13}]^-$, $[B_{11}H_{14}]^-$, $[B_{10}H_{13}]^-$, $[B_{12}H_{11}SH]^{2-}$ and $[B_{12}H_{10}(SCN)_2]^-$ give the parent ion, and hence the molecular weight.[40] The salts $Me_4N[B_3H_8]$ and $Cs[B_3H_8]$ have been examined in positive-ion mode in MeCN and methanol solvent systems, which exhibit differences. In MeCN, cationic cluster ions containing the cation and anion were observed, whereas in methanol solution only $[B(OMe)_4]^-$ cluster ions were observed, as a result of electrochemical processes operating in the ESI capillary.[41] A negative-ion study of $Me_4N[B_3H_8]$, $Cs[B_3H_8]$, $^nBu_4N[B_3H_8]$, $Cs_2[B_{10}H_{10}]$ and $Na_2[B_{12}H_{12}]$ in MeCN also yielded cluster ions formed between cations and anions.[42] Reactions between $RHSiCl_2$ (R = H, Me, Ph) and *nido*-$[B_{10}H_{12}]^{2-}$ in ether produce $B_{10}H_{14}$ and the silaborane anions $[RSiB_{10}H_{12}]^-$; the R = H compound was characterised by negative-ion ESI MS.[43]

Fullerenes

While MALDI is probably the premier technique for the analysis of simple fullerenes, ESI MS analysis can also be directly achieved as a result of either electrochemical oxidation or reduction in the electrospray capillary. A mixture of C_{60} and C_{70} (termed fullerite) has been analysed in dichloromethane solution and gave radical anions $[C_{60}]^{\bullet-}$ and $[C_{70}]^{\bullet-}$ in negative-ion mode, though no corresponding radical cations were observed in positive-ion mode.[44] C_{60} and C_{70} have also been analysed directly in toluene solution, and gave the radical ions $[C_{60}]^{\bullet-}$ and $[C_{60}]^{\bullet+}$ in negative- and positive-ion modes respectively; the presence of minor ions due to $C_{60}O^-$ and $C_{60}(CH_2Ph)^-$ were also observed.[45] Higher fullerenes such as C_{84} and C_{90} have been found to give dianions by electrochemical reduction in ESI MS; these fullerenes have higher electron affinities compared to C_{60} and C_{70}.[46] Oxidation or reduction using chemical reagents has also been used to generate fullerene ions for mass spectrometric analysis.[47] A number of indirect approaches have also been undertaken which provide easily ionisable fullerene derivatives for mass spectrometric analysis, including incorporation of crown ether groups[48,49] and amination.[50]

The addition of cyanide ions to fullerenes results in the formation of anionic fullerene-cyano adducts, such as $[C_{60}(CN)_m]^{2-}$ which can be detected by mass spectrometry.[51] When a 100-fold molar excess of NaCN is added to C_{60}, numerous polyanionic species $[(C_{60})_n(CN)_m]^{x-}$ are formed, where n = 1 – 3, m = 1 – 7 and x = 1 – 3, e.g. $[(C_{60})_3(CN)_7]^{3-}$, which, by CID studies is found to be $[C_{60}(CN)_2 \cdots C_{60}(CN)_3 \cdots C_{60}(CN)_2]^{3-}$.[52] For C_{70}, the most abundant anions are $[C_{70}(CN)_n]^-$ (n = 1, 2, 3) and $[C_{70}(CN)_n]^{2-}$ (n = 2, 4, 6) with $[C_{70}(CN)]^-$ the dominant ion at low concentrations.[53] The addition of cyanide ions to a mixture of higher fullerenes results in formation of adducts $[C_mCN]^-$ and $[C_m(CN)_2]^{2-}$ (m = 76, 78, 84, 86, 90, 92, 96 and 98); the relative abundances of the initially-formed mono-cyano adducts agreed well with percentages obtained by HPLC analysis of the fullerene mixture used, suggesting that this method could become a simple method for analysis of higher fullerene mixtures by ESI MS.[54] In a similar manner to cyanide addition, reaction of a solution of C_{60} in toluene with NaOMe/MeOH solution leads to C_{60}^- and $[C_{60}(OMe)_n]^-$ (n = 1, 3, 5, 7) as major ions.[48]

The fullerene oxides have themselves been the subjects of a number of studies using ESI MS. Using negative-ion ESI MS, intact molecular ions were generated for $C_{60}O$ and isomers of $C_{60}O_2$ without fragmentation.[55] It was found that the higher oxides showed enhanced sensitivity in ESI as a result of the increased electron affinity on oxidation.

Inorganic Phosphorus Compounds: Phosphoranes and Cyclophosphazenes

Pentacoordinated phosphoranes **4.7** (R = H, CH_3, ^iPr, ^tBu, CH_2Ph) give $[M + H]^+$ and $[M + Na]^+$ ions in their ESI mass spectra, which underwent different fragmentation paths upon CID.[56] Only one report appears to describe the application of ESI MS to the analysis of simple cyclotriphosphazenes of the type $[P_3N_3R_6]$, **4.8**. When R = OEt or OPh, good positive-ion ESI spectra were obtained by protonation of one of the ring nitrogens. CID of the ethoxy derivative proceeded readily by loss of C_2H_4 from the ethoxy groups, ultimately giving $P_3N_3(OH)_6$ or its tautomer metaphosphimic acid, $[NHP(O)OH]_3$. However, if R was more strongly electron-withdrawing (e.g. Cl or OC_6F_5) then negative-ion spectra containing intense $[P_3N_3R_5O]^-$ ions can be obtained.[57] These ions are formed by hydrolysis of one of the OR groups, giving an OH group which is then deprotonated to give the negative-ion species.

4.7   **4.8**

Summary

- Salts of singly-charge ions will generally give simple spectra, often of solvated ions at low cone voltages. Aggregate ions extending over a wide m/z range may also occur.
- Expect increasingly complex spectra as charges on ions increase.
- Negative-ion spectra are often simpler than corresponding positive-ion spectra, especially if there is a coordinating anion (e.g. NO_3^-, Cl^-).
- Reduction of metal ions such as copper(II), mercury(II) may occur readily. For all multiply-charged cations at high cone voltages, expect reduction to M^+ or MO^+.

References

1. As a general reference to this area, see G. R. Agnes and G. Horlick, *Applied Spectroscopy*, 1994, **48**, 655.
2. N. B. Cech and C. G. Enke, *Mass Spectrom. Rev.*, 2001, **20**, 362.
3. D. A. Barnett and G. Horlick, *J. Anal. Atom. Spectrom.*, 1997, **12**, 497.
4. C. Hao, R. E. March, T. R. Croley, J. C. Smith and S. P. Rafferty, *J. Mass Spectrom.*, 2001, **36**, 79; X. Zhao and J. Yinon, *Rapid Commun. Mass Spectrom.*, 2002, **16**, 1137; S. Zhou and M. Hamburger, *Rapid Commun. Mass Spectrom.*, 1996, **10**, 797.
5. C. E. C. A. Hop, *J. Mass Spectrom.*, 1996, **31**, 1314.

6. J. Dupont, R. F. de Souza, and P. A. Suarez, *Chem. Rev.*, 2002, **102**, 3667; '*Ionic liquids in synthesis*', P. Wasserscheid and T. Welton (Eds), Wiley-VCH, 2003; 'Room-temperature ionic liquids. Solvents for synthesis and catalysis', T. Welton, *Chem. Rev.*, 1999, **99**, 2071.

7. P. J. Dyson, J. S. McIndoe and D. Zhao, *Chem. Commun.*, 2003, 508.

8. D. Zhao, *Aust. J. Chem.*, 2004, **57**, 509.

9. G. P. Jackson and D. C. Duckworth, *Chem. Comm.*, 2004, 522.

10. P. J. Dyson, I. Khalaila, S. Luettgen, J. S. McIndoe and D. Zhao, submitted to *Chem. Comm.*

11. A. T. Blades, P. Jayaweera, M. G. Ikonomou and P. Kebarle, *J. Chem. Phys.*, 1990, **92**, 5900; P. Jayaweera, A. T. Blades, M. G. Ikonomou and P. Kebarle, *J. Am. Chem. Soc.*, 1990, **112**, 2452.

12. M. Peschke, A. T. Blades and P. Kebarle, *Int. J. Mass Spectrom.*, 1999, **185 – 187**, 685.

13. C. Hao and R. E. March, *J. Mass Spectrom.*, 2001, **36**, 509.

14. A. T. Blades, P. Jayaweera, M. G. Ikonomou, and P. Kebarle, *Int. J. Mass Spectrom. Ion Proc.*, 1990, **101**, 325.

15. A. T. Blades, P. Jayaweera, M. G. Ikonomou, and P. Kebarle, *Int. J. Mass Spectrom. Ion Proc.*, 1990, **102**, 251.

16. I. I. Stewart and G. Horlick, *Anal. Chem.*, 1994, **66**, 3983.

17. A. A. Shvartsburg, *J. Am. Chem. Soc.*, 2002, **124**, 12343.

18. Z. L. Cheng, K. W. M. Siu, R. Guevremont and S. S. Berman, *Org. Mass Spectrom.*, 1992, **27**, 1370.

19. S. Mollah, A. D. Pris, S. K. Johnson, A. B. Gwizdala III and R. S. Houk, *Anal. Chem.*, 2000, **72**, 985.

20. M. C. B. Moraes, J. G. A. Brito Neto and C. L. do Lago, *Int. J. Mass Spectrom.*, 2000, **198**, 121.

21. H. Lavanant, H. Virelizier and Y. Hoppilliard, *J. Am. Soc. Mass Spectrom.*, 1998, **9**, 1217.

22. L. Gianelli, V. Amendola, L. Fabrizzi, P. Pallavicini and G. G. Mellerio, *Rapid Commun. Mass Spectrom.*, 2001, **15**, 2347.

23. A. R. S. Ross, M. G. Ikonomou, J. A. J. Thompson, and K. J. Orians, *Anal. Chem.*, 1998, **70**, 2225.

24. C. L. Gatlin, F. Tureček and T. Vaisar, *Anal. Chem.*, 1994, **66**, 3950.

25. H. Deng and P. Kebarle, *J. Am. Chem. Soc.*, 1998, **120**, 2925.

26. O. Schramel, B. Michalke and A. Kettrup, *J. Chromatogr. A*, 1998, **819**, 231.

27. J. Shen and J. Brodbelt, *J. Mass Spectrom.*, 1999, **34**, 137.

28. D. Schroder, T. Weiske and H. Schwarz, *Int. J. Mass Spectrom.*, 2002, **219**, 729.

29. Y. Xu, X. Zhang and A. L. Yergey, *J. Am. Soc. Mass Spectrom.*, 1996, **7**, 25.

30. C. L. Gatlin, F. Tureček and T. Vaisar, *J. Mass Spectrom.*, 1995, **30**, 775.

31. G. J. Van Berkel, *J. Mass Spectrom.*, 2000, **35**, 773.

32. C. C. Raymon, D. L. Dick and P. K. Dorhout, *Inorg. Chem.*, 1997, **36**, 2678.

33. J. S. McIndoe and D. G. Tuck, *J. Chem. Soc., Dalton Trans.*, 2003, 244.

34. A. T. Blades and P. Kebarle, *J. Am. Chem. Soc.*, 1994, **116**, 10761.

35. I. I. Stewart, D. A. Barnett, and G. Horlick, *J. Anal. At. Spectrosc.*, 1996, **11**, 877.

36. W. Henderson, unpublished observations.

37. L. Charles and D. Pépin, *Anal. Chem.*, 1998, **70**, 353; L. Charles, D. Pépin and B. Casetta, *Anal. Chem.*, 1996, **68**, 2554.

38. B. K. Choi, D. M. Hercules and M. Houalla, *Anal. Chem.*, 2000, **72**, 5087.

39. P. Bussian, F. Sobott, B. Brutschy, W. Schrader and F. Schüth, *Angew. Chem., Int. Ed. Engl.*, 2000, **39**, 3901.

40. C. E. C. A. Hop, D. A. Saulys, A. N. Bridges, D. F. Gaines and R. Bakhtiar, *Main Group Metal Chem.*, 1996, **19**, 743.

41. C. E. C. A. Hop, D. A. Saulys and D. F. Gaines, *J. Am. Soc. Mass Spectrom.*, 1995, **6**, 860.

42. C. E. C. A. Hop, D. A. Saulys and D. F. Gaines, *Inorg. Chem.*, 1995, **34**, 1977.

43. J. A. Dopke, A. N. Bridges, M. R. Schmidt and D. F. Gaines, *Inorg. Chem.*, 1996, **35**, 7186.

44. A. Dupont, J.-P. Gisselbrecht, E. Leize, L. Wagner and A. Van Dorsselaer, *Tetrahedron Lett.*, 1994, **35**, 6083.

45. T.-Y. Liu, L.-L. Shiu, T.-Y. Luh and G.-R. Her, *Rapid Commun. Mass Spectrom.*, 1995, **9**, 93.

46. G. Khairallah and J. B. Peel, *Chem. Phys. Lett.*, 1998, **296**, 545.

47. G. J. Van Berkel and K. G. Asano, *Anal. Chem.*, 1994, **66**, 2096; J. F. Anacleto, M. A. Quilliam, R. K. Boyd, J. B. Howard, A. L. Lafleur and T. Yadav, *Rapid Commun. Mass Spectrom.*, 1993, **7**, 229; K. Hiraoka, I. Kudaka, S. Fujimaki and H. Shinohara, *Rapid Commun. Mass Spectrom.*, 1992, **6**, 254; S. Fujimaki, I. Kudaka, T. Sato, K. Hiraoka, H. Shinohara, Y. Saito and K. Nojima, *Rapid Commun. Mass Spectrom.*, 1993, **7**, 1077.

48. S. R. Wilson and Y. Wu, *J. Am. Chem. Soc.*, 1993, **115**, 10334.

49. S. R. Wilson and Y. Wu, *J. Am. Soc. Mass Spectrom.*, 1993, **4**, 596; S. R. Wilson and Q. Lu, *Tetrahedron Lett.*, 1993, **34**, 8043; S. R. Wilson, N. Kaprinidis, Y. Wu and D. I. Schuster, *J. Am. Chem. Soc.*, 1993, **115**, 8495; S. R. Wilson and Y. Wu, *J. Chem. Soc., Chem. Commun.*, 1993, 784.

50. S. R. Wilson and Y. Wu, *Org. Mass Spectrom.*, 1994, **29**, 186.

51. G. Khairallah and J. B. Peel, *Chem. Phys. Lett.*, 1997, **268**, 218.

52. A. A. Tuinman and R. N. Compton, *J. Phys. Chem. A*, 1998, **102**, 9791.

53. G. Khairallah and J. B. Peel, *J. Phys. Chem. A*, 1997, **101**, 6770.

54. G. Khairallah and J. B. Peel, *Int. J. Mass Spectrom.*, 2000, **194**, 115.

55. J.-P. Deng, C.-Y. Mou and C.-C. Han, *J. Phys. Chem.*, 1995, **99**, 14907.

56. Z. Liu, L. Yu, Y. Chen, N. Zhou, J. Chen, C. Zhu, B. Xin and Y. Zhao, *J. Mass Spectrom.*, 2003, **38**, 231.

57. C. V. Depree, E. W. Ainscough, A. M. Brodie, A. K. Burrell, C. Lensink and B. K. Nicholson, *Polyhedron*, 2000, **19**, 2101.

5 The ESI MS Behaviour of Coordination Complexes

Introduction

Coordination chemistry is a major field within the discipline of inorganic chemistry. A variety of techniques are commonly used in the characterisation of coordination complexes; organometallic complexes are often similar and are covered separately in Chapters 6 and 7. Widely used techniques include spectroscopy [nuclear magnetic resonance (NMR), infrared (IR) etc.], but the application of any particular technique is situation-dependent. In cases of paramagnetic metal centres NMR is of little use and other techniques must be employed. X-ray crystallography is often touted as the ultimate technique for substance characterisation. While this is often true, limitations need to be recognised. A crystal is usually needed and since the analysis is carried out on a single crystal rather than the bulk sample, the crystal may not be representative of the bulk material. For example, FAB MS has been used to analyse the organometallic compound $[(C_5Me_5)_4Cr_4]$, where the crystal used for X-ray analysis did not have the same composition as the bulk material.[1]

The power of the electrospray technique in the study of coordination complexes is due to its ability to sample *solution* systems (i.e. reaction mixtures of metal plus one or more ligands may be monitored), and paramagnetic substances may be analysed as easily as diamagnetic ones. Of course, mass spectrometry has its own limitations, so that generally a combination of a range of techniques often provides the most complete picture of the chemistry of the system under study.

The aim of this chapter is to describe the behaviour of different classes of coordination complexes by ESI MS. With this information in hand, the general ionisation and fragmentation processes that apply to coordination compounds should be understood and it should be possible to predict the likely behaviour of other, specific systems. Applications of ESI MS in coordination chemistry have developed enormously in recent years and so this chapter cannot attempt to give comprehensive coverage of all the individual systems reported. Instead, some of the major classes of coordination complex have been highlighted, with emphasis on preformed coordination complexes; metal salts, which may generate coordination complexes by (for example) dissolution in a solvent, have been covered in Chapter 4.

A number of detailed reviews have summarised recent developments in ESI MS analysis of coordination complexes;[2] these should be referred to for more detailed surveys.

Mass Spectrometry of Inorganic, Coordination and Organometallic Compounds W. Henderson and J. S. McIndoe
© 2005 John Wiley & Sons, Ltd   ISBNs: 0-470-85015-9 (HB); 0-470-85016-7 (PB)

Charged, 'Simple' Coordination Complexes

This class of compound often behaves straightforwardly in ESI MS analysis, particularly when the charge density of the complexes is relatively low. In these cases, the ES ionisation process (Section 3.9) serves to transfer the pre-existing solution ions into the gas phase. Providing that low fragmentation conditions are chosen (i.e. low cone voltage or equivalent) intact parent ions (or simple derivatives thereof) will generally be observed. However, for more highly charged systems, charge reduction is likely to occur; this can either occur by reduction of the metal centre, by ion-pairing with a counter-cation, or by fragmentation involving loss of a ligand. The examples chosen illustrate the different ionisation and fragmentation pathways for this class of compound.

5.2.1 Cationic Coordination Complexes

To illustrate the key points for simple, charged coordination complexes, the monocationic complexes $[Rh(py)_4Cl_2]^+Cl^-$ (py = pyridine) and *trans*-$[Co(en)_2Cl_2]^+Cl^-$ (en = NH$_2$CH$_2$CH$_2$NH$_2$) illustrate the similarities and differences between complexes with two different ancillary ligands. ESI MS spectra at a range of cone voltages are given in Figure 5.1 for the rhodium complex and Figure 5.2 for the cobalt complex.

Under low cone voltage conditions, spectrum (a) in both figures, the parent cation is seen as effectively the sole species. Upon increasing the cone voltage to 40 V, the complexes begin to undergo fragmentation, by differing pathways, that parallel those of other ionisation techniques such as FAB (Section 3.6.2). For the rhodium complex

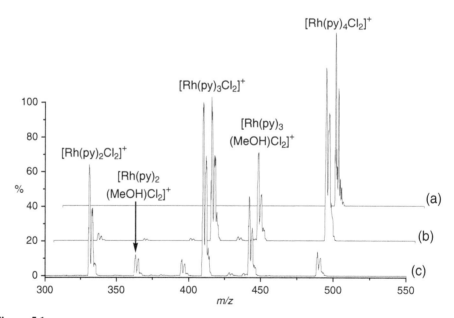

Figure 5.1
Positive-ion ESI MS spectra of the complex [Rh(pyridine)$_4$Cl$_2$]Cl in methanol solvent, at cone voltages of (a) 20 V, (b) 40 V and (c) 50 V, showing the fragmentation by loss of pyridine ligands, and solvation of the resulting fragments by methanol solvent molecules

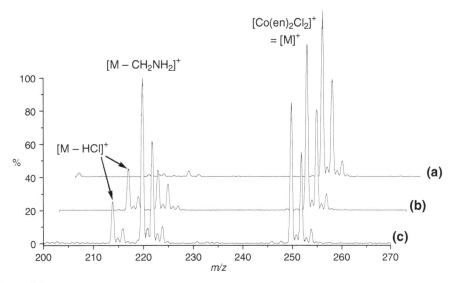

Figure 5.2
Positive-ion ESI MS spectra of the complex *trans*-[Co(NH$_2$CH$_2$CH$_2$NH$_2$)$_2$Cl$_2$]Cl in methanol solvent, at cone voltages of (a) 10 V, (b) 40 V and (c) 60 V, showing the intact cation at 10 V, and fragmentation by loss of HCl, and loss of CH$_2$NH$_2$ from the organic ligand at higher cone voltages

(Figure 5.1b), a neutral pyridine ligand is lost, giving [Rh(py)$_3$Cl$_2$]$^+$ (*m/z* 410), and its counterpart solvated by a methanol, [Rh(py)$_3$(MeOH)Cl$_2$]$^+$ (*m/z* 442); for these ions an entire neutral pyridine ligand is lost. In contrast, the cobalt complex contains bidentate ethylenediamine (en) ligands that form stable five-membered chelate rings. Instead of loss of an entire ligand, fragmentation involving the ligand itself occurs. The two observed ions at *m/z* 213 and 219, Figure 5.2b, are formed by loss of HCl and CH$_2$NH$_2$ from the parent [Co(en)$_2$Cl$_2$]$^+$ ion. For the en ligand, the C—C, C—N and N—H bonds are weaker than the aromatic C—C bonds in pyridine and so the ligand itself undergoes fragmentation. The presence of two chlorine atoms in the *m/z* 219 ion but only one in the *m/z* 213 ion can be seen when the isotope patterns for the two ions are examined. When the cone voltage is further increased to 60 V, Figure 5.2c, the intensities of the cobalt fragment ions increases. However, for the rhodium complex at 50 V, there is further loss of pyridine, giving the ions [Rh(py)$_2$(MeOH)$_x$Cl$_2$]$^+$ (x = 0, 1 and 2), and [Rh(py)$_3$Cl$_2$]$^+$ is now the base peak in the spectrum. Again, for this system containing only monodentate ligands, ligand loss occurs in preference to ligand fragmentation.

For more highly-charged species, the ESI MS behaviour again depends very strongly on the ligand type and on the charge. When the ligand is a stable, chelating donor such as bipyridine (bipy), then the resulting complex is stabilised towards fragmentation, relative to a complex containing easily fragmentable ligands (such as ethylenediamine). [Ru(bipy)$_3$]$^{2+}$ is an example of the former type and this complex gives the parent cation, with no observed fragment ions, at a cone voltage of 40 V, (Figure 5.3). Indeed, because of the stability of this complex, and the importance of complexes of this general type, it was one of the very first coordination compounds to be investigated using ESI MS.[3] Under gentle ionisation conditions in acetonitrile solvent, [Ru(bipy)$_3$]$^{2+}$ was the base peak together with low intensity solvated ions [Ru(bipy)$_3$ + nMeCN]$^{2+}$ (n = 1–4). At

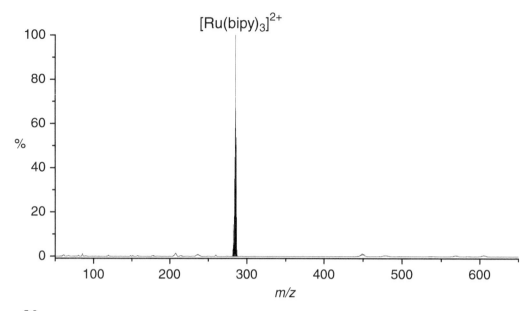

Figure 5.3
Positive-ion ESI MS spectrum of [Ru(bipy)$_3$]Cl$_2$ in methanol at a cone voltage of 40 V, showing exclusively the [Ru(bipy)$_3$]$^{2+}$ cation

higher cone voltages, [Ru(bipy)$_3$]$^{2+}$ undergoes fragmentation, giving [Ru(bipy)$_2$]$^{2+}$, [Ru(bipy)]$^{2+}$, [bipy + H]$^+$ and other minor ions. The bipyridine ligand is robust and so the parent dication fragments by loss of entire bipyridine ligands. The LSIMS spectrum of the same complex is shown in Figure 3.16 and the difference between the two spectra is dramatic. While the ESI mass spectrum gives solely the parent dication, and is easy to assign, the LSIMS spectrum shows only a weak [Ru(bipy)$_3$]$^{2+}$ anion, and a range of fragment ions, many of which are singly-charged. This comparison clearly demonstrates why the ESI technique has become so popular, and the LSIMS/FAB techniques less popular, for the analysis of coordination and organometallic compounds in recent years.

A tricationic complex such as the [Co(en)$_3$]$^{3+}$ cation (obtained easily as its chloride salt) provides a marked contrast. This is well known to be a highly stable coordination complex; the d^6 electronic configuration coupled with the presence of three bidentate chelating H$_2$NCH$_2$CH$_2$NH$_2$ (en) ligands means that this is a robust complex in solution. However, in the gas phase the high charge density on the cation results in a complex ESI MS spectrum, even at a very low cone voltage. At 10 V (Figure 5.4a), a considerable number of ions are seen; assignments for a number of them are given in the figure. At this low cone voltage, the assigned ions are derived from the parent trication by either loss of a proton (from the acidic NH$_2$ groups), or by gain of chloride; both of these processes serve to reduce the charge density, giving *dications* that are more stable in the gas phase. Hence, ions [M−H + nMeOH]$^{2+}$ (n = 0, 1, 2; *m/z* 119, 135 and 151) are formed by proton loss, [M + 2Cl]$^+$ (*m/z* 309) is formed by association with chloride, and [M−H + Cl]$^+$ (*m/z* 273) is formed as a result of both processes. When the cone voltage is raised to a modest value of 30 V (Figure 5.4b) more severe fragmentation occurs,

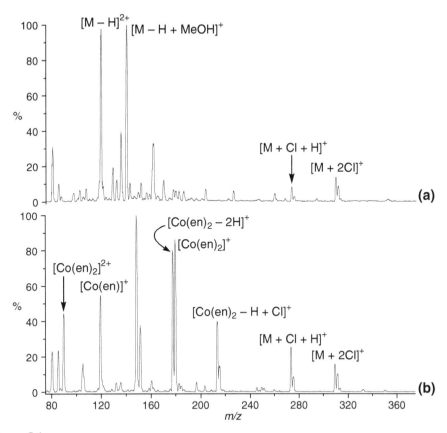

Figure 5.4
Positive-ion ESI MS spectra of the complex $[Co(NH_2CH_2CH_2NH_2)_3]Cl_3$ in methanol solvent, at cone voltages of (a) 10 V and (b) 30 V. The observation of a large number of ions in these spectra is a consequence of the high charge density of the cation

involving loss of ethylenediamine ligands giving ions such as $[Co(en)_2-H+Cl]^+$ (m/z 213) and $[Co(en)_2-2H]^+$ (m/z 177), as well as reduction of the cobalt from the $+3$ oxidation state to the $+1$ or $+2$ state, giving $[Co(en)_2]^+$ (m/z 179), $[Co(en)_2]^{2+}$ (m/z 89.5) and possibly $[Co(en)]^+$ (m/z 119; overlaps with $[M-H]^{2+}$ ion of parent cation).

The occurrence of chloride-containing ions in the $[Co(en)_3]Cl_3$ case is noteworthy, since it involves no actual fragmentation of the parent complex, but instead association, undoubtedly through a combination of electrostatic and hydrogen bond (N-H$\cdots$Cl) interactions. This phenomenon becomes increasingly important and common when the charge on the cation is $2+$ or higher. For example, analysis of the technetium complex $[Tc(MeCN)_4(PPh_3)_2](PF_6)_2$ by positive-ion ESI MS gave the aggregate ion $[Tc(MeCN)_4(PPh_3)_2(PF_6)]^+$ as the most intense peak, together with $[Tc(MeCN)_4(PPh_3)_2]^+$ and $[Tc(MeCN)_4(PPh_3)_2]^{2+}$.[4]

Ion association also occurs in the stabilisation of highly charged anions, such as cyanometallate anions (Section 5.2.3), and especially for large supramolecular self-assembled structures (Chapter 8).

5.2.2 Anionic Metal Halide Complexes

Many metals, particularly the transition metals, form anionic halide complexes. These are often the basic starting points in the chemistry of these elements. A good example comes from platinum chemistry, which forms the chloro complexes $[PtCl_4]^{2-}$, and $[PtCl_6]^{2-}$, which are both commonly-used for the synthesis of other platinum complexes. The ESI MS behaviour of these two complexes and some related systems illustrates the general behaviour of this class of coordination complex.

Negative-ion ESI MS spectra for $K_2[PtCl_6]$ in methanol solution, recorded at a range of cone voltages, are shown in Figure 5.5. At a low cone voltage (20 V, Figure 5.5a) the spectrum is dominated by the parent $[PtCl_6]^{2-}$ ion at m/z 204. There are also low intensity ions that also signal the onset of fragmentation even at this low cone voltage. Thus, in addition to the ion-pair $[KPtCl_6]^-$ (m/z 446), there are ions $[PtCl_5]^-$ (m/z 373), $[PtCl_4]^-$ (m/z 336) and $[PtCl_3]^-$ (m/z 301). As the cone voltage is further increased, the intensity of these ions increases, and the ion $[PtCl_2]^-$ (m/z 266) is also seen, until at 100 V (Figure 4.5e) $[PtCl_2]^-$ is the base peak. These ions contain platinum in lower oxidation states than the $+4$ state in $K_2[PtCl_6]$. These species are not present in the initial solution, but instead are formed by gas-phase reduction processes (refer also to Chapter 4). While the $+2$ state is common for platinum(II) complexes, the $+1$ state and $+3$ state – in the ions

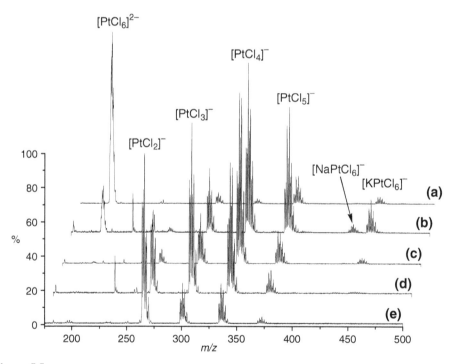

Figure 5.5
The ESI MS behaviour of $K_2[PtCl_6]$ in methanol solution, at a range of cone voltages. The parent $[PtCl_6]^{2-}$ ion (m/z 204) observed at low cone voltages, undergoes successive reduction/fragmentation at elevated cone voltages, giving the ions $[PtCl_5]^-$ [platinum(IV)], $[PtCl_4]^-$ [platinum(III)], $[PtCl_3]^-$ [platinum(II)], and $[PtCl_2]^-$ [platinum(I)]. The broad envelope of peaks for each species is a consequence of the isotopic richness of both platinum and chlorine (Appendix 1)

[PtCl$_2$]$^-$ and [PtCl$_4$]$^-$ respectively–are much rarer, though they are known. Thus, the ESI process is able to furnish gas phase chloroanions, which can then be easily converted into unusual, lower oxidation state chloroanions; a similar example of a main group element analogue comes from iodine chemistry, where the iodine(III) species [ICl$_4$]$^-$ forms [ICl$_3$]$^-$ in the mass spectrometer.[5]

The platinum(II) complex K$_2$[PtCl$_4$] behaves similarly to K$_2$[PtCl$_6$], except that the higher oxidation state (+3 or +4) platinum species are not seen. Thus, at a low cone voltage (10 V), [PtCl$_4$]$^{2-}$ is seen, together with lower intensity ions [PtCl$_3$]$^-$, [PtCl$_3$(H$_2$O)]$^-$, [PtCl$_3$(MeOH)]$^-$, [HPtCl$_4$]$^-$ and [KPtCl$_4$]$^-$. In all of these ions, the oxidation state of the platinum is in its starting +2 state. At a cone voltage of 30 V, the [PtCl$_3$]$^-$ ion is the base peak. Further increasing the cone voltage results in reduction, giving [PtCl$_2$]$^-$ as the base peak at 70 V.

Other transition metal halide anions behave in a similar fashion. As an example, for [AuCl$_4$]$^-$ (which is isoelectronic with [PtCl$_4$]$^{2-}$) the parent ion (*m/z* 339) dominates at low cone voltages (e.g. 30 V, Figure 5.6a), while the reduction product [AuCl$_2$]$^-$ (*m/z* 267) dominates at higher cone voltages (e.g. 70 V, Figure 5.6b). The ion [AuCl$_3$]$^-$, containing gold(II), has a very low intensity, consistent with the instability of the Au(II) oxidation state compared to Au(I), which is very stable.

The osmium salt K$_2$[OsCl$_6$] gives the [OsCl$_6$]$^{2-}$ ion as the base peak at 5 V, but at 20 V the base peak is [OsCl$_5$]$^-$ (still containing osmium(IV)) and at 100 V this ion together with the osmium(III) species [OsCl$_4$]$^-$ are the major ions observed. In this case, [OsCl$_3$]$^-$ and [OsCl$_2$]$^-$ are not seen. The different behaviour of K$_2$[PtCl$_6$] versus K$_2$[OsCl$_6$] at higher cone voltages probably reflects differences in the stabilities of the lower oxidation states of these two elements. Solutions of K$_2$[ReCl$_6$] have also been studied by ESI MS. The principal species observed were [ReCl$_5$]$^-$, [ReCl$_5$(H$_2$O)]$^-$, [ReCl$_5$(H$_2$O)$_2$]$^-$ and

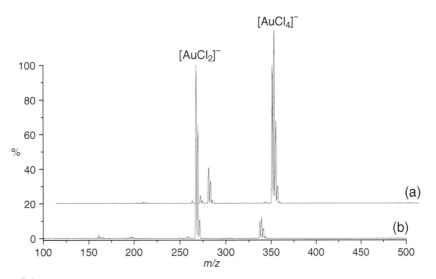

Figure 5.6
Negative-ion ESI mass spectra for Me$_4$N[AuCl$_4$] in acetonitrile solution, at cone voltages of (a) 30 V and (b) 70 V. The parent [AuCl$_4$]$^-$ ion dominates at the low cone voltage, but the reduced gold(I) species [AuCl$_2$]$^-$ dominates at the high cone voltage. In this system, the gold(II) ion [AuCl$_3$]$^-$ (*m/z* 302) has very low intensity, but is detectable

$[KReCl_6]^-$, together with some $[ReO_4]^-$. The intensity of the latter increased on standing the solution, confirming the instability of the rhenium chloro-anions in solution.[6]

Two studies have concerned the analysis of halocuprate anions by negative-ion ESI MS. Mixtures of CuY with two mole equivalents of LiX (X,Y = Cl, Br, I) were studied and gave homo- and hetero-halocuprate anions $[CuX_2]^-$, $[CuXY]^-$ and $[CuY_2]^-$. Mixtures of CuCN with LiX gave in addition, mixed halo-cyanocuprate anions, including species containing several copper(I) centres and lithium ions, which, on the basis of theoretical calculations, are believed to contain linear bridging cyanide ions, e.g. $[X-Cu-CN-Li-X]^-$.[7] The second study in this area investigated the ESI MS spectra of solutions from which various tetraalkylammonium and tetraalkylphosphonium chloro- and bromo-cuprates crystallise.[8] Aggregates containing more than one copper were typically observed, in some cases ion-paired with a countercation that serves to reduce the overall charge on the ion and thus stabilise it. As an example, the filtrate from which $(PEt_4)_2[Cu_2Br_4]$ crystallises gave $[Cu_2Br_4(PEt_4)]^-$, $[CuBr_2]^-$ and $[CuBr_3(PEt_4)]^-$ as well as several non-copper containing ions. When solutions of CuBr dissolved in acetonitrile were analysed, in the absence of any added countercation, a series of copper-bromide anions $[Cu_nBr_{n+1}]^-$ (n = 1 to 6) were observed.

5.2.3   Highly-Charged, Anionic Transition Metal Complexes – Cyanometallate Anions

As observed previously for highly-charged *cations* (Section 4.2.1) the high charge density of multiply-charged anions also poses interesting issues with respect to the observation or non-observation of intact parent ions in their ESI mass spectra. Here, the countercation can play a major role. The electrospray behaviour of small, highly-charged ions such as $[Fe(CN)_6]^{3-}$ (ferricyanide) and $[Fe(CN)_6]^{4-}$ (ferrocyanide) is of considerable interest because these complexes are very stable and inert, with ferrocyanide being particularly stable.

The general ESI MS features for these highly charged anions is illustrated in Figure 5.7, which gives negative-ion ESI mass spectra of $K_3[Fe(CN)_6]$ at cone voltages of 10, 20 and 50 V. A study on $K_4[Fe(CN)_6]$ and $K_3[Fe(CN)_6]$ showed very similar ions for both species.[9] The 10 V spectrum (Figure 5.7a) is relatively complex; the parent $[Fe(CN)_6]^{3-}$ ion is not observed and instead a range of charge-reduced ions are seen, formed by reduction of the iron centre, loss of CN^- or aggregation with K^+ ions. Major species observed are $[KFe(CN)_5]^-$, $[Fe(CN)_3]^-$, $[KFe(CN)_6]^{2-}$ and $[Fe(CN)_6]^{2-}$; the latter formally contains iron(IV) and is formed by loss of an electron (i.e. oxidation) from the parent $[Fe(CN)_6]^{3-}$ ion. The observation of these iron(II), (III) and (IV) oxidation state species clearly indicates the instability of the highly charged ions in this particular system, even at a low cone voltage. When the cone voltage is raised, to 20 V (Figure 5.7b), the spectrum simplifies, with $[KFe(CN)_5]^-$, $[Fe(CN)_3]^-$ and $[Fe(CN)_4]^-$ being the major observed species. Increasing the cone voltage further, to 50 V (Figure 5.7c) gives the species $[Fe(CN)_4]^-$, $[Fe(CN)_3]^-$ and $[Fe(CN)_2]^-$, containing iron in the +3, +2 and +1 oxidation states respectively. This high cone voltage spectrum thus resembles those of other metal cationic and anionic species [e.g. $PtCl_6^{2-}$, oxoanions] that give lower oxidation state species at elevated cone voltages.

Highly charged ions such as $[Fe(CN)_6]^{3-}$ are, however, stable when the excess charge can be reduced by favourable associations with countercations, such as quaternary phosphonium ions, R_4P^+. Hence, ESI MS analysis of $[Fe(CN)_6]^{3-}$ with the

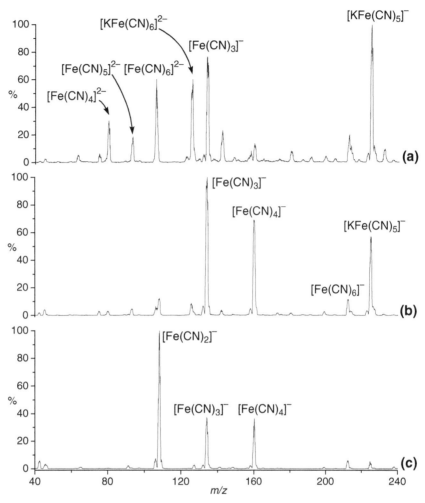

Figure 5.7
The instability of highly charged ions in the gas phase is clearly shown by the negative-ion ESI mass spectra of $K_3[Fe(CN)_6]$ in methanol solution at cone voltages of (a) 10 V, (b) 20 V and (c) 50 V. The various observed ions are formed by cyanide loss, changes in the iron oxidation state, and aggregation with a potassium counterion

countercations $MePh_3P^+$ or Ph_4P^+ led to a rich array of mainly negatively-charged aggregate ions.[10] For example, $(MePh_3P)_3[Fe(CN)_6]$ gave 23 aggregates ranging from $[(MePh_3P)_2Fe(CN)_6]^-$ to $[(MePh_3P)_{37}\{Fe(CN)_6\}_{14}]^{5-}$. The charge density of these aggregates tends towards -0.15 per component ion. Structural features that are likely to be responsible for the stability of these aggregates were gleaned from examination of several X-ray crystal structures, and included electrostatic interactions (consistent with an ionic material), $C–H \cdots NC–Fe$ and $C–H \cdots O–N–Fe$ hydrogen bonds, and to a lesser extent, interactions (sextuple phenyl embraces) involving the cations. The key feature is that *intact* $[Fe(CN)_6]$ groups are observed in the mass spectrum, albeit in an aggregated state.

The nitroprusside anion $[Fe(CN)_5NO]^{2-}$ gives the parent dianion in ESI MS at a low cone voltage of 10 V, together with minor ions $[Fe(CN)_5]^{2-}$ (by loss of NO), $[Fe(CN)_3]^-$, and the ions $[HFe(CN)_5NO]^-$ and $[NaFe(CN)_5NO]^-$ formed by ion pairing. Additionally, a range of ion-paired species $\{M_x[Fe(CN)_5NO]_y\}^{z-/z+}$ up to $x = 8$ and $y = 5$ were also observed, with the greatest amount of ion pairing occurring when M is a small cation such as Li^+ or Me_4N^+.[11]

A number of new, anionic complexes containing both CO and cyanide ligands, namely $(PPN)_3[Fe(CN)_5(CO)]$ and $(Ph_4P)_2[Fe(CN)_4(CO)_2]$ have also been characterised using ESI MS. The first complex gave the parent-derived ion $[(PPN)_2Fe(CN)_5(CO)]^-$ plus $[(PPN)Fe(CN)_3(CO)]^-$ and $[Fe(CN)_3]^-$, while the latter gave the parent-derived ion $[(Ph_4P)Fe(CN)_4(CO)_2]^-$ plus $[(Ph_4P)Fe(CN)_4]^-$ and $[Fe(CN)_3]^-$. The ruthenium complex $(Ph_4P)[Ru(CN)_3(CO)_3]$ gave the expected $[M]^-$ ion in negative-ion ESI MS, together with CO-loss fragment ions $[Ru(CN)_3(CO)_2]^-$ and $[Ru(CN)_3(CO)]^-$.[12] Again, this indicates the tendency for highly charged anions to aggregate with one or more cations to reduce the charge, as well as the tendency towards facile loss of neutral CO compared to anionic CN^- in these species, especially when more than one CO ligand is present.

(Neutral) Metal Halide Coordination Complexes

Neutral complexes containing halide ligands are very common in transition metal chemistry. One facile method by which metal halide complexes (and transition metal halide complexes in particular) are able to undergo ionisation is by loss of a halide anion. This is described in general terms by:

$$L_nMX \longrightarrow [L_nM]^+ \quad \text{and/or} \quad [L_nM(solvent)]^+$$

where M is the metal atom, X is the halide (or other anionic ligand) and L_n are the ancillary ligands.

The observed cationic species $[L_nM]^+$ may or may not be solvated, depending on the donor ability of the solvent used, and the cone voltage. At elevated cone voltages, loss of the most weakly bonded neutral ligand generally occurs, which is typically the coordinated solvent molecule. By adding to the analyte solution a small amount of a donor ligand (with better coordinating ability than the solvent), displacement of the halide ligand and complexation of the resulting cation can be enhanced. This has the effect of enhancing signal intensity and simplifying spectra. Pyridine is generally an excellent choice, due to its volatility and good coordinating properties towards most transition metal centres.

As an example to illustrate the general behaviour, Figure 5.8 shows positive-ion ESI MS spectra of the complex *trans*-$[PtMeI(PPh_3)_2]$. Although formally an organometallic complex (containing a metal-carbon bond), this complex displays the classic features of this particular ionisation route.

At a low cone voltage (5 V) with added pyridine (Figure 5.8a), the base peak in the spectrum is $[PtMe(PPh_3)_2(pyridine)]^+$ (*m/z* 813), formed by replacement of an anionic chloride by a neutral pyridine. A low intensity ion at *m/z* 630 is identified as the bis(pyridine) species $[PtMe(PPh_3)(pyridine)_2]^+$. When the cone voltage is increased to

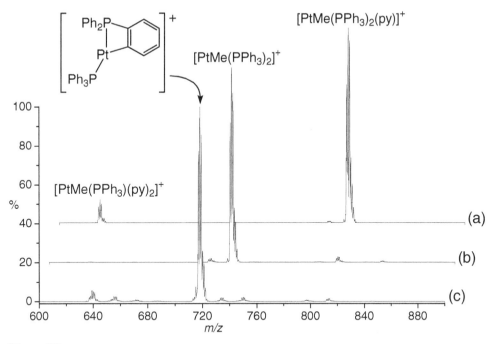

Figure 5.8
Positive-ion ESI mass spectra of the complex *trans*-[PtMeI(PPh$_3$)$_2$] in methanol solution with added pyridine (py). Spectrum (a) shows the formation of the solvated species [PtMe(PPh$_3$)$_2$(pyridine)]$^+$ at a cone voltage of 5 V, which loses pyridine at a cone voltage of 40 V (b). At even higher cone voltages (spectrum (c) 80 V) CH$_4$ is eliminated and a cyclometallated species is formed

40 V (Figure 5.8b), the labile pyridine ligand is lost, and the base peak in the spectrum becomes the 'bare' [PtMe(PPh$_3$)$_2$]$^+$ ion at *m/z* 734, giving a very simple and easily assignable spectrum, where the observed species can be correlated easily with the parent complex. The choice of cone voltage is, as with many other systems, critical. Figure 5.8c shows the ESI MS spectrum of this system at a cone voltage of 80 V, where it can be seen that the base peak is now at *m/z* 718. This is assigned to a species that contains a cyclometallated triphenylphosphine ligand, as shown in the Figure. Thus, in the fragmentation of [PtMe(PPh$_3$)$_2$]$^+$, CH$_4$ is lost. This cyclometallated species is frequently observed when platinum-triphenylphosphine complexes are subjected to elevated cone voltages (of the order of 80 V). By comparison, the LSIMS spectrum of the same complex (Figure 3.17) is more complex, and shows only a weak [PtMe(PPh$_3$)$_2$]$^+$ ion.

The halide-loss mechanism described above appears to be quite general for a large number of transition metal halide complexes, which may have a range of ancillary ligands including organometallic ligands (alkyl, aryl, cyclopentadienyl, CO etc.) and other ligands such as phosphines. Other examples of neutral metal-halide complexes that have been shown to behave in this way include *trans*-[IrCl(CO)(PPh$_3$)$_2$], [RuCl(η^5-C$_5$H$_5$)(PPh$_3$)$_2$], *cis*-[PtCl$_2$(PPh$_3$)$_2$], *cis*-[PtCl(CH$_2$COCH$_2$R)(PPh$_3$)$_2$] (R = chlorine, hydrogen),[13] neutral ruthenium(II) complexes of macrocyclic selenoethers of the type $[RuX_2(macrocycle)]$ (X = Cl, Br, I),[14] ruthenium(II) complexes of ditelluroethers of the type $[RuX_2(R-TeCH_2CH_2TeR)]$ (X = Cl, Br, I; R = Me, Ph),[15] and rhenium nitrido complexes containing functionalised diphosphine ligands such as $Re(\equiv N)Cl_2\{(Ph_2PCH_2CH_2)_2O\}$.[16] For a dinuclear chloride-bridged complex such as $[Ru(CO)_3Cl_2]_2$, addition of pyridine results in cleavage of the dinuclear unit, and displacement of a chloride ion, giving $[Ru(CO)_3Cl(pyridine)]^+$ as the base peak in the spectrum.

An example of a more complicated coordination complex is the polymethylene-bridged tungsten-platinum complex $(\eta^5\text{-}C_5Me_5)W(CO)_3\text{-}(CH_2)_4\text{-}PtMe_2(I)(L)$ (L = 4, 4'-dimethyl-2,2'-bipyridine), where the iodide bound to platinum is lost in ESI MS analysis in MeCN solution, giving $[M-I+MeCN]^+$ at low cone voltage (15 V) and $[M-I]^+$ at higher cone voltage (50 V).[17] In this respect, the complex is behaving as a transition metal halide coordination complex, with the organometallic portion of the complex behaving merely as a spectator.

As expected, the ionisation route is not restricted to halides and transition metal complexes, but also extends to other metals (e.g. main group elements) and other labile leaving groups, such as carboxylate and related ligands. As an example, the ESI MS analysis of the mercury trifluoroacetate complex $[Hg(O_2CCF_3)_2L_2]$ (L = tri-o-methoxyphenylphosphine) gave $[Hg(O_2CCF_3)L_2]^+$ as the base peak but the corresponding perchlorate complex $[Hg(ClO_4)_2(PCy_3)_2]$ gave a strong peak due to $[Hg(O_2CCH_3)(PCy_3)_2]$ and weaker peaks $[Hg(ClO_4)(PCy_3)_2]^+$ and $[Hg(PCy_3)_2]^{2+}$. In this case the perchlorate anion is a much poorer ligand towards mercury(II) and is readily lost or replaced by acetate present in the solvent used.[18] Main group element halides such as R_3SnCl and Ph_3PbCl also ionise by loss of the halide ligand and are described in Section 6.2.

Even in the presence of readily protonatable groups, the halide loss appears to be a facile ionisation pathway for neutral halide complexes, since the platinum(II) chloride complex of the readily protonatable ('electrospray-friendly') ligand $P(p\text{-}C_6H_4NMe_2)_3$ (Section 6.3.1) gave $[M-Cl]^+$ and $[M-Cl+solvent]^+$ ions instead of the $[M+H]^+$ ion.[19] However, in a study of ruthenium(II) complexes of the highly basic phosphatriaza-adamantane (pta, **5.1**) ligand $[Ru(\eta^6\text{-}p\text{-}cymene)(pta)X_2]$, the behaviour was found to be dependent on the strength of binding of the anion X, with weakly-bonding iodide and thiocyanate complexes giving the $[M-X]^+$ cation, but more strongly-bound chloro and bromo complexes giving protonated $[M+H]^+$.[20]

5.1

Some exceptions to the above general principles have however been observed. For example, the ruthenium-nitrosyl complexes $[Ru(NO)Cl_3(EPh_3)_2]$ complexes (E = P, As, Sb) did not give simple positive ions by halide loss, but in the presence of alkali metal ions gave $[M+cation]^+$.[21] A number of platinum(II) anticancer drugs, which contain NH_3 or amine ligands, also frequently ionise by addition of a sodium ion rather than loss of a chloride and are discussed in greater detail in Section 5.12.

Metal Complexes of Polydentate Oxygen Donor Ligands: Polyethers, Crown Ethers, Cryptands and Calixarenes

The general structural feature of polyether-containing molecules is the presence of $-CH_2CHRO-$ (R = H, Me) repeating units; the oxygen atoms of these polyether chains are able to coordinate a wide range of metal ions. Polyethers are simple acyclic compounds, often synthesised as oligomeric mixtures with molecules having differing chain lengths as a result of different numbers of CH_2CH_2O (EO) repeat units. Figure 5.9 shows the positive-ion ESI mass spectrum of polyethylene glycol 600 (PEG600), having

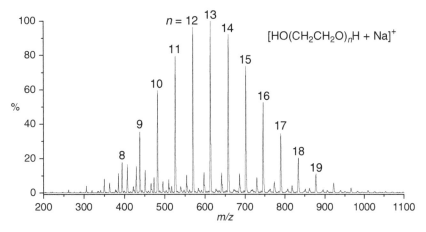

Figure 5.9
Positive-ion ESI mass spectrum of polyethylene glycol 600, in methanol solution with added sodium iodide, at a cone voltage of 50 V, showing the presence of $[M + Na]^+$ adduct ions for a range of oligomer chain lengths

the structure $HO(CH_2CH_2O)_nH$; the average relative molar mass of this material is 600 Da. The spectrum clearly shows a Gaussian-like distribution of $[M + Na]^+$ ions, separated by 44 Daltons (the mass of a CH_2CH_2O unit), formed from individual oligomers with different molar masses. The base peak in the spectrum, at *m/z* 614, is due to the n = 13 oligomer and thus the average molar mass of the material is around 600 Da. This demonstrates the usefulness of ESI MS in determining molecular weight distributions in oligomeric materials. However, some caution must be exercised when looking at absolute ion intensities, since it has been found that molecules containing longer EO chains often display a preference for binding to the larger K^+ ion over the Na^+ ion.

Crown ethers are related to polyethylene glycols, but are cyclic molecules, with no terminal hydroxyl groups, and are typically obtained as pure materials as opposed to oligomeric mixtures. They typically display a high selectivity for coordination to particular metal cations, dependent on the fit between the cation size and the size of the cavity in the crown ether.

ESI MS can be readily used to probe the selectivity of different crown ether ligands towards alkali metal cations. Because signal intensities in ESI MS are dependent on the desolvation of the ion and its transmission efficiency into the detector, similar complexes formed between a metal and several ligands (or between a ligand and several similar metals) will have similar ES ionisation efficiencies. Hence ESI MS can be used to determine relative binding selectivities. If a ligand-metal combination of known binding strength is incorporated into the experiment, the results can be calibrated. Various studies have concerned binding of metal ions to various host molecules such as crown ethers and

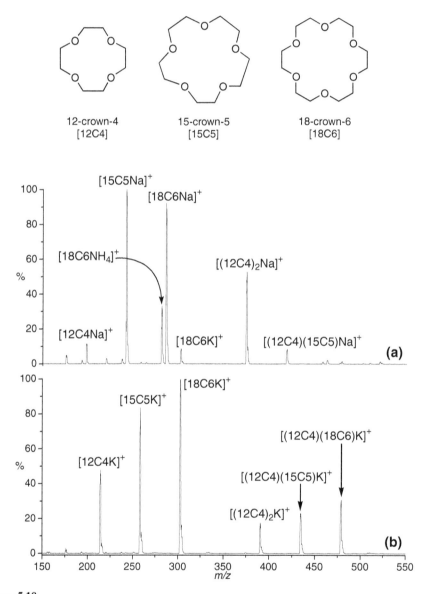

Figure 5.10

Positive-ion ESI mass spectra for a 1:1:1 molar mixture of 12-crown-4 [12C4], 15-crown-5 [15C5] and 18-crown-6 [18C6] in methanol at a cone voltage of 20 V. In spectrum (a) an excess of NaCl is added, while in spectrum (b) an excess of KBr is added. The differences between the two spectra indicate the preferences of the different crown ethers for binding cations of different sizes

caged crown ethers,[22] crown ethers with pendant ether groups,[23] and calixarenes.[24] Reviews by Brodbelt[25] and Vincenti[26] summarise the general features of this area.

A simple ESI MS experiment illustrates how metal-ligand binding affinities can be investigated using ESI MS. The spectra of a 1:1:1 molar mixture of the three crown ethers 12-crown-4 [12C4], 15-crown-5 [15C5], and 18-crown-6 [18C6], with added sodium chloride (Figure 5.10a) and added, potassium bromide (Figure 5.10b) show various metal-ligand adduct ions. Even though the amount of metal cation is not closely controlled, the spectra still illustrate the general features. With added Na^+ ions, the highest intensity peak is due to $[15C5Na]^+$ (*m/z* 243), closely followed by $[18C6Na]^+$ (*m/z* 287), indicating that these two crown ethers easily form complexes with the Na^+ ion. However, the $[12C4Na]^+$ ion at *m/z* 199 is of low intensity, since the Na^+ ion is not a good fit for the cavity in this crown ether. Instead, the 12C4 forms a 'sandwich' type adduct $[(12C4)_2Na]^+$ (*m/z* 375) and a similar weaker $[(12C4)(15C5)Na]^+$ ion at *m/z* 419. Also noteworthy is the presence of an ion $[18C6K]^+$ at *m/z* 303, demonstrating the preference for the 18C6 crown ether to bind the larger (adventitious) potassium ion, even in the presence of excess added Na^+ ions. In the presence of excess K^+ ions (Figure 5.10b), the crown ether mixture is dominated by the ions $[18C6K]^+$ (*m/z* 303) and $[15C5K]^+$ (*m/z* 259), together with $[12C4K]^+$ (*m/z* 215) and various sandwich type ions involving 12-crown-4, as indicated.

Cryptands, typified by cryptand[2.2.2] are even more selective metal-binding ligands; the metal ion binds in the cavity provided by the ether and amine groups. In the case of this molecule, the cavity is very well matched to the potassium ion. The positive ion ESI mass spectrum of cryptand[2.2.2] in methanol solution at a cone voltage of 10 V is shown in Figure 5.11a. With no added metal cations, the cryptand 'scavenges' adventitious cations from solution $- H^+$, NH_4^+, Na^+ and K^+. When the cone voltage is raised to 50 V, the $[M + NH_4]^+$ ion disappears (Figure 5.11b). When KBr is added to the solution, and the spectrum is recorded at a cone voltage of 10 V, the only species observed is the $[M + K]^+$ ion (Figure 5.11c). The strong preference for binding potassium ions can be shown by the ESI mass spectrum of cryptand[2.2.2] with an excess of an equimolar mixture of K^+, Rb^+ and Cs^+ ions. The base peak is the $[M + K]^+$ ion, with a very small $[M + Rb]^+$ ion, and no $[M + Cs]^+$ observed.

Porphyrins and Metalloporphyrins

The behaviour of porphyrins and phthalocyanins, and their metal complexes, is rather dependent on the exact nature of the system. ESI MS analysis of a range of porphyrins yielded $[M + H]^+$ ions for the free ligands.[27]

octaethylporphyrin, H_2OEP tetraphenylporphyrin, H_2TPP

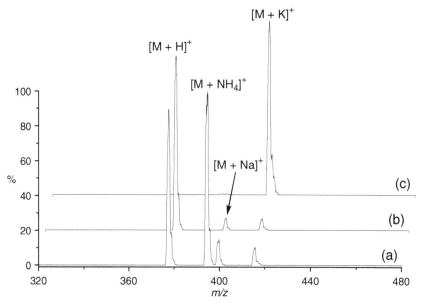

Figure 5.11

Positive-ion ESI mass spectrum of cryptand[2.2.2] in methanol solution: (a) at a cone voltage of 10 V with no added cations, showing adducts with various adventitious cations, (b) at a cone voltage of 50 V and (c) at 10 V with added KBr

The ESI MS behaviour of porphyrin complexes of trivalent or tetravalent metals containing ancillary anionic ligands (such as halide) also tends to be relatively straightforward, giving cations formed by loss of the anion. The positive-ion ESI MS spectrum of the iron(III) porphyrin complex TPPFeCl (H_2TPP = 5,10,15,20-tetraphenylporphyrin) at a cone voltage 100 V is shown in Figure 5.12. The spectrum is dominated by the $[TPPFe]^+$ cation, with the solvated species $[TPPFe(MeOH)]^+$ as a low intensity ion at both cone voltages, together with $[(TPPFe)_2OH]^+$, which is undoubtedly a species with a bridging hydroxy group; such species are very well-known in porphyrin chemistry.

Other related complexes behave the same way, for example $Mn^{III}(OEP)Cl$ (H_2OEP = octaethylporphyrin) gives $[Mn(OEP)]^+$,[27] the tin(IV) hydroxy complex $Sn(TPP)(OH)_2$ gives $[Sn(TPP)(OH)]^+$ [28] and the gallium complex $[Ga(OEP)Cl]$ gives $[Ga(OEP)]^+$ and the dimeric hydroxy-bridged $[\{Ga(OEP)\}_2OH]^+$.[29] The zirconium complex $[Zr(OEP)Cl_2]$ gave more complex spectra, with the most abundant ions being $[Zr(OEP)OH]^+$, $[Zr(OEP)Cl]^+$ and dimeric $[\{Zr(OEP)OH\}_2]^{2+}$.[29] This behaviour, involving ionisation by loss of an anionic ligand (usually halide), is reminiscent of that shown by other metal-halide and related complexes (Section 5.3).

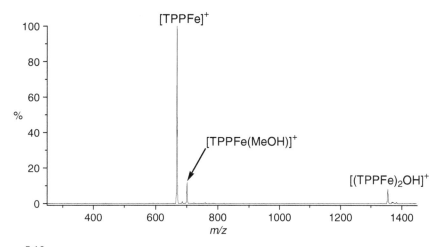

Figure 5.12
Positive-ion ESI mass spectrum of the iron(III) porphyrin complex TPPFeCl (TPP = 5,10,15, 20-tetraphenylporphyrin), at a cone voltage of 100 V

The ESI MS behaviour of neutral metalloporphyrin complexes of divalent metals can be more complex, giving either radical $[M]^{\bullet+}$, protonated $[M + H]^+$ or metallated $[M + Na]^+$ ions depending on the metal and the solvent. Other parameters such as flow rate, temperature, and capillary needle voltage can also influence the ionisation. Initial studies in this area by Van Berkel and coworkers investigated the ESI MS spectra of alkyl-substituted metalloporphyrins, and found that they gave molecular radical cations, $[M]^{\bullet+}$.[27] A subsequent study confirmed this behaviour with M(OEP) complexes (M = magnesium, nickel, copper, zinc) and concluded that the origin of the radical cations was due to an electrochemical oxidation process occurring in the ESI source, as discussed in Section 3.9.1.[30] The use of a chemical oxidant in the solvent system, such as 2,3-dichloro-5,6-dicyano-1,4-benzoquinone (DDQ) or SbF_5 in a CH_2Cl_2-CF_3CO_2H solvent system was found to promote the formation of the $[M]^{\bullet+}$ cations from Co(OEP) and Ni(OEP), as well as forming the dications $[M]^{2+}$.[31] A study of the magnesium, nickel, copper, zinc and vanadyl complexes of OEP, etioporphyrin I and TPP was carried out using ESI MS, and the ion abundance of each complex monitored as a function of concentration, and found to be dependent on the redox potential of the complex. Binary mixtures of complexes with redox potentials differing by < 0.1 V were found to give ion abundances proportional to concentration, but when the redox potential difference was > 0.1 V, ionisation of the complex with lowest oxidation potential preferentially occurred.[32]

Several studies have concerned lanthanide complexes. The lanthanide(III) sandwich complexes ML_2 (L = substituted phthalocyaninate ligand; M = europium, gadolinium) have been characterised using ESI MS,[33] where the parent molecular radical cations $[M]^{\bullet+}$ were observed. A number of mixed-ligand complexes of the type Ln(TPP)(Pc), $Ln_2(TPP)(Pc)_2$ and $Ln_2(TPP)_2(Pc)$ (Ln = samarium, europium, gadolinium) were also studied using ESI FTICR MS, and in all cases intense signals due to the radical cations $[M]^{\bullet+}$ were observed, together with overlapping $[M + H]^+$ signals.[34] Signals due to multiply-charged ions were also observed, and it was proposed that this was due to successive oxidation of the ligand(s).

Metal Alkoxides – Highly Moisture-Sensitive Coordination Compounds

The analysis of metal alkoxides by ESI MS presents a challenge due to the high moisture-sensitivity of this type of compound. However, by using 'superdry' solvents, this problem is easily overcome. Using this modification, ESI mass spectra were initially recorded for a wide range of moisture-sensitive alkoxides, including, $Zr(OEt)_4$, $Ti(OEt)_4$, $Al(OEt)_3$, $Si(OEt)_4$ and $(EtO)_3SiCH_2CH_2CH_2NH_2$.[35] $Zr(OEt)_4$ gave $[nZr(OEt)_4 + OEt]^-$ ions as the major observed species, the intensity of which could be enhanced by addition of a small amount of sodium ethoxide. Cationic niobium and tantalum alkoxide clusters have been generated in solution by addition of an equimolar quantity of $B(C_6F_5)_3$ to the metal ethoxide $Nb(OEt)_5$ or $Ta(OEt)_5$ and studied using ESI MS.[36] The dominant species observed were $[M(OEt)_4]^+$, $[M_2O(OEt)_7]^+$, $[M_2(OEt)_9]^+$ and $[M_3O(OEt)_{12}]^+$. Mixed niobium-tantalum species were also observed, and CID experiments furnished structural information on the various cations.

ESI MS has subsequently been used to study the initial stages of polycondensation reactions of various main group and transition element alkoxides such as the characterisation and hydrolysis of $Si(OEt)_4$.[37,38] Marshall and co-workers have used the power of the ESI FTICR MS technique to characterise hydrolysis and condensation intermediates using alkali and transition metal ions to ionise the various species; Na^+ ions bind non-specifically to silicates, and Ni^{2+} ions forms the tris(tetraethylorthosilicate) complex $[Ni\{Si(OEt)_4\}_3]^{2+}$. Other studies have investigated condensation reactions on $Ti(OR)_4$ ($R = {}^iPr$ and ^nBu),[39] $Ge(OEt)_4$[40] and $Sn(O^tBu)_4$.[41] The condensation of mixtures of $Ge(OEt)_4$ with $Si(OEt)_4$[40] and $Ti(O^nBu)_4$ with $Si(OEt)_4$[42] have also been studied.

Various organosilicon alkoxide compounds have been studied by ESI MS including the *tert*-butoxysilanols ($^tBuO)_3SiOH$, $(^tBuO)_2Si(OH)_2$, $HO[(^tBuO)_2SiO]_2H$, and $HO[(^tBuO)_3$-$SiO]_3H$.[43] In positive-ion mode, the compounds gave $[M + Na]^+$ and $[2M + Na]^+$ ions, through aggregation with adventitious sodium cations. In negative-ion mode $(^tBuO)_3$-$SiOH$ and $HO[(^tBuO)_3SiO]_3H$ give $[M-H]^-$ ions, while $(^tBuO)_2Si(OH)_2$, $HO[(^tBuO)_2$-$SiO]_2H$ and $HO[(^tBuO)_3SiO]_3H$ gave adducts with chloride ions $[M + Cl]^-$ or $[2M + Cl]^-$, presumably through hydrogen bonding with the silanol hydrogen atoms. Protonated $[M + H]^+$ ions were observed when hydrochloric acid (1M) was added to the samples and the ESI mass spectra immediately recorded. It is possible that the ionisation occurs at the silanol groups, since attempts at obtaining ESI MS spectra of a low molecular weight siloxane, Me_3Si-O-$SiMe_3$ have been unsuccessful. This indicates that the Si-O-Si group is not readily protonated or metallated, due to involvement of the oxygen lone pairs in bonding with the silicon. ESI MS has also been used in a study of the silatrane reaction products between geothermal silica and triethanolamine, giving silicon-alkoxide products of the type $N(CH_2CH_2OH)_n(CH_2CH_2OR)_{3-n}$ where R is the silatranyl moiety $-Si(OCH_2CH_2)_3N$.[44]

Related to alkoxides are silsesquioxane materials, formed by hydrolysis and condensation of $RSiX_3$, giving molecular materials of the composition $[RSiO_{3/2}]$. Since these materials are often soluble in organic solvents, they are often used as model compounds for silica. A number of silsesquioxanes have been analysed in hexane solution by APCI MS and turbo ionspray MS techniques, and gave $[M + H]^+$ ions. In some cases, addition of ammonium acetate, sodium chloride or sodium acetate allowed the detection of $[M + NH_4]^+$ or $[M + Na]^+$ ions.[45] Heterosilsesquioxanes, formally derived by substitution of a silicane atom by a heterometal centre (antimony, tin, vanadium, tungsten, aluminium or boron) were also characterised. A considerable number of other studies have concerned the analysis of silsesquioxanes by ESI MS and MALDI-TOF.[46]

β-Diketonate Complexes

Surprisingly few ESI MS studies have been carried out with diketonate complexes, containing the monoanionic bidentate acetylacetonate (acac) ligand $CH_3COCHCOCH_3^-$, and substituted analogues thereof. The acac ligand forms complexes with a wide range of metals, including main group metals, transition metals, lanthanides and actinides. It is found that the best ESI mass spectra are obtained when the complexes are analysed in the presence of NH_4^+ ions. However, the actual ions observed are very dependent on the metal centre. The general properties are best illustrated by means of a number of examples.

The complex $Co(acac)_3$ is an example of a very stable cobalt(III) derivative, and ESI mass spectra for this complex are given in Figure 5.13. In the presence of added NH_4^+

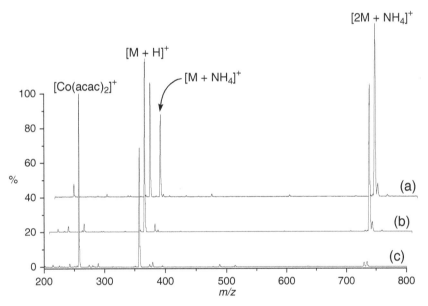

Figure 5.13
Positive-ion ESI mass spectra for $[Co(acac)_3]$ (acac $= CH_3COCHCOCH_3$) in methanol with added ammonium formate, at cone voltages of (a) 5 V, (b) 15 V and (c) 40 V, showing the formation of ammonium adduct ions, and fragmentation by loss of acac$^-$ at the highest cone voltage

ions, this complex gives $[M + H]^+$, $[M + NH_4]^+$ and $[2M + NH_4]^+$ ions at low cone voltages (Figure 5.13a), the latter being the base peak in the spectrum. At higher cone voltages (e.g. 15 V, Figure 5.13b) the $[M + NH_4]^+$ ion fragments, giving the $[M + H]^+$ ion as the base peak, but the $[2M + NH_4]^+$ ion is still significant. At a cone voltage of 40 V (Figure 5.13c) a new species $[Co(acac)_2]^+$ is formed by loss of an acac ligand.

Spectra for the chromium(III) complex $Cr(acac)_3$ are fairly similar to those of the cobalt complex, except that at higher cone voltages, solvated ions $[Cr(acac)_2(MeOH)_x]^+$

(x = 1, 2) are also seen in addition to $[Cr(acac)_2]^+$. Both the cobalt(III) and chromium(III) derivatives are characterised by inert d-electron subshells–d^6 and d^3 respectively–and the stability of these configurations is well known in transition metal chemistry. Loss of an acac ligand therefore only occurs at moderate cone voltages.

A main group metal complex such as $Al(acac)_3$ behaves in broadly the same way as the cobalt and chromium cases, with the ions $[M + H]^+$, $[M + NH_4]^+$ and $[2M + NH_4]^+$ being observed at a low cone voltage (5 V) with added ammonium formate. The aggregate ion $[2M + NH_4]^+$ is again the base peak in the spectrum of this complex. Upon increasing the cone voltage to 15 V, fragmentation by loss of $acac^-$ occurs more readily than for the cobalt and chromium complexes; at 40 V $[Al(acac)_2]^+$ is the base peak in the spectrum, and $[M + H]^+$ has only very low intensity. The $Al(acac)_3$ therefore shows a greater tendency to undergo fragmentation that the cobalt and chromium counterparts, reflecting weaker metal-ligand binding, but otherwise the general mass spectral features are quite similar.

The complexes $Fe(acac)_3$ and $Mn(acac)_3$–containing d^5 iron(III) and d^4 manganese(III)–are less substitutionally inert than their cobalt and chromium counterparts, and this is reflected in their ESI MS behaviour. Both complexes behave similarly and the manganese complex will be used to illustrate the general features. Spectra at three cone voltages (5, 15 and 40 V) are shown in Figure 5.14 and can be compared with the cobalt example in Figure 5.13. At 5 V (Figure 5.14a) the base peak is a solvolysis ion, $[Mn(acac)_2(MeOH)]^+$ at m/z 285. Parent-derived ions $[M + H]^+$, $[M + NH_4]^+$ and $[2M + NH_4]^+$ have a surprisingly low intensity for this complex. In addition, there are ions $[Mn(acac)_2]^+$ (m/z 253), $[Mn_2(acac)_5]^+$ (m/z 605) and $[Mn_2(acac)_4(OMe)]^+$ (m/z 537) that indicate significant lability of the complex. At 15 V (Figure 5.14b) the base

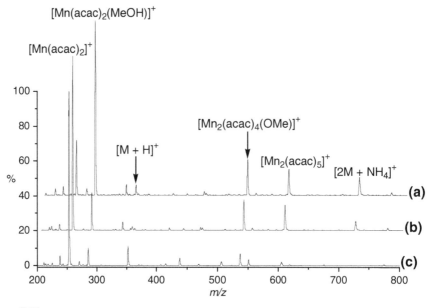

Figure 5.14
Positive-ion ESI mass spectra for $[Mn(acac)_3]$ ($acac = CH_3COCHCOCH_3$) in methanol with added ammonium formate, at cone voltages of (a) 5 V, (b) 15 V and (c) 40 V

peak is $[Mn(acac)_2]^+$, which remains at 40 V. Thus, for the five different $M(acac)_3$ complexes (of Co, Cr, Fe, Mn and Al), their ESI MS behaviour broadly correlates with their known chemical properties.

For a divalent metal complex such as $Pd(acac)_2$ in the presence of ammonium formate, $[M + H]^+$, $[M + NH_4]^+$ and $[2M + NH_4]^+$ were all observed as major ions at a cone voltage of 10 V, together with a lower intensity $[3M + NH_4]^+$ ion. In this case, the greater tendency to aggregate is probably a result of a less bulky metal complex. However, not all $M(acac)_2$ complexes show enhanced aggregation; positive-ion ESI mass spectra of the fluorinated analogue $Cu(CF_3COCHCOCF_3)_2$ under the same conditions gave a strong $[M + NH_4]^+$ ion, but no $[2M + NH_4]^+$.

Lanthanide complexes behave in a similar manner, with added ammonium ions an effective ionisation aid that do not cause any breakdown reactions of the diketonate complex. For example $Eu(fod)_3$ (fod = 6,6,7,7,8,8,8-heptafluoro-2,2-dimethyl-3,5-octanedionato) gives $[M + NH_4]^+$ as the base peak at m/z 1057 when analysed in methanol solution with added ammonium formate (Figure 5.15a).

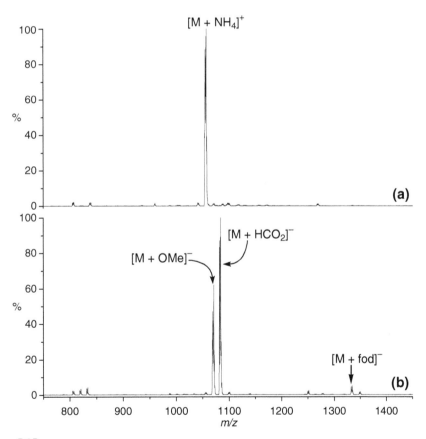

Figure 5.15
ESI mass spectra of the europium *β*-diketonate complex $Eu(fod)_3$ (fod = 6,6,7,7,8,8,8-heptafluoro-2,2-dimethyl-3,5-octanedionato, $CF_3CF_2CF_2COCHCOBu^t$) in methanol solution with added ammonium formate. (a) Positive-ion spectrum at a cone voltage of 20 V, showing exclusive formation of the $[M + NH_4]^+$ ion (m/z 1057). (b) Negative-ion spectrum at a cone voltage of 20 V, showing formation of $[M + OMe]^-$, $[M + HCO_2]^-$ and $[M + fod]^- \equiv [Eu(fod)_4]^-$ ions

It is worth noting that an early study on related lanthanide complexes [Ln(^{t}BuCOCH-CO^{t}Bu)$_3$] (Ln = europium, gadolinium, ytterbium) in methanol-water solutions with added acetic acid gave [M + H]$^+$ ions, together with ligand exchange products containing acetate ligands in place of diketonate ligands.[47] Similar results were also obtained for the complexes [Ln(tfc)$_3$] (tfc = D-3-trifluoroacetylcamphorate).[48] These studies indicated that organic acids, which are commonly added to electrospray solvents for the analysis of 'robust' organic compounds, are not appropriate for coordination complexes if ligand-exchanged species are not desired. Tandem mass spectrometry of the protonated europium and ytterbium ions [M(^{t}BuCOCHCO^{t}Bu)$_3$ + H]$^+$ gave divalent ions [M(^{t}BuCOCHCO^{t}Bu)]$^+$, consistent with the known tendency for these metals to form divalent complexes.

A number of metal-diketonate complexes also give negative ions. The negative-ion ESI mass spectrum of Eu(fod)$_3$ is illustrated in Figure 5.15b and shows various anionic adducts with OMe$^-$ (from the solvent), HCO$_2^-$ (added as ammonium formate, NH$_4^+$HCO$_2^-$) and fod$^-$ itself, either from ligand redistribution reactions or from traces of free fod in the sample. Lanthanide β-diketonate complexes have been widely used as paramagnetic 'contact shift' reagents in NMR spectroscopy, although their use has greatly declined as a result of increasingly powerful superconducting magnets. A Ln(β-diketonate)$_3$ complex (Ln = lanthanide element) is able to interact with donor atoms on organic molecules, readily increasing its coordination number from six in the original complex, thus accounting for the formation of adducts with various anionic species.

[Cu(CF$_3$COCHCOCF$_3$)$_2$] also gives [Cu(CF$_3$COCHCOCF$_3$)$_3$]$^-$ (*m/z* 684) as the base peak at a cone voltage of 20 V, together with weak signals due to [CF$_3$COCHCOCF$_3$]$^-$ (*m/z* 207) and an ion at *m/z* 477 that is assigned to the copper(I) reduction product [Cu(CF$_3$COCHCOCF$_3$)$_2$]$^-$. In contrast, no such ions were observed by an example of a tris(acac) complex of a 3d transition metal, namely Co(acac)$_3$, presumably because of its inability to expand upon six-coordination.

Metal Complexes of Carbohydrates

Metal ions are known to interact with carbohydrates, which (to the coordination chemist) are polydentate oxygen donor ligands; such interactions are of biochemical importance for the transport and storage of metals, the function of metalloenzymes, and the metabolism of toxic metal ions.[49]

Various different aspects of metal-carbohydrate complexes have been investigated using ESI MS:

(a) *Characterisation of preformed metal-carbohydrate complexes.* The MS technique provides additional characterisation data, for example in a study of cationic trimethylplatinum(IV) derivatives of carbohydrates, of the type [Me$_3$PtL]$^+$ (L = isopropylidene-protected carbohydrate).[50] In this case, the cationic nature of the complexes results in observation of the expected parent [M]$^+$ ions.

(b) *Using metal coordination of carbohydrates to provide structural information.* Distinguishing isomeric structures by mass spectrometry is often difficult, and this can be particularly problematical in the field of carbohydrate chemistry, because of the possible existence of many stereoisomers for any carbohydrate, especially oligosaccharides. Metal ion complexation has been used to distinguish isomeric carbohydrate structures by selectively coordinating a metal ion to the carbohydrate, coupled with MS^n studies. Types of carbohydrates investigated include branched trisaccharides (distinguished using Ca^{2+} and Mg^{2+} coordination)[51], aminosaccharides (distinguished using cobalt complexes[52] or nickel complexes[53]) and underivatised hexose sugars (e.g. D-glucose, D-galactose, D-mannose etc, distinguished using Pb^{2+} complexation[54]).

(c) *Inclusion complexes formed by cyclodextrins.* Cyclodextrins (CDs) are cyclic (toroidal) molecules comprised of linked carbohydrate units, with a central hydrophobic cavity and a hydrophilic exterior. CDs are able to selectively include a range of guest species in their cavities, including organometallic molecules such as ferrocene derivatives (Section 7.4.1). For example, ESI MS showed formation of a 2:1 complex between α-CD and ferrocene, while other ferrocene derivatives formed a 1:1 adduct.[55]

Metal Complexes of Amino Acids, Peptides and Proteins

A summary of the interactions of metal ions with proteins, peptides and amino acids has appeared;[56] in this section some of the highlights of this area are provided.

5.9.1 Amino Acids

Amino acids, of the general structure $H_2NCHRCOOH$, contain amine and carboxylate functions that are capable of coordination to metal centres; additionally, the R group may contain metal-coordinating groups such as thioether (methionine), thiolate (cysteine) or imidazole (histidine). Amino acids would therefore be expected to form complexes with the majority of metallic or metalloid elements in the periodic table. A number of studies have investigated the complexes formed from binary mixtures (metal ion + amino acid), using ESI MS.[57] A specific recent example involving the main group element lead has been carried out, prompted by the high toxicity of lead and the absence of good characterisation data on lead-amino acid complexes. Complexes between Pb^{2+} ions and all 20 of the common amino acids were detected by ESI MS, and the study culminated in the first crystal structure of a lead(II)-amino acid complex.[58] Complexes formed between bismuth(III) and sulfur containing amino acids (cysteine, homocysteine, methionine) and the tripeptide glutathione have also been studied by ESI MS.[59]

The majority of studies have concerned ternary mixtures (metal + amino acid + co-ligand L, e.g. 2,2′-bipyridine), involving the metal ions nickel(II), cobalt(II), copper(II) and zinc, which form an abundance of amino acid complexes.[60] Such ternary complexes have high stability constants in solution and the ESI mass spectra of these systems not surprisingly show abundant gas phase ions of the type $[Cu^{II}(H_2NCHRCOO)(L)]^+$. As an example, Figure 5.16 shows the positive-ion ESI mass spectrum of a mixture of copper(II) sulfate, 2,2′-bipyridine and L-β-phenylalanine [$H_2NCH(CH_2Ph)CO_2H$, Phe], which shows the ternary ion $[Cu^{II}(bipy)(Phe-H)]^+$ as the base peak.

The majority of the common α-amino acids form this ion, though the thiol-amino acid cysteine undergoes oxidation to the disulfide cystine with copper(II), and the basic amino acids lysine and arginine form doubly-charged ions $[Cu^{II}(H_2NCHRCOO + H)(L)]^+$, due to protonation on the amino acid side chain. The role of the auxiliary neutral ligand in these complexes is to firstly occupy two of the copper coordination sites (thereby blocking coordination of a second amino acid anion, which would give a product species

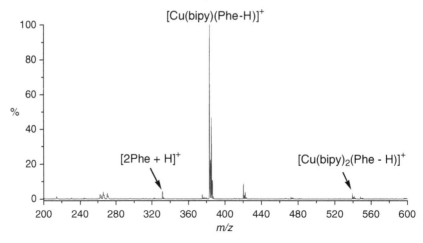

Figure 5.16
The positive-ion ESI mass spectrum of a mixture of $CuSO_4$, 2,2'-bipyridine and phenylalanine (Phe) at a cone voltage of 10 V in methanol-water (1:1 v/v), showing the $[Cu^{II}(bipy)(Phe-H)]^+$ ion as the base peak, together with several minor ions

with no charge), and secondly to act as an electron donor for copper. Extensive CID studies of these metal amino acid complex ions have been carried out.

5.9.2 Proteins and Peptides

Large biomolecules such as proteins can easily be studied by ESI MS. Detailed discussions of the application of ESI MS to protein analysis are beyond the scope of this text but are readily available elsewhere.[61] The example given in Section 3.9.2 illustrates the behaviour of a protein by means of a simple example, horse heart myoglobin, with a molecular weight of 16,951.49 Da. Multiple charging furnishes ions in the m/z range 800–1600 (Figure 3.40), giving the characteristic pattern of ions observed for a protein.

Because metal ions often interact with proteins (which are built up from numerous amino acid residues, linked through amide bonds) and may have a specific role in the function of the protein, there have been multiple studies of protein-metal interactions by ESI MS. Specific systems that have been studied include iron-sulfur proteins,[62] metal ion binding of horse heart cytochrome c,[63] and proteins that bind calcium,[64] copper(II),[65] and zinc.[66] Many of these studies have concerned the determination of the metal binding stoichiometry, which can be obtained from the increase in mass of the protein upon metal ion binding. In cases where the metal may have variable oxidation states (such as copper) FTICR data may be necessary in order to correctly identify the observed species. As an example of this, the protein ubiquitin forms a complex with copper ions. Analysis of the 12+ ion by FTICR MS allowed resolution of the isotope pattern for the ion, which gave a better fit–in terms of m/z and intensities–to the theoretical parameters for the copper(II)-containing ion $[ubiquitin + 10H + Cu^{II}]^{12+}$. An alternative possibility that was considered and discounted – the $[ubiquitin + 11H + Cu^{I}]^{12+}$ ion – differs only by 1/12 m/z, and therefore high resolution data were obviously required to distinguish the two possibilities.[65]

As with proteins, there have been various studies that investigate the interactions of peptides with metal centres; in some cases metal ion adduction has been investigated to provide sequence information on the peptide. For example, the peptides angiotensin I and angiotensin II form multiply-charged complexes with zinc ions in their ESI mass spectra. From CID of the $[M + Zn]^{2+}$ ion for angiotensin II, the observation of certain zinc-containing fragment ions indicates that the histidine residues His-6 and His-9 are involved in zinc coordination.[67] It is worth noting in this context that zinc-complexing peptides and proteins typically utilise histidine residues at the binding site. Other examples of peptide-metal complexes that have been studied by ESI MS include those of copper(II)[68], zinc[69] palladium[70] and alkali metals.[71]

Ternary complexes, formed by self assembly of (for example) copper(II), with 2,2'-bipyridine (bipy) and a dipeptide[72] or a tripeptide[73] have also been studied by ESI MS. Abundant complexes of the type $[Cu(peptide-H)(bipy)]^+$ and $[Cu(peptide)(bipy)]^{2+}$ were observed for a range of tripeptides. CID studies provided fragment ions that were indicative of the amino acid sequence of the peptide and was able to distinguish tripeptides containing isomeric leucine and isoleucine residues.

Metal complexes have also been used as 'labels' for the MS analysis of proteins. For example, adduction with silver ions has been found to be useful in labelling peptides and proteins, especially ones that contain the amino acid residue methionine, which is a thioether with a strong binding affinity for Ag^+. However, in the gas phase the silver-peptide complex rearranges so that the silver also binds to nitrogen and oxygen atoms on the peptide backbone. The silver-sulfur bond is often broken upon CID, producing silver-peptide product ions that do not contain methionine groups.[74] Organometallic dienyl-iron cations of the type $[(\eta^5\text{-}RC_6H_6)Fe(CO)_3]^+$ have also been effectively used for labelling peptides and proteins, with cysteine and histidine residues being particularly targeted by these reagents.[75] The addition of silver[76] and organomercury derivatives[77] has also assisted analysis of peptides and proteins by MALDI-TOF MS.

Oxoanions, Polyoxoanions and Related Species

5.10.1 Simple Transition Metal Oxoanions

Studies in this area have centred on both simple mononuclear oxyanions and on polyoxometallate anions; oxyanions of main group non-metallic elements have been previously discussed (Section 4.4).

Large, multiply-charged anions are able to delocalise and hence stabilise the charge within, whereas smaller multiply-charged anions require a hydration sphere for stability and often undergo protonation to reduce the charge density. Removal of the hydration sphere under CID conditions may then generate unstable ions that fragment to separate the charges. Under higher cone voltage conditions (dependent on the oxoanion), oxygen loss (i.e. reduction) occurs, giving oxoanions with the metal in a lower oxidation state.

In early studies in this area, simple oxoanions such as $[MnO_4]^-$, $[Cr_2O_7]^{2-}$, $[CrO_3Cl]^-$, $[RuO_4]^-$ and $[RuO_2Cl_3]^-$ were studied by ESI MS in acetonitrile solution, and gave the parent anion as the base peak with minimal fragmentation.[78] In the case of $[MnO_4]^-$ and $[Cr_2O_7]^{2-}$, solvated ions $[MnO_4(MeCN)_x]^-$ (x = 1,2) and $[Cr_2O_7(MeCN)_x]^{2-}$ (x = 1, 2) were observed, together with the aggregate ion $[K(MnO_4)_2]^-$ and protonated $[HCrO_4]^-$, formed from the equilibrium:

$$[Cr_2O_7]^{2-} + H_2O = 2[HCrO_4]^-$$

The parent dianions $[MO_4]^{2-}$ (M = chromium, molybdenum, tungsten) are **not** observed in ESI MS analysis because of the high charge density, which is reduced by protonation to give the $[HMO_4]^-$ ion.[79] The same behaviour is even observed in alkaline solution, where tungstate gives the $[HWO_4]^-$ ion, with a small amount of $[HW_2O_7]^-$ or $[NaWO_4]^-$.[80,81]

When oxoanions are subjected to CID, for example by applying a cone voltage, fragment ions, having the metal in a lower oxidation state, are typically observed. Thus, for $[MnO_4]^-$, $[Cr_2O_7]^{2-}$ and $[RuO_4]^-$, fragment ions $[MO_x]^-$ (x = 2,3) (M = manganese, chromium, ruthenium) are observed.

To illustrate these principles, Figure 5.17 shows ESI MS spectra of a solution of potassium permanganate at cone voltages of 10 and 50 V. At 10 V the only species

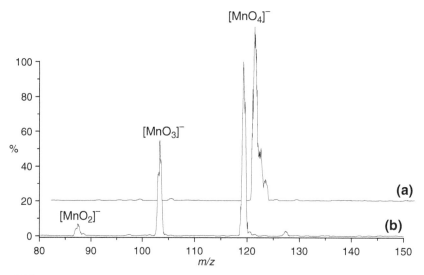

Figure 5.17
Negative-ion ESI mass spectra for an aqueous solution of $KMnO_4$ at cone voltages of (a) 10 V and (b) 50 V, showing breakdown of the parent $[MnO_4]^-$ ion to $[MnO_3]^-$ and $[MnO_2]^-$

observed is the parent $[MnO_4]^-$ ion, while at 50 V, the fragment ions $[MnO_3]^-$ and $[MnO_2]^-$ are also seen. This represents a reduction of the metal centre and is a commonly observed feature in CID studies of higher oxidation state metal complexes. The cone voltage that is required to achieve this is very dependent on the metal centre; for MnO_4^-, a relatively low cone voltage causes fragmentation, while for a stable oxoanion such as ReO_4^-, much more forcing conditions are required (Figure 5.18). These observations correlate well with the chemical properties of MnO_4^- (a strong oxidant that is easily reduced) and ReO_4^- (which is rather stable).

Several studies have concerned the ESI MS analysis of perrhenate systems, $[ReO_4]^-$, in both positive- and negative-ion modes. In negative-ion mode, and using a low cone voltage, the expected $[ReO_4]^-$ ion dominates the negative-ion spectrum,[80] as illustrated in Figure 5.18. However, at very high cone voltages such as 160 V (Figure 5.18b),

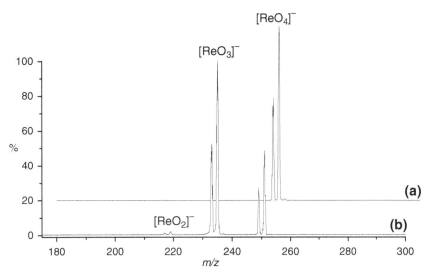

Figure 5.18
Negative-ion ESI mass spectra of ammonium perrhenate, NH_4ReO_4 in methanol solution at cone voltages of (a) 40 V and (b) 160 V. The robustness of the ReO_4^- ion is clearly displayed by the very high cone voltage required to effect fragmentation, when compared with the isoelectronic MnO_4^- ion (Figure 5.17)

fragmentation results in observation of $[ReO_2]^-$ and $[ReO_3]^-$ ions.[82] The high cone voltage required contrasts with the ease of reduction of the MnO_4^- ion (Figure 5.17). The presence of two major rhenium isotopes (^{185}Re 59.7 %, ^{187}Re 100 % relative abundance) can easily be seen for these ions.

In a positive-ion ESI MS study of ammonium and alkali metal (Na, K) perrhenates, a wide range of previously unknown polyoxorhenate ions were detected, including the series $[M_{x+1}Re^V_xO_{4x}]^+$ (x = 1 – 5, M = NH_4, Na, K), and the series $[K_{x+2}Re^VRe^{VII}O_{4x+3}]^+$ (x = 0 – 4); all series have $\{MReO_4\}$ as the polymerisation unit.[83] ESI MS has also been used in a kinetic study of the base-catalysed hydrolysis of CH_3ReO_3 to $[ReO_4]^-$. The $[ReO_4]^-$ ion gave only the parent ion in solution and the tosylate anion (p-$CH_3C_6H_4SO_3^-$) was used as an internal standard. Using gentle ionisation conditions the intermediate species $[CH_3ReO_3(OH).4H_2O]^-$ was observed.[84] The $[ReO_4]^-$ ion has also been observed as a product of atmospheric oxidation of a rhenium colloid; in this study, the ions $[ReO_4 + nH_2O]^-$ (n = 1 – 3) were observed in addition to the $[ReO_4]^-$ ion.[6]

The analysis of a $[Cr_2O_7]^{2-}$ solution in *alcohol* produces rather different results to acetonitrile, due to the reactivity of the solvent. Negative-ion ESI mass spectra for a solution of $Na_2Cr_2O_7$ in methanol are shown in Figure 5.19. At a low cone voltage (10 V, Figure 5.19a), the base peak is due to the species $[MeCrO_4]^-$, together with $[HCrO_4]^-$ and a trace of $[HCrO_3]^-$. Confirmation of the assignment of the methylated species is easily obtained by recording a spectrum of $Na_2Cr_2O_7$ in ethanol solution, whereby the corresponding $[EtCrO_4]^-$ species is observed at *m/z* 145. These species are likely to be chromate esters of the type $ROCrO_3^-$, and may represent intermediates in the well known oxidation of alcohols by chromium(VI) species. At a higher cone voltage (30 V, Figure 5.19b) $[HCrO_3]^-$ becomes the base peak in the spectrum, while at 90 V (Figure 5.19c), the chromium(III) species $[CrO_2]^-$ (*m/z* 84) is also observed.

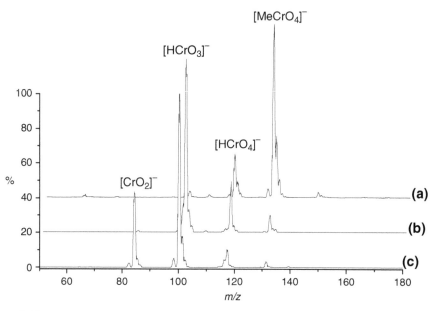

Figure 5.19
The formation of the chromium(VI) ester species [MeCrO$_4$]$^-$ in a methanolic solution of Na$_2$Cr$_2$O$_7$ is shown in (a) (cone voltage 10 V). At higher cone voltages (b) 30 V, (c) 90 V, charge reduction occurs, giving species such as [HCrO$_3$]$^-$ and [CrO$_2$]$^-$

A detailed study of the species present in alkali metal (lithium, sodium, potassium) dichromate solutions (using both positive- and negative-ion modes) has been recently reported.[85] In negative-ion mode, the protonated chromate ion [HCrO$_4$]$^-$ was the most abundant ion at a cone voltage of 20 V; other ions observed were singly-charged anions. A range of higher nuclearity polyoxochromate anions were observed, with up to Cr$_6$ species observed in some cases. Fragment ions such as [CrO$_2$]$^-$, [CrO$_3$]$^-$ were also observed at a higher cone voltage, consistent with the other studies described above. In positive-ion mode, a substantial number of aggregate clusters were observed, with up to twelve chromium atoms. It was concluded that the preservation of a tetrahedral geometry at chromium strongly influences the degree of polymerisation in the chromate system, compared to related molybdate and tungstate systems, where the metal is typically six coordinate.

A number of related nitrido complexes have also been characterised by ESI MS, with compounds containing [OsO$_3$N]$^-$, [OsNCl$_4$]$^-$, [RuNCl$_4$]$^-$ and [Cr(NBu^t)(mnt)$_2$]$^-$ (mnt = maleonitriledithiolato, S$_2$C$_2$(CN)$_2$) all giving the expected parent anion.[86] [OsO$_3$N]$^-$ yielded [OsO$_2$N]$^-$ and [OsO$_2$]$^-$ upon CID, while [OsO$_3$]$^-$ was not observed, suggesting that the Os≡N triple bond is stronger than the Os=O bond. CID of [MNCl$_4$]$^-$ (M = ruthenium, osmium) showed fragment ions due to loss of chlorine atoms, with the ruthenium complex undergoing greater fragmentation than the osmium analogue.

5.10.2 Reactivity Studies Involving Molybdate and Tungstate Ions

Reactions between molybdate ions and polyhydroxy compounds have been known for many years and recently ESI MS has been applied to the study of these reactions.

Products from the reaction between molybdate and D-lyxose and D-mannitol were identified using negative-ion ESI MS.[87] A wide range of species was detected, including ones with up to three molybdenum atoms. A reactivity study between molybdate ions and the 2,3- or 2,5-isomers of dihydroxybenzoic acid has also been carried out using ESI MS.[88]

Molybdate and tungstate have been reacted with galactaric acid and galacturonic acid and found by ESI MS studies to confirm and extend previous solution-state studies using NMR spectroscopy. Species with 1:1, 1:2 and 2:1 stoichiometries were observed. Although some dehydration and decarboxylation processes occurred during the ES ionisation process, the ESI MS method was proposed to be a rapid and simple method for identifying coordination complexes between oxoanions and carbohydrates.[89]

5.10.3 Polyoxoanions

From the very earliest days of the application of ESI MS to inorganic systems, the utility of the technique for the characterisation of polyoxoanions was realised with the report of the ESI MS spectrum of the heteropolymolybdate ion $[S_2Mo_{18}O_{62}]^{4-}$.[90] Since then, a wide range of polyoxoanion systems has been studied by ESI MS. The high formal charge present on many polyoxoanion species is delocalised, the charge density is therefore generally low, and parent ions are typically observed for many species. In some cases, charge reduction can occur by the formation of an ion pair, for example with a counter cation such as a tetraalkylammonium. In other cases, it has been concluded that some isopolyoxo-molybdate and -tungstate species are formed by a condensation process occurring during ES ionisation.[91,92]

The general behaviour of a stable heteropolyanion system is shown in Figure 5.20, for phosphotungstic acid, $H_3[PW_{12}O_{40}]$, which contains the Keggin-type $[PW_{12}O_{40}]^{3-}$ polyoxoanion, with an encapsulated PO_4 group. At a cone voltage of 40 V, the base peak is the trianion $[PW_{12}O_{40}]^{3-}$, together with a lower intensity protonated ion

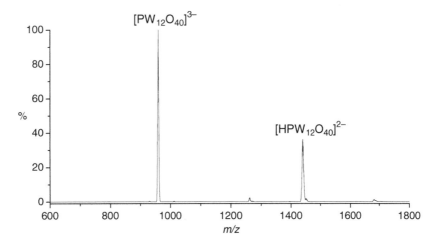

Figure 5.20
Negative-ion ESI mass spectrum of phosphotungstic acid, $H_3[PW_{12}O_{40}]$ in methanol solution at a cone voltage of 40 V, showing the $[PW_{12}O_{40}]^{3-}$ (*m/z* 959) and $[HPW_{12}O_{40}]^{2-}$ (*m/z* 1440) ions

$[HPW_{12}O_{40}]^{2-}$.[81] The observation of the intact parent trianion, even at moderately high cone voltages, is because the charge density on the species is relatively low.

Overall, a wide range of polyoxoanions have been investigated using ESI MS; an early study in this field by Lau and co-workers investigated a series of polyoxomolybdates and polyoxotungsates by ESI MS.[93] Parent ions were observed with minimal fragmentation, for doubly-, triply- and quadruply-charged polyoxoanions, for example $[M_6O_{19}]^{2-}$ (M = molybdenum, tungsten), and the Keggin-type heteropolyanions $[PW_{12}O_{40}]^{3-}$ and $[SiM_{12}O_{40}]^{4-}$ (M = molybdenum, tungsten). CID spectra of $[W_6O_{19}]^{2-}$, $[W_{10}O_{32}]^{4-}$, $[Mo_6O_{19}]^{2-}$ and $[Mo_8O_{26}]^{4-}$ all show fragment ions due to multiple loss of WO_3 or MoO_3 units; for example $[Mo_6O_{19}]^{2-}$ fragments to give the ions $[Mo_5O_{16}]^{2-}$, $[Mo_4O_{13}]^{2-}$, $[Mo_3O_{10}]^{2-}$ and $[Mo_2O_7]^{2-}$. The strong tendency of molybdates to undergo polycondensation reactions is nicely illustrated by the ESI MS behaviour of $(^nBu_4N)_2$-$[Mo_2O_7]$ in methanol solution in the range m/z 250 – 1200, which showed the presence of various molybdate species, $[(^nBu_4N)Mo_2O_7]^-$, $[Mo_2O_4(OMe)_5]^-$, $[H_3Mo_2O_7(OMe)_2]^-$, $[(^nBu_4N)Mo_3O_{10}]^-$, $[(^nBu_4N)Mo_4O_{13}]^-$, $[(^nBu_4N)Mo_5O_{16}]^-$ and $[HMo_6O_{18}(OMe)_2]^-$.[94] In addition, in a separate study, an acidic solution of tungstate ions (pH 3.45) gave $[H_4W_{12}O_{40}]^{4-}$, $[H_5W_{12}O_{40}]^{3-}$ and $[H_6W_{12}O_{40}]^{2-}$ as major ions.[80]

A study on highly-charged Keggin-type polyoxotungstate ions of the type $[XW_{11}O_{39}]^{n-}$ (X = Si, P, n = 8 or 7 respectively) showed the presence of doubly- and triply-charged pseudomolecular ions with associated alkali metal cations.[95] Spectra were found to be dependent on the counterion and on the initial concentration of the polyoxotungstate. Using lithium salts, the dilution process was found to effect rearrangement to a Keggin-type dodecatungstate, induced by the ESI process. Mixed-metal polyoxomolybdates, which also contain organometallic RhCp* (Cp* = η^5-C_5Me_5) groups as part of the framework, have been studied by ESI MS, with the intention of detecting unstable solution species.[94] Thus, the reaction of $[RhCl_2Cp^*]_2$ with $(^nBu_4N)_2[Mo_2O_7]$ in methanol gave evidence for the species $[Cp^*RhMo_3O_8(OMe)_5]^-$ (m/z 809) which dimerises to give the ion $[(Cp^*Rh)_2Mo_6O_{20}(OMe)_2(^nBu_4N)]^-$.

In a related area, peroxotungstate species have been identified in reaction solutions prepared by dissolution of tungsten metal in aqueous hydrogen peroxide. These species are formed by replacement of O^{2-} in a normal polyoxometallate, by the peroxo ion, O_2^{2-}. Upon dissolution of tungsten in hydrogen peroxide, species based on the $[W_6O_{19}]^{2-}$ ion, with up to four successive replacements of O^{2-} by O_2^{2-} were observed. In addition, the peroxotungstate species $[HWO_3(O_2)]^-$ and $[HWO_2(O_2)_2]^-$ were observed, together with $[HWO_4]^-$.[81]

In contrast to the considerable amount of work that has been done on polyoxomolybdates and -tungstates, much less work has been done with other metal polyoxoanions. Polyvanadate systems appear to be relatively complex, with species interconverting in solution and during the ESI process. What is clear, however, is that the conditions under which spectra of labile systems are recorded need to be closely controlled. The protonated decavanadate ion $[H_4V_{10}O_{28}]^{2-}$ has been studied by negative-ion ESI MS, and as expected gave the parent dianion (m/z 481) together with a weaker signal assigned to diprotonated $[H_2V_{10}O_{28}]^{4-}$ (m/z 240).[96] Aqueous vanadate solutions have been found to give various ions in ESI MS analysis, dependent on the pH and the concentration. At a pH near 10, vanadate solution was found to give only $[H_2VO_4]^-$, with no evidence for $[HVO_4]^{2-}$ or $[VO_4]^{3-}$. Under acidic conditions, (pH 4.68) the well-known decavanadate species in various protonation states, i.e. $[H_5V_{10}O_{28}]^-$ and $[H_4V_{10}O_{28}]^{2-}$ are observed, but not more highly charged analogues.[80]

Significantly different results were reported by other workers, who were able to identify many previously unknown polyvanadate anions and cations, with evidence that many of these species form as a result of evaporation effects in the ESI MS process.[97] As an example, for ammonium metavanadate, two series of ions were observed, namely $[H_xV_yO_z]^-$ (x = 0 to 1; y = 1 to 10; z = 3 to 26) and $[H_xV_yO_z]^{2-}$ (x = 0,1; y = 3 to 17; z = 9 to 44), together with the anions $[H_2VO_4]^-$, $[V_{10}O_{28}]^{6-}$ and $[H_5V_{10}O_{28}]^-$. At high cone voltages, mixed-valence polyoxovanadate ions were observed, as a result of reduction processes. A detailed ESI MS study on isopolyoxo-niobates and -tantalates revealed complex speciation, but yielded primarily ions based on the common $\{M_6O_{19}\}$ moiety, with some protonation of most species observed.[98]

5.10.4 Miscellaneous Oxo Complexes

A series of multiply-charged oxo-bridged complexes containing μ_2-oxo bridges have been characterised by ESI MS, with no fragmentation observed, and in most cases the base peak was due to the intact cation with two to four charges.[99] Cationic trinuclear μ_3-oxo complexes $[M_3O(RCO_2)_6(H_2O)_3]^+$ (M = chromium, iron; R = alkyl) give strong ESI MS spectra as befits their cationic nature, and exchange of the coordinated water molecules with solvent molecules (methanol and pyridine) was observed.[100] Dinuclear bis(μ-oxo) complexes of manganese(IV),[101] iron(III, IV)[102] and copper(II)[103] have also been studied.

A study of the reaction products between Al_2O_3 or $AlOH(O_2CCH_3)_2$ and acetic or propionic acids has been carried out using ESI MS to investigate the nature of the cationic species produced. A wide range of oxonium-bridged tri-aluminium species, containing the putative $Al_3(\mu_3$-O) core, were observed.[104]

Studies on neutral transition metal oxo compounds have also been carried out. The complex $[Cp*_2Mo_2O_5]$ has been studied using a combined electrochemical flowcell, and on-line ESI MS.[105] In positive-ion mode, this species gives ions with from one to four molybdenum atoms, with protonated $[Cp*_2Mo_2O_5 + H]^+$ the most intense ion, together with $[Cp*MoO_2]^+$, $[Cp*MoO_3H_2]^+$ and $[Cp*_3Mo_3O_7]^+$, while in negative-ion mode, $[Cp*MoO_3]^-$ is observed as the base peak, together with $[Cp*_2Mo_2O_6]^-$. The on-line electrochemical study of this compound revealed the formation of previously unknown mono-, di-, tri- and tetra-nuclear Mo(V), Mo(IV) and mixed-valence compounds, that were identified by their characteristic isotope patterns and fragment ions.

Metal Clusters

While mass spectrometric techniques have been widely applied to the analysis of metal *carbonyl* clusters, with a core of metal atoms surrounded by CO ligands (Chapter 7), very little work has been done with non-carbonyl clusters. Many metal *aggregates* (e.g. metal thiolate or sulfido-bridged multi-metallic assemblies of metals) have been studied but since these do not contain metal-metal bonds they are strictly not classified as clusters and are discussed in Section 5.12.

A rare example of a true, non-carbonyl metal cluster that has been characterised by ESI MS is the anionic cluster $[W_6Cl_8\{(\mu\text{-NC})Mn(CO)_2Cp\}_6]^{2-}$. This has a central $\{W_6Cl_8\}$ cluster core, with tungsten-tungsten bonds and each tungsten coordinated by the cyanide nitrogen of a $CpMn(CO)_2CN$ group. The ESI mass spectrum gave the parent ion, together with a series of fragments $[W_6Cl_8(NC)_n\{Mn(CO)_2Cp\}_{6-n}]^{2-}$ (n = 1 to 5) formed by loss of the $CpMn(CO)_2$ groups, with the cyanide groups remaining coordinated to the cluster.[106]

Compounds with Anionic Sulfur and Selenium Donor Ligands

5.12.1 Metal Sulfide, Selenide and Related Complexes

The sulfide ligand, S^{2-}, is a soft donor ligand, with a strong propensity to bridge metal centres, and the ability to form complexes with a diverse range of metal centres.

As a result of the importance of sulfide-bridged Fe-S clusters in the enzyme nitrogenase, the model cubane derivative $[Fe_4S_4(SH)_4]^{2-}$ (as its $^nPr_4N^+$ salt) has been studied by negative-ion ESI MS.[107] The parent ion $[Fe_4S_4(SH)_4]^{2-}$ was observed at m/z 242, together with the ion-paired species $[Fe_4S_4(SH)_4(Pr_4N)]^-$ (m/z 670); such ion-paired species are very commonly observed for multiply-charged anions. In addition, a low intensity peak at m/z 484 was assigned to the oxidised species $[Fe_4S_4(SH)_4]^-$; the substituted cluster $[Fe_4S_4(SEt)_4]^{2-}$ behaved similarly, and it was suggested that the oxidation was caused by traces of residual oxygen in the nitrogen nebulising gas used. Substitution reactions of $[Fe_4S_4(SH)_4]^{2-}$ with $p\text{-}CF_3C_6H_4SH$ (RSH) were also carried out *in situ*, with reaction mixtures monitored by ESI MS, resulting in detection of the species $[Fe_4S_4(SH)_{4-x}(SR)_x]^{2-}$ and $[Fe_4S_4(SH)_{4-x}(SR)_x(Pr_4N)]^-$ (x = 0 to 4). At low cone voltages, the intensities of the dianions was found to parallel the intensities observed by NMR spectroscopy. Other cubane-type metal-sulfide aggregates which have been studied by ESI MS are the cyclic tricubane cluster $[Cp*_2Mo_2Fe_2S_4]_3(\mu\text{-}S_4)_3$, which gives the protonated species $[M+H]^+$ and $[M+2H]^{2+}$ when analysed in tetrahydrofuran with acetic acid added,[108] and double cubanes of the type $Fe_4E_4\text{-}E\text{-}Fe_4E_4$ (E = sulfur or selenium) linked by a bridging sulfur or selenium atom. Since such linked cubanes are considered to be fragile, the gentle nature of ES ionisation was employed to provide evidence of the existence of such linked cubanes in solution. Thus, $(Bu_4N)_4$ $[(Fe_4S_4Cl_3)_2(\mu\text{-}S)]$ gave $(Bu_4N)_4[(Fe_4S_4Cl_3)_2(\mu\text{-}S)(Bu_4N)_2]^{2-}$ in the negative-ion ESI spectrum, while the related neutral cuboidal cluster $\{[VFe_4S_6(PEt_3)_4]_2(\mu_2\text{-}S)\}$ yielded an intense $[M+H]^+$ ion. Protonation of these sulfide clusters probably occurs at the electron-rich sulfide atoms.[109]

The iron-sulfide-nitrosyl anions $[Fe_2S_2(NO)_4]^{2-}$ and $[Fe_4S_3(NO)_7]^-$ have been known since 1858, and ESI FTICR MS has recently been applied to the study of solution speciation of these complexes.[110] The negative-ion ESI MS spectrum of a freshly prepared solution of $Na_2[Fe_2S_2(NO)_4]$ in methanol showed the protonated ion $[Fe_2S_2(NO)_4H]^-$, which rapidly converts to a species of m/z 647, identified as $[Fe_5S_4(NO)_8]^-$. Subsequently, this converts in solution to a mixture of the known $[Fe_4S_3(NO)_7]^-$ together with the new species $[Fe_7S_6(NO)_{10}]^-$. Solutions of $Na[Fe_4S_3(NO)_7]$ also show the slow formation of $[Fe_5S_4(NO)_8]^-$ and $[Fe_7S_6(NO)_{10}]^-$. It was considered that the ESI spectra were reflective of the solution species present, due to the strong dependence on spectra on the history of the solution being analysed, and the loss of NO ligands (but not disruption of the FeS cores) upon introduction of fragmentation (through the electrospray capillary potential).

Various large selenide clusters have also been investigated using ESI MS. The large gold cluster $[Au_{16}Se_8(dppe)_6]^{2+}$ (dppe = $Ph_2PCH_2CH_2PPh_2$) gives solely the parent dication when analysed in the m/z range 200–7000 using ESI FTICR MS; this is an excellent example of the power of the ESI technique for the analysis of high mass compounds, in this case 6568 g mol^{-1}.[111] Related compounds characterised by ESI MS are $[Au_{10}Se_4(dppm)_4]^{2+}$ (dppm = $Ph_2PCH_2PPh_2$) and $[Au_{18}Se_8(dppe)_6]^{2+}$.[112]

ESI MS has also been used to confirm the identity of the novel tungsten complex, $[(\eta^5\text{-}C_5Me_5)WOSSe]^-$, containing three different terminal chalcogenide atoms (oxygen, sulfur and selenium); the complex gave the expected $[M]^-$ ion as the base peak in the negative-ion spectrum and the isotopic richness of tungsten and selenium aided the confirmation by examination of observed and calculated isotope distribution patterns.[113] The related tris(sulfido) complex $[\eta^5\text{-}(C_5Me_5)WS_3]$ forms coordination complexes with other metal centres, and ESI MS has been used to characterise $[\eta^5\text{-}C_5Me_5)WS_3Cu(PPh_3)_3(NO_3)]^+$, $[\eta^5\text{-}C_5Me_5)WS_3Ag_3(PPh_3)_3]^+$, and $[\eta^5\text{-}C_5Me_5)WS_3Pd(dppe)]^+$. Parent ions, together with fragment species were observed, with isotopic distribution patterns again being used in the assignment. The related neutral species $[\eta^5\text{-}C_5Me_5)WS_3Au(PPh_3)]$ proved more difficult to study, but addition of lithium chloride to a tetrahydrofuran (THF) solution yielded $[\eta^5\text{-}C_5Me_5)WS_3Au(PPh_3) + Li(THF)_n]^+$ ions ($n = 0, 1$), with the Li$^+$ ion probably attaching at a sulfide centre.[114]

The dinuclear platinum(II) sulfide complex $[(Ph_3P)_2Pt(\mu\text{-}S)_2Pt(PPh_3)_2]$ contains highly reactive sulfide ligands and this complex shows a very rich reactivity, which has been probed extensively using ESI MS. These studies are summarised separately in a case study in Chapter 8. Sulfide and selenide aggregates of cadmium and zinc, that also contain thiolate ligands, are described in Section 5.12.3.

5.12.2   Metal Dithiocarbamate and Dithiophosphate Complexes

The dithiocarbamate ligand, $R_2NCS_2^-$ typically acts as a bidentate chelating ligand through both sulfur atoms and complexes have been obtained with a wide range of metals. One of the interesting features of this type of ligand is that it is electron-rich and able to stabilise metal centres in high oxidation states, such as iron(IV), nickel(IV) and copper(III).[115]

A number of neutral diethyl dithiocarbamate complexes of manganese(III), copper(II), nickel(II) and cobalt(III) of the type $[M(S_2CNEt_2)_n]$ ($n = 2, 3$) give oxidised ions formed by electrochemical oxidation in the metal capillary of an ESI spectrometer. In addition, ions formed by loss of the dithiocarbamate anion are also observed.[116,117] Alternatively, external oxidising agents such as $NOBF_4$[118] can be used to oxidise neutral metal-dithiocarbamate complexes of iron(III), cobalt(III), nickel(II) and copper(II), and $AgNO_3$[119] oxidises iron(III) and nickel(II) dithiocarbamates. However, in some cases, dithiocarbamate complexes of platinum(II), mercury(II), cobalt(III), rhodium(IIII), and iridium(III) can act as ligands towards Ag$^+$ ions, resulting in the formation of various heterobimetallic dithiocarbamate complexes.[119] Electrochemical oxidation of dithiocarbamate complexes either off-line, or coupled on-line to an ESI mass spectrometer also furnishes oxidised product ions.[120] It is worth contrasting the ease of oxidation of the above dithiocarbamate complexes with a series of molybdenum(VI) and tungsten(VI) tris-dithiolene complexes **5.2**, which give parent anions in negative-ion spectra, due to the ease of reduction of these hexavalent complexes in the ESI capillary.[121]

$$\left[\begin{array}{c} R \\ R' \end{array} \underset{C-S}{\overset{C-S}{\underset{|}{C}}} \right]_3 M$$

5.2

Other systems that have been studied by ESI MS are dithiocarbamate complexes of cadmium and mercury,[122] and the cationic platinum(II) complex $[Pt(\eta^1\text{-}S_2CNEt_2)L]^+$ (L = η^3-Ph$_2$PCH$_2$CH$_2$PPhCH$_2$CH$_2$PPh$_2$) that can be chemically oxidised to the platinum(III) cation $[Pt(S_2CNEt_2)L]^{2+}$, detected by ESI MS.[123] Related dithiophosphate $[(RO)_2PS_2]^-$ complexes have also been studied. The analysis of platinum(II) dithiophosphate complexes was among one of the earliest applications of the ESI MS technique in inorganic chemistry;[124,125] more recent investigations of dithiophosphate complexes of cadmium and mercury,[126] and zinc[127] have also been reported.

5.12.3 Metal Thiolate Complexes

The thiolate ligand RS$^-$ is an important ligand in coordination chemistry; the soft sulfur atom has a strong preference for forming stable complexes particularly with soft metal centres (e.g. mercury), but thiolate complexes of the majority of transition and main group metals are known. The presence of additional lone pairs of electrons in the M-SR moiety facilitates the formation of metal centres bridged by thiolate ligands. The thiol group also occurs in biological systems (in the amino acid cysteine), so that peptides and proteins that contain the cysteinyl residue are typically able to act as thiolate ligands towards a range of metal centres. Given their importance, many studies have been carried out on this class of coordination complex. The discussion below serves to illustrate some of the more important features of their mass spectrometric behaviour.

$$HS-CH_2-\underset{\underset{NH_2}{\big\backslash}}{\overset{\overset{CO_2H}{\big/}}{CH}} \qquad\qquad HS-CH_2-CH_2-\underset{\underset{NH_2}{\big\backslash}}{\overset{\overset{CO_2H}{\big/}}{CH}}$$

cysteine   homocysteine

$$\underset{H_2N}{\overset{HO_2C}{\big\backslash}}CH-CH_2-CH_2-CONH-\underset{\underset{CONH-CH_2-CO_2H}{\big\backslash}}{\overset{\overset{CH_2SH}{\big/}}{CH}}$$

glutathione

A neutral metal-thiolate complex containing basic groups behaves analogously to other such neutral coordination complexes. As an example, the antiarthritic gold(I) thiolate drug *Auranofin* **5.3** [R$_3$PAuSR, R = glucose tetra-acetate] gives $[M + H]^+$, $[M + NH_4]^+$ and $[2M + NH_4]^+$ ions at a low cone voltage (10 V) with added ammonium formate (Figure 5.21a). When the cone voltage is raised to 30 V (Figure 5.21b), the intensities of the ammoniated ions decrease dramatically, with $[M + H]^+$ becoming the base peak. In addition, an ion at *m/z* 993, which is of low intensity in the 10 V spectrum, becomes more intense at 30 V and is the base peak at 50 V (Figure 5.21c). This species is assigned as $[(Et_3PAu)_2SR]^+$, containing a thiolate ligand bridging between two gold(I) phosphine units; such species are very well known in gold(I) thiolate chemistry, due to the excellent

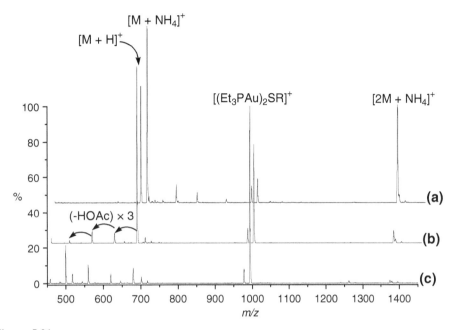

Figure 5.21
Positive-ion ESI mass spectra for the antiarthritic gold-thiolate drug *Auranofin* with added ammonium formate, at cone voltages of (a) 10 V, (b) 30 V, and (c) 50 V

bridging properties of thiolate ligands. At elevated cone voltages, fragmentation by loss of HOAc is also observed as a series of ions separated by m/z 60 from the $[M + H]^+$ ion. For comparison, the LSIMS spectrum of *Auranofin* (Figure 3.18), shows a range of ions and only a very weak $[M + H]^+$ ion, so that identification of the (pseudo)molecular ion, and hence the parent molecule, is more difficult with this technique.

Auranofin **5.3**

The antiarthritic gold(I) thiolate drug *Myochrisine* is known to be a mixture of components and various oligomeric structures $[AuSR]_n$ have been proposed, based on various ring and chain forms of the gold-thiolate repeat unit **5.4**. Analysis of *Myochrisine* and an ammonium salt analogue by ESI MS indicated that the principal component is a tetrameric species, probably containing an eight-membered Au_4S_4 ring.[128] A related oligomeric silver(I) thiomalate material $\{Na[Ag\{SCH(CO_2)CH_2CO_2\}].0.5H_2O\}_n$ (n = 24 to 34) has been analysed by negative-ion ESI MS which demonstrated a similar stability of a tetrameric silver-thiomalate unit.[129]

$$\left[\begin{array}{c} \quad\quad CO_2Na \\ \quad\quad | \\ Au-S-CH \\ \quad\quad \backslash \\ \quad\quad\quad CH_2 \\ \quad\quad\quad | \\ \quad\quad\quad CO_2Na \end{array} \right]_n$$

5.4

Various bismuth compounds such as bismuth subcitrate and bismuth subsalicylate have been used for many years for the treatment of ulcers, and because of the affinity of bismuth(III) for sulfur-based ligands a number of MS studies have recently been carried out in this area. Bismuth thiosalicylate species have been detected using ESI MS and have been proposed as models for bismuth subsalicylate.[130] The reactions of $BiCl_3$, bismuth subsalicylate and $Bi(NO_3)_3$ with the thiol-containing molecules cysteine, homocysteine and the tripeptide glutathione have been monitored using ESI MS and provided evidence for bismuth complexes of these biomolecules. In contrast, no complexes were observed between the bismuth compounds and methionine (a thioether donor) which indicates the importance of the 'thiolate anchor' in complex formation between bismuth and such molecules.[59] The reaction of $Bi(NO_3)_3$ with $HSCH_2CH_2OH$ has been found to give $[Bi(SCH_2CH_2OH)_2]NO_3$, containing a weakly-bound bidentate nitrate in the solid state. As expected, this material gives the parent $[Bi(SCH_2CH_2OH)_2]^+$ cation in APCI MS as the dominant peak at a low cone voltage of $10\,V$. At elevated cone voltages, the ion $[BiSCH_2CH_2O]^+$ is formed.[131] Related studies have been carried out using APCI MS on bismuth(III) complexes of aminoethanethiolate ligands derived from $H_2NCH_2CH_2SH$ and $Me_2NCH_2CH_2SH$[132] and using ESI MS on bismuth(III) complexes of thiol esters, of the type $Bi(SCH_2CO_2Et)Cl_2$, $Bi(SCH_2CO_2Me)_2Cl$ and $Bi(SCH_2CO_2Me)_3$.[133] Antimony(III) also shows a preference for binding to soft thiolate ligands and a study on the reaction of the antileishmanial drug potassium antimony(III) tartrate with glutathione (RSH) using NMR spectroscopy and ESI MS revealed exclusive formation of the complex $[Sb(SR)_3]$.[134]

The reaction of chromium(VI) – a well-known carcinogen – with glutathione is believed to be the first step in metabolism of chromium(VI) and thus several studies have concerned the ESI MS analysis of chromium thiolate complexes. Thus, the reaction of $Cr_2O_7^{2-}$ with oxidized glutathione (RSSR) yielded negative-ion signals for $[RSSR-H]^-$, $[RS]^-$ and the chromium(VI) thiolate complex $[CrO_3(SR)]^-$.[135] A chromium(V) glutathione complex, $[Cr^VO(LH_2)_2]^{3-}$ (LH_5 = glutathione), isolated from the reaction of Cr(VI) with glutathione at pH 7, has been characterised by ESI MS, together with chromium(III) reduction products of the chromium(VI) plus glutathione reaction.[136] Other chromium(V) amino acid complexes have also been characterised using ESI MS.[137]

Several studies have investigated the ESI MS behaviour of zinc, cadmium and mercury thiolates containing the thiophenolate (SPh^-) ligand. For the dinuclear thiolate bridged anion $[Hg_2(SPh)_6]^{2-}$, the only species observed in the ESI MS spectrum, even at a low cone voltage, was the $[Hg(SPh)_3]^-$ ion, indicating that the dinuclear unit is unstable with respect to dissociation.[138] By comparison, the anion $[Cd_4(SPh)_{10}]^{2-}$ gives the parent dianion at low cone voltages, but at slightly higher cone voltages undergoes dissociation to the $[Cd_2(SPh)_5]^-$ monoanion;[139] this is illustrated in Figure 1.18 and is a good example of the necessity of recording high resolution isotope patterns in order confirm ion assignments.

ESI MS has also been applied to the study of thiophenolate-capped metal sulfide aggregates, such as $[E_4Cd_{10}(SPh)_{16}]^{4-}$ (E = S, Se), $[S_4Zn_{10}(SPh)_{16}]^{4-}$ and $[S_4Cd_{17}(SPh)_{28}]^{2-}$.[140] Fragmentation of the metal chalcogenide core occurs only at high cone voltages, and fragmentation proceeds by loss of SPh^- or $M(SPh)_3^-$ units, that decrease the charge. For the tetraanionic species, it was possible to record the isotope pattern of the parent tetraanion on a quadrupole mass analyser; the expected 0.25 *m/z* separation of adjacent peaks in the isotope pattern was observed. ESI MS was also used to investigate exchange reactions, for example a mixture of $[S_4Cd_{10}(SPh)_{16}]^{4-}$ and $[S_4Zn_{10}(SPh)_{16}]^{4-}$, after equilibration, gave ions deriving from the species $[S_4Cd_{10-x}Zn_x(SPh)_{16}]^{4-}$, where x = 0 to 10. Chemical oxidation of SPh^- ligands from $[S_4Cd_{10}(SPh)_{16}]^{4-}$ by iodine has also been monitored using ESI MS, leading to the formation of a CdS nanocluster material.[140]

Characterisation of Metal-Based Anticancer Drugs, Their Reaction Products and Metabolites

5.13.1   Characterisation of Anticancer-Active Platinum Complexes

Platinum coordination complexes are widely used in the clinic for the treatment of many types of cancer. The archetypal complex *cisplatin*, cis-[PtCl$_2$(NH$_3$)$_2$], has been widely used with much success, but because of side effects other platinum(II) and platinum(IV) compounds have been developed as second-generation drugs. Several of these are in clinical use and others are undergoing clinical trials. Because of their importance, various studies have been carried out in the mass spectrometric characterisation of these drugs, their hydrolysis products, reactions with biological molecules, and detection of metabolites. The structures of various platinum complexes of this class are shown in Scheme 5.1.

Several ESI MS studies have been carried out with *cisplatin;* this somewhat surprisingly gives a strong $[M + Na]^+$ ion, together with lower intensity ions $[M + K]^+$ and $[M + Na - NH_3]^+$. The complex *tetraplatin* also gave the $[M + Na]^+$ ion as the base peak, but in contrast, the hydroxo complex *iproplatin* gave $[M + H]^+$ as the base peak,

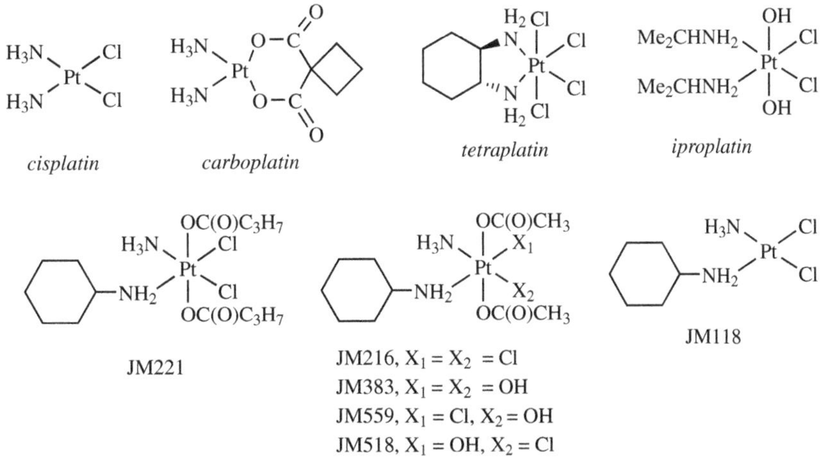

Scheme 5.1
The structures of various anticancer active platinum complexes that have been studied by ESI MS

cisplatin   *carboplatin*   *tetraplatin*   *iproplatin*

JM221

JM216, X$_1$ = X$_2$ = Cl
JM383, X$_1$ = X$_2$ = OH
JM559, X$_1$ = Cl, X$_2$ = OH
JM518, X$_1$ = OH, X$_2$ = Cl

JM118

presumably through protonation of a coordinated hydroxyl.[141] In a separate study, *cisplatin*, and its corresponding *trans*-isomer, *trans*-[PtCl$_2$(NH$_3$)$_2$], have been analysed in a mixed water-acetonitrile-acetic acid solution by positive-ion ESI MS; the formation of the [M + Na]$^+$ ion for *cisplatin* was confirmed, while *transplatin* gave the [M−Cl + MeCN]$^+$ ion, with no sodiated ion observed. Ionisation by loss of a halide ligand is typical behaviour for many other neutral transition metal halide complexes, as discussed in Section 4.3. Another study on *cisplatin* investigated the hydrolysis reaction in water and, after 24 hours incubation, the dominant ion was the monohydrated complex [PtCl(H$_2$O)(NH$_3$)$_2$]$^+$.[142] In a separate study, the same monohydrated complex in an acetonitrile solvent showed ions [M−H$_2$O]$^+$, [M]$^+$, and [M−H$_2$O + MeCN]$^+$, the latter formed by substitution of the water ligand by a more strongly coordinating MeCN.[143]

Studies of a number of second-generation anticancer drugs have also been carried out. *Carboplatin* has been characterised by ESI MS, giving ions of the type [(*carboplatin*)$_n$ + cation]$^+$ (cation = hydrogen or sodium; n = 1 to 4).[144] Figure 5.22a shows part of the positive-ion ESI mass spectrum of carboplatin, indicating the formation of the [*carboplatin* + Na]$^+$ ion. In this complex, the carboxylate oxygens clearly provide a site for ionisation.

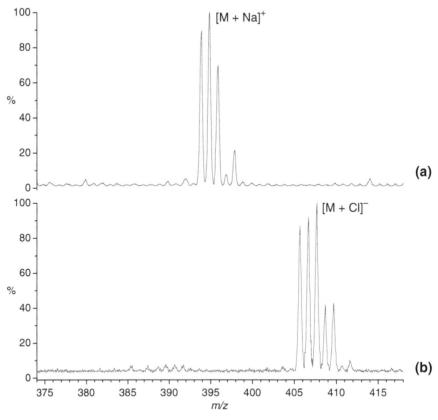

Figure 5.22
ESI mass spectra for the platinum anticancer drug *carboplatin*, recorded in methanol with added NaCl: (a) Positive-ion spectrum at a cone voltage of 10 V; (b) negative-ion spectrum at a cone voltage of 5 V

The platinum(IV) complexes JM221 and JM216 have been studied using positive-ion ESI MS, with both compounds giving strong $[M + Na]^+$ and $[M + K]^+$ ions, together with $[M + H]^+$ ions, the intensity of which could be increased by addition of a small amount of acetic acid. Aggregate ions of the type $[2M + cation]^+$ (cation = hydrogen, sodium, potassium) were also observed.[145]

Both *cisplatin* and *carboplatin* also give negative ions $[M + Cl]^-$, which are enhanced in intensity if chloride is added to the solution.[146] Figure 5.22b shows part of the negative-ion ESI mass spectrum of carboplatin with added NaCl, showing formation of the $[M + Cl]^-$ ion. The chloride ion presumably interacts with the N—H protons through hydrogen bonding, and thus negative-ion spectra (which are often less 'cluttered' than positive-ion spectra) might prove to be useful for many other platinum anticancer drugs that contain N—H protons.

The '*trans* effect'[147] has been found to play a role in the fragmentation of dinuclear platinum(II) anticancer complexes of the types **5.5** and **5.6**.[148] Using ESI coupled with surface-induced dissociation (SID) tandem MS, the same fragment ions were observed for the *cis* and *trans* isomers, but the relative intensities were dependent on the actual isomer. The position of the chloride, having a higher *trans*-effect, plays a significant role in determining fragmentation by labilising the Pt—N bond *trans* to the chloride.

Various other platinum complexes with potential use as antitumour agents have been characterised by ESI MS.[149]

5.13.2 Reactions of Platinum Anticancer Drugs with Biomolecules and Detection of Metabolites

Cisplatin and other platinum anticancer drugs are known to interact with biological molecules and, not surprisingly, various studies have concerned the detection and characterisation of such species by ESI MS. The thiol-containing tripeptide glutathione (GSH) is an example of such a material that is known to reduce the toxic side effects of *cisplatin*. *Cisplatin* and glutathione interact in phosphate buffer to form two conjugates, *cis*-[Pt(NH$_3$)$_2$Cl(SG)] and *cis*-{[Pt(NH$_3$)$_2$Cl]$_2$(μ-SG)}$^+$, which were characterised by LC MS. However, when rats and human patients were given *cisplatin* preceded by an intravenous administration of GSH, the *cisplatin*-glutathione conjugates were found to be absent in the plasma.[150] Interaction of *cisplatin* with the chemopreventative agent selenomethionine (SeMetH) has been studied using ESI MS and NMR spectroscopy.[151] Reaction intermediates identified were *cis*-[PtCl(SeMetH)(NH$_3$)$_2$]$^+$, *cis*-[Pt(SeMet)(NH$_3$)$_2$]$^+$, [PtCl(SeMetH)(NH$_3$)]$^+$, and [Pt(SeMet)(SeMetH)]$^+$, which are formed by displacement of coordinated chloride by the amino acid, in some cases with loss of a proton. In a related study, also using ESI MS and NMR spectroscopy, reactions of *carboplatin* with selenomethionine were investigated.[145] The species [Pt(NH$_3$)$_2$(CBDCA-*O*)(SeMetH-*Se*), [Pt(NH$_3$)(CBDCA-*O*)(SeMet-*Se,N*)], [Pt(CBDCA-*O,O*)(SeMetH-*Se,N*) and [Pt(SeMet-*Se,N*)$_2$], where CBCDA is cyclobutane-1,1-dicarboxylate, and italicised atoms indicate those coordinated to platinum. These studies confirmed that selenomethionine reacts by ring opening of the platinum-dicarboxylate ring system.

DNA is known to be the primary active target site of *cisplatin*. Accordingly, studies of the interactions of DNA, together with model oligonucleotides and nucleosides with anticancer platinum complexes have been studied using ESI MS (and related ionspray) techniques, and this has been summarised in a detailed review.[152]

The binding of *cisplatin* with the metal-ion transport protein transferrin has also been studied using ESI MS, which indicates that the *cisplatin* initially docks with, and then covalently bonds to the hydroxyl functional group of threonine 457, with loss of HCl then giving a transferrin-O-PtCl$(NH_3)_2$ adduct.[153]

LC-MS has also been used to identify metabolites of JM216 in human plasma ultrafiltrates. The metabolites JM118, JM383, JM518 and JM559 were detected, again primarily as their $[M + Na]^+$ ions.[154]

5.13.3   Other Non-Platinum Anticancer Agents

A range of coordination complexes of other metals also show good anticancer activity and much current research is focused on this field. Not surprisingly, studies on some of these systems by ESI MS are starting to appear in the literature. As an illustrative example, the complexes *cis*- and *trans*-[RuCl$_2$(SOMe$_2$)$_2$] which have very promising anticancer activity, have been studied by positive-ion ESI MS.[155] Reactions of these isomers with deoxynucleosides were monitored using ESI MS and ^{1}H NMR spectroscopy. Thus, both the *cis* and *trans* isomers react with 2′-deoxyguanosine to give two species containing a single coordinated nucleoside, and a *bis* adduct with two coordinated nucleosides. This suggests that both complexes may react *in vivo* with adjacent guanine bases in DNA, forming crosslinks analogous to *cisplatin*. By comparison, reactions with other nucleosides (2′-deoxycytidine, thymidine) were not significant, suggesting that the interactions with these DNA bases might be of lesser importance.

In Situ Formation of Coordination Complexes as an Ionisation Technique

It will be apparent, from many of the chapters of this book, that the ionisation of a neutral inorganic complex can proceed through a range of processes, including oxidation, protonation, loss of an anionic ligand etc. Ionisation through metallation is also a relatively common occurrence, especially involving adventitious or added alkali metal salts. Much rarer, however, is ionisation by deliberate addition of other metal cations, particularly transition metals. The general technique of metal ion adduction has recently been termed **Coordination Ionspray Mass Spectrometry** (CIS MS) by Bayer and co-workers[156] who investigated silver adducts of various compounds. However, Bayer *et al.* demonstrated that the presence of the electric field, which is a prerequisite of the ESI technique, is not required for CIS MS, though a weak electric field can be helpful in spray stabilisation.

Silver(I) ions have a strong affinity for soft donor ligands; the use of silver ions in the ESI MS analysis of π-hydrocarbons is discussed separately in Chapter 7. Silver ions have also been employed for the analysis of phosphines, arsines and stibines; this is discussed in Chapter 6. Mixtures of the ligands PPh$_3$, AsPh$_3$ and SbPh$_3$ with added AgNO$_3$ or Cu$^+$ ions were studied by ESI MS, and the coordination number of the species observed increases from two in [Ag(PPh$_3$)$_2$]$^+$ to four for [Ag(SbPh$_3$)$_4$]$^+$, consistent with the decreasing ligand donor ability and increasing metal-ligand bond distance in the series PPh_3–$AsPh_3$–$SbPh_3$.[157] The use of competition experiments such as this allows determination of the binding characteristics of a series of related ligands. Similarly, a range of adducts of $AgNO_2$ with PPh_3, $AsPh_3$ and $SbPh_3$ have also been investigated by ESI MS in acetonitrile solution, giving ions such as $[Ag(ER_3)]^+$, $[Ag(MeCN)_n]^+$ ($n = 1$, 2), $[AgCl_2]^-$, $[Ag(NO_2)_2]^-$, $[Ag(ER_3)(MeCN)]^+$ and $[Ag(ER_3)_2]^+$ as well as less intense higher aggregates $[Ag_2(NO_2)(ER_3)_2]^+$, $[Ag_2(NO_2)_3]^-$ and $[Ag_2Cl_2(NO_2)]^-$.[158]

Pyridyl and polyether derivatives of divalent transition metal ions have been investigated as alternative ionisation agents for pharmaceutical compounds, in place of protonation. The method involves electrospraying a ternary solution mixture consisting of the pharmaceutical compound of interest, L, a transition metal salt (of copper, nickel or cobalt), and the auxiliary pyridyl or polyether complexing agent (aux), giving ions of the type $[(L - H)M(aux)]^+$. This method resulted in improved ionisation efficiency for several of the pharmaceuticals studied, and 2,2′-bipyridine was found to give the most abundant metal containing ions. Polyethers were instead found to preferentially complex sodium ions.[159] The method has also been used for the analysis of amino acids and peptides,[160] quinolone antibiotics and other pharmaceutical compounds,[161] flavonoids[162] and an amino acid ester.[163]

Interestingly, the addition of ruthenium(III) chloride to a range of complex organic analytes such as tetronasin **5.7** (an *ionophore* that strongly binds alkali metal ions) surprisingly results in an increase of the $[M + H]^+$ ion at the expense of the $[M + Na]^+$ ion, and no ruthenium adducts were detected. In the absence of added ruthenium(III) chloride only a very low intensity $[M + H]^+$ ion was seen, and the enhancement of the $[M + H]^+$ ion was found to increase with added ruthenium(III) chloride, which is thus acting as an 'alkali metal sponge'. Although the exact mechanism is unknown, it was proposed that the formation of highly insoluble alkali metal salts of the ions $RuCl_4^-$ and $RuCl_5^{2-}$ removes alkali metal ions from solution. Compounds such as monensin A, valinomycin (a cyclic peptide), and dimethyl β-cyclodextrin behaved in a similar manner, and the method might show promise for simplifying spectra formed by metallation with multiple metal ions.[164]

Tetronasin
5.7

Summary

- For charged species with a low charge density, expect simple spectra at low cone voltages.

- Expect fragmentation and more complex spectra for more highly charged complexes (charge per metal centre ≥ 2).
- For metal halide complexes, loss of halide (promoted by addition of a good donor ligand such as pyridine) is a facile ionisation route, simplifying spectra.
- For neutral complexes, ionisation may proceed via cationisation (e.g. H^+, Na^+, NH_4^+, Ag^+), or less frequently, anionisation (Cl^-), or oxidation/reduction.
- ESI MS is ideal for investigating speciation in mixtures of metal ions and ligands *in solution*.

References

1. B. J. Thomas, J. F. Mitchell, K. H. Theopold and J. A. Leary, *J. Organomet. Chem.,* 1988, **348**, 333.

2. I. I. Stewart, *Spectrochim. Acta*, 1999, **B54**, 1649; R. Colton, A. D'Agostino and J. C. Traeger, *Mass Spectrom. Rev.*, 1995, **14**, 79; J. C. Traeger and R. Colton, in *Advances in Mass Spectrometry*, vol 14 (1998) Ch 29, 637; C. L. Gatlin and F. Tureček, in *Electrospray Ionisation Mass Spectrometry: Fundamentals, Instrumentation and Applications*, R. B. Cole (Ed.), Wiley, 1997, Ch.15; R. E. Shepherd, *Coord. Chem. Rev.*, 2003, **247**, 147.

3. V. Katta, S. K. Chowdhury and B. T. Chait, *J. Am. Chem. Soc.*, 1990, **112**, 5348.

4. E. Freiberg, W. M. Davis, A. Davison and A. G. Jones, *Inorg. Chem.*, 2002, **41**, 3337.

5. J. S. McIndoe and D. G. Tuck, *J. Chem. Soc., Dalton Trans.*, 2003, 244.

6. M. R. Mucalo and C. R. Bullen, *J. Colloid Interface Sci.*, 2001, **239**, 71.

7. B. H. Lipshutz, K. L. Stevens, B. James, J. G. Pavlovich and J. P. Snyder, *J. Am. Chem. Soc.*, 1996, **118**, 6796.

8. C. Hasselgren, G. Stenhagen, L. Ohrstrom and S. Jagner, *Inorg. Chim. Acta*, 1999, **292**, 266.

9. M. C. B. Moraes, J. G. A. Brito Neto and C. L. do Lago, *Int. J. Mass Spectrom.*, 2000, **198**, 121.

10. P. A. W. Dean, K. Fisher, D. Craig, M. Jennings, O. Ohene-Fianko, M. Scudder, G. Willett and I. Dance, *J. Chem. Soc., Dalton Trans.*, 2003, 1520.

11. G. A. Lawrance, M. Maeder, Y.-M. Neuhold, K. Szaciłowski, A. Barbieri and Z. Stasicka, *J. Chem. Soc., Dalton Trans.*, 2002, 3649.

12. S. M. Contakes, S. C. N. Hsu, T. B. Rauchfuss and S. R. Wilson, *Inorg. Chem.*, 2002, **41**, 1670.

13. W. Henderson and C. Evans, *Inorg. Chim. Acta*, 1999, **294**, 183.

14. W. Levason, J. J. Quirk, G. Reid and W. Levason, *J. Chem. Soc., Dalton Trans.*, 1997, 3719.

15. W. Levason, S. D. Orchard, G. Reid and V.-A. Tolhurst, *J. Chem. Soc., Dalton Trans.*, 1999, 2071.

16. E. Marotta, A. M. Gioacchini, F. Tisato, A. Cagnolini, L. Uccelli and P. Traldi, *Rapid Commun. Mass Spectrom.*, 2001, **15**, 2046; M. Tubaro, E. Marotta, P. Traldi, C. Bolzati, M. Porchia, F. Refosco and F. Tisato, *Int. J. Mass Spectrom.*, 2004, **232**, 239.

17. X. Yin and J. R. Moss, *J. Organomet. Chem.*, 1998, **557**, 259.

18. R. Colton and D. Dakternieks, *Inorg. Chim. Acta*, 1993, **208**, 173.

19. C. Decker, W. Henderson and B. K. Nicholson, *J. Chem. Soc., Dalton Trans.*, 1999, 3507.

20. C. S. Allardyce, P. J. Dyson, D. J. Ellis, P. A. Salter and R. Scopelliti, *J. Organomet. Chem.*, 2003, **668**, 35.

21. S. Chand, R. K. Coll and J. S. McIndoe, *Polyhedron*, 1998, **17**, 507.

22. M. L. Reyzer, J. S. Brodbelt, A. P. Marchand, Z. Chen, Z. Huang and I. N. N. Namboothiri, *Int. J. Mass Spectrom.*, 2001, **204**, 133.

23. S. Williams, S. M. Blair, J. S. Brodbelt, X. Huang and R. A. Bartsch, *Int. J. Mass Spectrom.*, 2001, **212**, 389.

24. B. J. Goolsby, J. S. Brodbelt, E. Adou and M. Blanda, *Int. J. Mass Spectrom.*, 1999, **193**, 197.

25. J. S. Brodbelt, *Int. J. Mass Spectrom.*, 2000, **200**, 57.

26. M. Vincenti, *J. Mass Spectrom.*, 1995, **30**, 925.

27. G. J. Van Berkel, S. A. McLuckey and G. L. Glish, *Anal. Chem.*, 1991, **63**, 1098.

28. K. A. Murphy, A. M. Cartner, W. Henderson and N. D. Kim, *J. Forensic Ident.*, 1999, **49**, 269.

29. J. Witowska-Jarosz, L. Górski, E. Malinowska and M. Jarosz, *J. Mass Spectrom.*, 2002, **37**, 1236.

30. G. J. Van Berkel, S. A. McLuckey and G. L. Glish, *Anal. Chem.*, 1992, **64**, 1586.

31. G. J. Van Berkel and F. Zhou, *Anal. Chem.*, 1994, **66**, 3408.

32. V. E. Vandell and P. A. Limbach, *J. Mass Spectrom.*, 1998, **33**, 212.

33. J. Jiang, R. C. W. Liu, T. C. W. Mak, T. W. D. Chan and D. K. P. Ng, *Polyhedron*, 1997, **16**, 515.

34. R. L. C. Lau, J. Jiang, D. K. P. Ng and T.-W. D. Chan, *J. Am. Soc. Mass Spectrom.*, 1997, **8**, 161.

35. T. Løver, W. Henderson, G. A. Bowmaker, J. M. Seakins and R. P. Cooney, *J. Mater. Chem.*, 1997, **7**, 1553.

36. K. A. Zemski, A. W. Castleman, Jr. and D. L. Thorn, *J. Phys. Chem. A*, 2001, **105**, 4633.

37. R. E. Bossio, S. D. Callahan, A. E. Stiegman and A. G. Marshall, *Chem. Mater.*, 2001, **13**, 2097.

38. S. Cristoni, L. Armelao, E. Tondello and P. Traldi, *J. Mass Spectrom.*, 1999, **34**, 1380.

39. S. Cristoni, L. Armelao, S. Gross, E. Tondello and P. Traldi, *Rapid Commun. Mass Spectrom.*, 2000, **14**, 662.

40. S. Cristoni, L. Armelao, S. Gross, R. Seraglia, E. Tondello and P. Traldi, *Rapid Commun. Mass Spectrom.*, 2002, **16**, 733.

41. L. Armelao, G. Schiavon, R. Seraglia, E. Tondello, U. Russo and P. Traldi, *Rapid Commun. Mass Spectrom.*, 2001, **15**, 1855.

42. S. Cristoni, P. Traldi, L. Armelao, S. Gross and E. Tondello, *Rapid Commun. Mass Spectrom.*, 2001, **15**, 386.

43. J. Beckmann, D. Dakternieks, A. Duthie, M. L. Larchin and E. R. T. Tiekink, *Appl. Organomet. Chem.*, 2003, **17**, 52.

44. T. Kemmitt and W. Henderson, *Aust. J. Chem.*, 1998, **51**, 1031.

45. R. Bakhtiar, *Rapid Commun. Mass Spectrom.*, 1999, **13**, 87; R. Bakhtiar and F. J. Feher, *Rapid Commun. Mass Spectrom.*, 1999, **13**, 687.

46. D. R. Bujalski, H. Chen, R. E. Tecklenburg, E. S. Moyer, G. A. Zank and K. Su, *Macromolecules*, 2003, **36**, 180; R. E. Tecklenburg, W. E. Wallace, and H. Chen, *Rapid Commun. Mass Spectrom.*, 2001, **15**, 2176; R. J. J. Williams, R. Erra-Balsells, Y. Ishikawa, H. Nonami, A. N. Mauri and C. C. Riccardi, *Macromol. Chem. Phys.*, 2001, **202**, 2425; D. P. Fasce, R. J. J. Williams, R. Erra-Balsells, Y. Ishikawa, and H. Nonami, *Macromolecules*, 2001, **34**, 3534; B. Hong, T. P. S. Thoms, H. J. Murfee and M. J. Lebrun, *Inorg. Chem.*, 1997, **36**, 6146; B. Devreese, P. Smet, F. Verpoort, L. Verdonck and J. Van Beeumen, *Rapid Commun. Mass Spectrom.*, 1998, **12**, 1204.

47. J. M. Curtis, P. J. Derrick, A. Schnell, E. Constantin, R. T. Gallagher and J. R. Chapman, *Inorg. Chim. Acta*, 1992, **201**, 197.

48. J. M. Curtis, P. J. Derrick, A. Schnell, E. Constantin, R. T. Gallagher and J. R. Chapman, *Org. Mass Spectrom.*, 1992, **27**, 1176.

49. B. Gyurcsik and L. Nagy, *Coord. Chem. Rev.*, 2000, **203**, 81; D. M. Whitfield, S. Stojkovski and B. Sarkar, *Coord. Chem. Rev.*, 1993, **122**, 171; S. Yano, *Coord. Chem. Rev.*, 1988, **92**, 113; D. Steinborn and H. Junicke, *Chem. Rev.*, 2000, **100**, 4283.

50. H. Junicke, C. Bruhn, R. Kluge, A. S. Serianni and D. Steinborn, *J. Am. Chem. Soc.*, 1999, **21**, 6232.

51. A. Fura and J. A. Leary, *Anal. Chem.*, 1993, **65**, 2805.

52. H. Desaire and J. A. Leary, *Anal. Chem.*, 1999, **71**, 4142.

53. G. Smith and J. A. Leary, *Int. J. Mass Spectrom.*, 1999, **193**, 153.

54. J.-Y. Salpin and J. Tortajada, *J. Mass Spectrom.*, 2002, **37**, 379.

55. R. Bakhtiar and A. E. Kaifer, *Rapid Commun. Mass Spectrom.*, 1998, **12**, 111.

56. C. L. Gatlin and F. Tureček in *Electrospray Ionisation Mass Spectrometry. Fundamentals, Instrumentation and Applications*, R. B. Cole (Ed), John Wiley & Sons Inc., New York, 1997, Chapter 15.

57. W. A. Tao, D. Zhang, F. Wang, P. D. Thomas and R. G. Cooks, *Anal. Chem.*, 1999, **71**, 4427; Y. Xu, X. Zhang and A.L. Yergey, *J. Am. Soc. Mass Spectrom.*, 1996, **7**, 25.

58. N. Burford, M. D. Eelman, W. G. LeBlanc, T. S. Cameron and K. N. Robertson, *Chem. Comm.*, 2004, 332.

59. N. Burford, M. D. Eelman, D. E. Mahony and M. Morash, *Chem. Comm.*, 2003, 146.

60. For selected examples see: C. L. Gatlin and F. Tureček, *J. Mass Spectrom.*, 2000, **35**, 172; H. Lavanant, E. Hecquet and Y. Hoppilliard, *Int. J. Mass Spectrom.*, 1999, **185 – 187**, 11; C. L. Gatlin, F. Tureček and T. Vaisar, *J. Am. Chem. Soc.*, 1995, **117**, 3637; S. R. Wilson, A. Yasmin and Y. Wu, *J. Org. Chem.*, 1992, **57**, 6941; C. L. Gatlin, F. Tureček and T. Vaisar, *J. Mass Spectrom.*, 1995, **30**, 775; C. L. Gatlin and F. Tureček, *J. Mass Spectrom.*, 1995, **30**, 1605; C. L. Gatlin, F. Tureček and T. Vaisar, *J. Mass Spectrom.*, 1995, **30**, 1617; C. L. Gatlin and F. Tureček, *J. Mass Spectrom.*, 1995, **30**, 1636.

61. See for example: J. A. Loo and R. R. Ogorzalek Loo, in *Electrospray Ionisation Mass Spectrometry. Fundamentals, Instrumentation and Applications*, R. B. Cole (Ed.), John Wiley & Sons Inc., New York, 1997, Chapter 11; A. P. Snyder, *Biochemical and Biotechnological Applications of Electrospray Ionisation Mass Spectrometry*, Oxford University Press, Oxford, 1998; *Experimental Mass Spectrometry*, D. H. Russell, (Ed.) Plenum Publishing Corp., New York, 1994.

62. S. Kazanis, T. C. Pochapsky, T. M. Barnhart, J. E. Penner-Han, U. A. Mirza and B. T. Chait, *J. Am. Chem. Soc.*, 1995, **117**, 6625; M. Jaquinod, E. Leize, N. Potier, A.-M. Albrecht, A. Shanzer and A. V. Dorsselaer, *Tetrahedron Lett.*, 1993, **34**, 2771.

63. A. M. Bond, R. Colton, A. D'Agostino, J. C. Traeger, A. J. Downard and A. J. Canty, *Inorg. Chim. Acta*, 1998, **267**, 281.

64. P. Hu, Q.-Z. Ye and J. A. Loo, *Anal. Chem.*, 1994, **66**, 4190; E. C. Dell'Angelica, C. H. Schleicher and J. A. Santome, *J. Biol. Chem.*, 1994, **269**, 28929; F. Hoffman, P. James, T. Vorherr and E. Carafoli, *J. Biol. Chem.*, 1994, **268**, 10252.

65. C. Q. Jiao, B. S. Freiser, S. R. Carr and C. J. Cassady, *J. Am. Soc. Mass Spectrom.*, 1995, **6**, 521.

66. M. H. Allen and T. W. Hutchens, *Rapid Commun. Mass Spectrom.*, 1992, **6**, 469; T. W. Hutchens, M. H. Allen, C. M. Li, and T.-T. Lip, *FEBS Lett.*, 1992, **309**, 170.

67. J. A. Loo, P. Hu and R. D. Smith, *J. Am. Soc. Mass Spectrom.*, 1994, **5**, 959.

68. P. Hu and J. A. Loo, *J. Am. Chem. Soc.*, 1995, **117**, 11314; T. Vaisar, C. L. Gatlin and F. Tureček, *Int. J. Mass Spectrom. Ion Proc.*, 1997, **162**, 77; P. Mineo, D. Vitalini, D. La Mendola, E. Rizzarelli, E. Scamporrino and G. Vecchio, *Rapid Commun. Mass Spectrom.*, 2002, **16**, 722.

69. L. Grøndahl, N. Sokolenko, G. Abbenante, D. P. Fairlie, G. R. Hanson and L. R. Gahan, *J. Chem. Soc., Dalton Trans.*, 1999, 1227.

70. X. Luo, W. Huang, Y. Mei, S. Zhou and L. Zhu, *Inorg. Chem.*, 1999, **38**, 1474.

71. J. Wang, R. Guevremont and K. W. M. Siu, *Eur. Mass Spectrom.*, 1995, **1**, 171; J. Wang, F. Ke, K. W. M. Siu and R. Guevremont, *J. Mass Spectrom.*, 1996, **31**, 159.

72. C. L. Gatlin, R. D. Rao, F. Tureček and T. Vaisar, *Anal. Chem.*, 1996, **68**, 263.

73. T. Vaisar, C. L. Gatlin, R. D. Rao, J. L. Seymour and F. Tureček, *J. Mass Spectrom.*, 2001, **36**, 306.

74. H. Li, K. W. M. Siu, R. Guevremont and J. C. Y. Le Blanc, *J. Am. Soc. Mass Spectrom.*, 1997, **8**, 781.

75. K. L. Bennett, J. A. Carver, D. M. David, L. A. P. Kane-Maguire and M. M. Sheil, *J. Coord. Chem.*, 1995, **34**, 351.

76. I. K. Chu, D. M. Cox, X. Guo, I. Kireeva, T.-C. Lau, J. C. McDermott and K. W. M. Siu, *Anal. Chem.*, 2002, **74**, 2072.

77. E. J. Zaluzec, D. A. Gage and J. T. Watson, *J. Am. Soc. Mass Spectrom.*, 1994, **5**, 359.

78. T.-C. Lau, J. Wang, K. W. M. Siu and R. Guevremont, *Chem. Commun.*, 1994, 1487.

79. S. Mollah, A. D. Pris, S. K. Johnson, A. B. Gwizdala III and R. S. Houk, *Anal. Chem.*, 2000, **72**, 985.

80. C. S. Truenbach, M. Houalla and D. M. Hercules, *J. Mass Spectrom.*, 2000, **35**, 1121.

81. M. J. Deery, O. W. Howarth and K. R. Jennings, *J. Chem. Soc., Dalton Trans.*, 1997, 4783.

82. F. Sahureka, R. C. Burns and E. I. Von Nagy-Felsobuki, *J. Am. Soc. Mass Spectrom.*, 2001, **12**, 1136.

83. F. Sahureka, R. C. Burns and E. I. Von Nagy-Felsobuki, *Inorg. Chem. Commun.*, 2002, **5**, 23.

84. J. H. Espenson, H. Tan, S. Mollah, R. S. Houk and M. D. Eager, *Inorg. Chem.*, 1998, **37**, 4621.

85. F. Sahureka, R. C. Burns and E. I. von Nagy-Felsobuki, *Inorg. Chim. Acta*, 2002, **332**, 7.

86. T.-C. Lau, Z. Wu, J. Wang, K. W. M. Siu and R. Guevremont, *Inorg. Chem.*, 1996, **35**, 2169.

87. V. Kovácik, L. Petrǔs, C. Versluis and W. Heerma, *Rapid Commun. Mass Spectrom.*, 1996, **10**, 1807.

88. A. Karaliota, V. Aletras, D. Hatzipanayioti, M. Kamariotaki and M. Potamianou, *J. Mass Spectrom.*, 2002, **37**, 760.

89. M. J. Deery, T. Fernandez, O. W. Howarth and K. R. Jennings, *J. Chem. Soc., Dalton Trans.*, 1998, 2177.

90. R. Colton and J. C. Traeger, *Inorg. Chim. Acta*, 1992, **201**, 153.

91. D. K. Walanda, R. C. Burns, G. A. Lawrence and E. I. von Nagy-Felsobuki, *J. Chem. Soc., Dalton Trans.*, 1999, 311.

92. D. K. Walanda, R. C. Burns, G. A. Lawrence and E. I. von Nagy-Felsobuki, *J. Cluster Sci.*, 2000, **11**, 5 – 28.

93. T.-C. Lau, J. Wang, R. Guevremont and K. W. M. Siu, *J. Chem. Soc., Chem. Commun.*, 1995, 977.

94. S. Takara, S. Ogo, Y. Watanabe, K. Nishikawa, I. Kinoshita and K. Isobe, *Angew. Chem. Int. Ed. Engl.*, 1999, **38**, 3051.

95. M. Bonchio, O. Bortolini, V. Conte and A. Sartorel, *Eur. J. Inorg. Chem.*, 2003, 699.

96. E. Chinea, D. Dakternieks, A. Duthie, C. A. Ghilardi, P. Gili, A. Mederos, S. Midollini and A. Orlandini, *Inorg. Chim. Acta*, 2000, **298**, 172.

97. D. K. Walanda, R. C. Burns, G. A. Lawrance and E. I. von Nagy-Felsobuki, *Inorg. Chim. Acta*, 2000, **305**, 118.

98. F. Sahureka, R. C. Burns, and E. I. Von Nagy-Felsobuki, *Inorg. Chim. Acta*, 2003, **351**, 69.

99. U. N. Andersen, C. J. McKenzie and G. Bojesen, *Inorg. Chem.*, 1995, **34**, 1435.

100. A. van den Bergen, R. Colton, M. Percy and B. O. West, *Inorg. Chem.*, 1993, **32**, 3408.

101. C. P. Horwitz, J. T. Warden and S. T. Weintraub, *Inorg. Chim. Acta*, 1996, **246**, 311.

102. Y. Dong, H. Fujii, M. P. Hendrich, R. A. Leising, G. Pan, C. R. Randall, E. C. Wilkinson, Y. Zang, L. Que Jr., B. G. Fox, K. Kauffmann and E. Münck, *J. Am. Chem. Soc.*, 1995, **117**, 2778.

103. S. Mahapatra, J. A. Halfen, E. C. Wilkinson, G. Pan, X. Wang, V. G. Young, Jr., C. J. Cramer, L. Que Jr., and W. B. Tolman, *J. Am. Chem. Soc.*, 1996, **118**, 11555.

104. M. Bartók, P. T. Szabó, T. Bartók, G. Szöllösi and K. Balázsik, *Rapid Commun. Mass Spectrom.*, 2001, **15**, 65.

105. J. Gun, A. Modestov, O. Lev, D. Saurenz, M. A. Vorotyntsev and R. Poli, *Eur. J. Inorg. Chem.*, 2003, 482.

106. C. S. Weinert, N. Prokopuk, S. M. Arendt, C. L. Stern and D. F. Shriver, *Inorg. Chem.*, 2001, **40**, 5162.

107. H. R. Hoveyda and R. H. Holm, *Inorg. Chem.*, 1007, **36**, 4571.

108. H. Kawaguchi, K. Yamada, S. Ohnishi and K. Tatsumi, *J. Am. Chem. Soc.*, 1997, **119**, 10871.

109. J. Huang, S. Mukerjee, B. M. Segal, H. Akashi, J. Zhou and R. H. Holm, *J. Am. Chem. Soc.*, 1997, **119**, 8662.

110. M. Lewin, K. Fisher and I. Dance, *Chem. Commun.*, 2000, 947.

111. D. Fenske, T. Langetepe, M. M. Kappes, O. Hampe and P. Weis, *Angew. Chem., Int. Ed. Engl.*, 2000, **39**, 1857.

112. V. W.-W. Yam and E. C.-C. Cheng, *Angew. Chem., Int. Ed. Engl.*, 2000, **39**, 4240.

113. H. Kawaguchi and K. Tatsumi, *Angew. Chem., Int. Ed. Engl.*, 2001, **40**, 1266.

114. J.-P. Lang, H. Kawaguchi and K. Tatsumi, *J. Organomet. Chem.*, 1998, **569**, 109.

115. A. M. Bond and R. L. Martin, *Coord. Chem. Rev.*, 1984, **54**, 23; D. Coucouvanis, Prog. *Inorg. Chem.*, 1970, **11**, 233.

116. D. F. Schoener, M. A. Olsen, P. G. Cummings and C. Basic, *J. Mass Spectrom.*, 1999, **34**, 1069.

117. A. R. S. Ross, M. G. Ikonomou, J. A. J. Thompson and K. J. Orians, *Anal. Chem.*, 1998, **70**, 2225.

118. A. M. Bond, R. Colton, A. D'Agostino, J. Harvey and J. C. Traeger, *Inorg. Chem.*, 1993, **32**, 3952.

119. A. M. Bond, R. Colton, Y. A. Mah and J. C. Traeger, *Inorg. Chem.*, 1994, **33**, 2548.

120. A. M. Bond, R. Colton, A. D'Agostino, A. J. Downard and J. C. Traeger, *Anal. Chem.*, 1995, **67**, 1691.

121. P. Falaras, C.-A. Mitsopoulou, D. Argyropoulos, E. Lyris, N. Psaroudakis, E. Vrachnou and D. Katakis, *Inorg. Chem.*, 1995, **34**, 4536.

122. A. M. Bond, R. Colton, J. C. Traeger and J. Harvey, *Inorg. Chim. Acta*, 1993, **212**, 233.

123. A. M. Bond, R. Colton, D. A. Fiedler, J. E. Kevekordes, V. Tedesco and T. F. Mann, *Inorg. Chem.*, 1994, **33**, 5761.

124. R. Colton, V. Tedesco and J. C. Traeger, *Inorg. Chem.*, 1992, **31**, 3865.

125. R. Colton, J. C. Traeger and V. Tedesco, *Inorg. Chim. Acta*, 1993, **210**, 193.

126. A. M. Bond, R. Colton, J. C. Traeger and J. Harvey, *Inorg. Chim. Acta*, 1994, **224**, 137; A. M. Bond, R. Colton, J. C. Traeger and J. Harvey, *Inorg. Chim. Acta*, 1995, **228**, 193; A. M. Bond, R. Colton, J. Harvey and R. S. Hutton, *J. Electroanalyt. Chem.*, 1997, **426**, 145.

127. T. J. Cardwell, R. Colton, N. Lambropoulos, J. C. Traeger and P. J. Marriott, *Anal. Chim. Acta*, 1993, **280**, 239.

128. H. E. Howard-Lock, D. J. LeBlanc, C. J. L. Lock, R. W. Smith and Z. Wang, *Chem. Commun.*, 1996, 1391; D. J. LeBlanc, R. W. Smith, Z. Wang, H. E. Howard-Lock and C. J. L. Lock, *J. Chem. Soc., Dalton Trans.*, 1997, 3263.

129. K. Nomiya, Y. Kondoh, H. Nagano and M. Oda, *J. Chem. Soc., Chem. Commun.*, 1995, 1679.

130. N. Burford, M. D. Eelman and T. S. Cameron, *Chem. Commun.*, 2002, 1402.

131. L. Agocs, G. G. Briand, N. Burford, T. S. Cameron, W. Kwiatkowski and K. N. Robertson, *Inorg. Chem.*, 1997, **36**, 2855.

132. G. G. Briand, N. Burford, T. S. Cameron and W. Kwiatkowski, *J. Am. Chem. Soc.*, 1998, **120**, 11374.

133. G. G. Briand, N. Burford, M. D. Eelman, T. S. Cameron and K. N. Robertson, *Inorg. Chem.*, 2003, **42**, 3136.

134. H. Sun, S. C. Yan and W. S. Cheng, *Eur. J. Biochem.*, 2000, 5450.

135. A. Levina and P. A. Lay, *Inorg. Chem.*, 2004, **43**, 324.

136. A. Levina, L. Zhang and P. A. Lay, *Inorg. Chem.*, 2003, **42**, 767.

137. H. A. Headlam, C. L. Weeks, P. Turner, T. W. Hambley and P. A. Lay, *Inorg. Chem.*, 2001, **40**, 5097.

138. G. A. Bowmaker, I. G. Dance, R. K. Harris, W. Henderson, I. Laban, M. L. Scudder and S.-W. Oh, *J. Chem. Soc., Dalton Trans.*, 1996, 2381.

139. T. Løver, W. Henderson, G. A. Bowmaker, J. M. Seakins and R. P. Cooney, *Inorg. Chem.*, 1997, **36**, 3711.

140. T. Løver, G. A. Bowmaker, J. M. Seakins, R. P. Cooney and W. Henderson, *J. Mater. Chem.*, 1997, **7**, 647.

141. G. K. Poon, P. Mistry and S. Lewis, *Biol. Mass Spectrom.*, 1991, **20**, 687.

142. M. Cui and Z. Mester, *Rapid Commun. Mass Spectrom.*, 2003, **17**, 1517.

143. H. C. Ehrsson, I. B. Wallin, A. S. Andersson and P. O. Edlund, *Anal.Chem.*, 1995, **67**, 3608.

144. Q. Liu, J. Lin, P. Jiang, J. Zhang, L. Zhu and Z. Guo, *Eur. J. Inorg. Chem.*, 2002, 2170.

145. G. K. Poon, G. M. F. Bisset and P. Mistry, *J. Am. Soc. Mass Spectrom.*, 1993, **4**, 588.

146. W. Henderson, unpublished observations.

147. T. G. Appleton, H. C. Clark and L. E. Manzer, *Coord. Chem. Rev.*, 1973, **10**, 335.

148. T. G. Schaaf, Y. Qu, N. Farrell and V. H. Wysocki, *J. Mass Spectrom.*, 1998, **33**, 436.

149. J. Zhang, Q. Liu, C. Duan, Y. Shao, J. Ding, Z. Miao, X.-Z. You and Z. Guo, *J. Chem. Soc., Dalton Trans.*, 2002, 591; D. N. Mason, G. B. Deacon, L. J. Yellowlees and A. M. Bond, *J. Chem. Soc., Dalton Trans.*, 2003, 890; A. Kung, M. Galanski, C. Baumgartner and B. K. Keppler, *Inorg. Chim. Acta*, 2002, **339**, 9.

150. A. Bernareggi, L. Torti, R. M. Facino, M. Carini, G. Depta, B. Casetta, N. Farrell, S. Spadacini, R. Ceserani and S. Tognella, *J. Chromatogr. B*, 1995, **669**, 247.

151. Q. Liu, J. Zhang, X. Ke, Y. Mei, L. Zhu and Z. Guo, *J. Chem. Soc., Dalton Trans.*, 2001, 911.

152. J. L. Beck, M. L. Colgrave, S. F. Ralph and M. M. Sheil, *Mass Spectrom. Rev.*, 2001, **20**, 61.

153. C. S. Allardyce, P. J. Dyson, J. Coffey and N. Johnson, *Rapid Commun. Mass Spectrom.*, 2002, **16**, 933.

154. G. K. Poon, P. Mistry, F. I. Raynaud, K. R. Harrap, B. A. Murrer and C. F. J. Barnard, *J. Pharmaceut. Biomed. Anal.*, 1995, **13**, 1493; G. K. Poon, F. I. Raynaud, P. Mistry, D. E. Odell, L. R. Kelland, K. R. Harrap, C. F. J. Barnard and B. A. Murrer, *J. Chromatogr. A*, 1995, **712**, 61.

155. J. M. Davey, K. L. Moerman, S. F. Ralph, R. Kanitz and M. M. Sheil, *Inorg. Chim. Acta*, 1998, **281**, 10.

156. E. Bayer, P. Gfrörer and C. Rentel, *Angew. Chem., Int. Ed. Engl.*, 1999, **38**, 992.

157. L. S. Bonnington, R. K. Coll, E. J. Gray, J. I. Flett and W. Henderson, *Inorg. Chim. Acta*, 1999, **290**, 213.

158. A. Cingolani, Effendy, M. Pellei, C. Pettinari, C. Santini, B. W. Skelton and A. H. White, *Inorg. Chem.*, 2002, **41**, 6633.

159. E. J. Alvarez and J. S. Brodbelt, *J. Am. Soc. Mass Spectrom.*, 1998, **9**, 463.

160. C. L. Gatlin, F. Tureček and T. Vaisar, *J. Mass Spectrom.*, 1995, **30**, 1605; C. L. Gatlin, F. Tureček and T. Vaisar, J. Mass Spectrom., 1995, **30**, 1617; C. L. Gatlin, R. D. Rao, F. Tureček and T. Vaisar, *Anal. Chem.*, 1996, **68**, 263.

161. E. J. Alvarez, V. H. Vartanian and J. S. Brodbelt, *Anal. Chem.*, 1997, **69**, 1147; J. Shen and J. S. Brodbelt, *Rapid Commun. Mass Spectrom.*, 1999, **13**, 1381.

162. M. Satterfield and J. S. Brodbelt, *Anal. Chem.*, 2000, **72**, 5989.

163. T. Bieńkowski, A. Brodzik-Bieńkowska and W. Danikiewicz, *J. Mass Spectrom.*, 2002, **37**, 617.

164. C. B. W. Stark, N. P. Lopes, T. Fonseca and P. J. Gates, *Chem. Commun.*, 2003, 2732.

6 The ESI MS Behaviour of Main Group Organometallic Compounds

Introduction

Organometallic compounds of the main group elements are dominated by alkyl and aryl derivatives. The emphasis will be placed on these compounds, where most studies have been carried out. The chapter is subdivided according to Periodic Table Groups. Two reviews specifically cover applications of ESI MS to organometallic chemistry.[1]

Organometallic Derivatives of Group 14 Elements

6.2.1 Organosilicon Compounds

For this class of compound, the behaviour is highly dependent on the nature of the substituents on the compound; the presence of hydrolysable halides will result in the observation of solvolysis products, while compounds with ancillary basic groups such as ethers will give $[M + H]^+$ or $[M + \text{cation}]^+$ (cation = e.g. sodium, potassium etc.). An example illustrating this is the acylsilanes $Me_3C\text{-}SiMe(COR^1)(CH_2OR^2)$ (R^1 or $R^2 =$ Me or Ph), which give $[M + H]^+$ when analysed with added CF_3CO_2H, but alkali metal adducts $[M + \text{cation}]^+$, when analysed in MeCN solution with added alkali metal salts.[2]

Studies include a range of oligo[(dimethylsilylene)phenylene]s of the type **6.1** and **6.2**, which give π-complexes with alkali metal cations,[3] and the analysis of F – endblocked polydiphenylsiloxanes with masses up to 3000 Da.[4] Poly(dimethylsiloxane) has also been detected by ESI MS as a component of the extract from medical silicone rubber.[5]

6.1   **6.2**

Mass Spectrometry of Inorganic, Coordination and Organometallic Compounds   W. Henderson and J. S. McIndoe
© 2005 John Wiley & Sons, Ltd   ISBNs: 0-470-85015-9 (HB); 0-470-85016-7 (PB)

By using a dry ethanol mobile phase, it was possible to obtain ESI MS data for the moisture-sensitive silicon alkoxide $(EtO)_3SiCH_2CH_2CH_2NH_2$.[6]

6.2.2 Organogermanium Compounds

Little work has been carried out on the analysis of organogermanium compounds by ESI MS. The germanium compound Ge-132 (bis-carboxyethyl germanium sesquioxide) has been studied by tandem ESI MS which led to the deduction of its structure.[7] In a later study, this germanium compound, together with other carboxylato-organogermanium compounds were studied by ESI MS.[8] Different hydrolysis products were observed, including cyclic oligomers.

6.2.3 Organotin Compounds

Tin has a large number of isotopes (10), which results in relatively complex but highly distinctive isotope patterns in mass spectra; Figure 6.1 gives the isotopic distribution of tin.

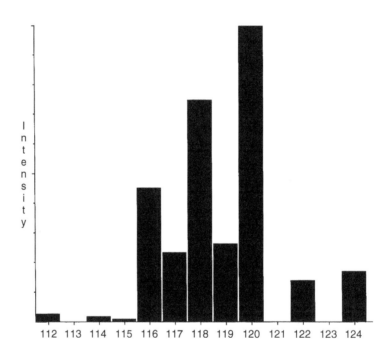

Figure 6.1
The isotopic distribution of tin

In general, organo-tin halides R_3SnX give solvated cations $[R_3Sn(solvent)_x]^+$ (x = 0, 1) in positive-ion mode, and anions $[R_3SnXY]^-$ in negative-ion mode. However, in the case of fluoride, the Sn-F bonds are strong, and since fluoride forms stable Sn-F-Sn bridges, a number of higher nuclearity species are observed, in both the solid and solution states. This is nicely illustrated by ESI MS analysis of the organotin fluorides $PhMe_2SnF$ and $(Me_3SiCH_2)_3SnF$ in acetonitrile. Both compounds give $[R_3Sn]^+$ and $[(R_3Sn)_2F]^+$, with $PhMe_2SnF$ also giving $[(PhMe_2Sn)_3F_2]^+$ and $[(PhMe_2Sn)_4F_3]^+$. In negative-ion mode, both compounds gave $[R_3SnF_2]^-$, with $PhMe_2SnF$ also giving $[(PhMe_2Sn)_2F_3]^-$ and $[(PhMe_2Sn)_3F_4]^-$. The greater tendency for $PhMe_2SnF$ to form more highly aggregated species correlated well with the higher Lewis acidity of this compound, which promotes the formation of stronger fluoride bridges between tin atoms, also observed in the solid state structures of these compounds. Thus, ESI MS provides indirect evidence on the extended structures of these materials.[9]

Hydrolysis products of a range of organotin halides Ph_3SnCl, Ph_3SnBr, Ph_2SnCl_2, Me_3SnCl and Bu_3SnCl have also been studied in both positive- and negative-ion modes in aqueous acetonitrile solvent.[10] Positive ions observed included $[(R_3Sn)_n(OH)_{n-1}]^+$ and solvated $[R_3Sn(MeCN)_x]^+$ ($x = 1$, 2); when pyridine was added, the species $[Ph_3Sn(pyridine)]^+$ was predominantly observed. Negative ions observed for Ph_3SnCl included the five-coordinate species $[Ph_3SnX_2]^-$, $[Ph_3SnX(OH)]^-$, $[Ph_3SnOSnPh_3X]^-$ ($X = Cl$, Br) and $[Ph_3SnO]^-$, while Ph_2SnCl_2 gave predominantly $[Ph_2SnCl_3]^-$. Mixed-halide species $[Bu_3SnXY]^-$ ($X = Y = Cl$ or Br) were also observed in solutions containing two halides. Similar results were obtained in a parallel study which investigated organotin compounds such as Ph_3SnCl, Ph_2SnCl_2, Bu_3SnCl, Bu_2SnCl_2 and $BuSnCl_3$ in methanol solution, where the $[M - Cl]^+$ or $[M - Cl + MeOH]^+$ ions were the principal species, along with various oxo-bridged species.[11] The organotin trifluoro-methanesulfonates $R_3Sn(OSO_2CF_3)$ ($R = Me$, Bu, Ph) similarly give $[R_3Sn(MeCN)]^+$ as the dominant ion in MeCN solution, together with $[R_3Sn]^+$.[12]

Other organotin compounds that have been characterised by ESI MS include dibutyltin perfluoroalkanecarboxylates (which give hydrolysis products in MeCN-H_2O solution),[13] a series of di- and tri-organotin 3,6-dioxaheptanoates and 3,6,9-trioxadecanoates,[14] a series of triphenyltin, tri-n-butyltin and di-n-butyltin derivatives containing 18-crown-6 and 15-crown-5 rings (which give the expected $[M + Na]^+$ and $[M + K]^+$ ions),[15] methylene-bridged tin species,[16] and some six-coordinate phosphonate-functionalised aryltin compounds.[17] A series of bis(imidazolyl)borate complexes of organotin(IV) have been studied by ESI MS; hydrolysed species were observed in methanol solution for the mono- and di-organotin(IV) derivatives, while non-hydrolysed aggregates were detected for triorganotin(IV) derivatives).[18]

A selection of larger organo-tin compounds have also been characterised by ESI MS. The distannoxane $^tBu_2ClSnOSnClMe_2$ has been studied in methanol-water solution, giving a series of species containing three tin atoms, as a result of hydrolysis reactions.[19] A carbohydrate-derived distannoxane[20] and spacer-bridged tetraorganodistannoxanes[21] have also been characterised using ESI MS. The large aggregates $[(RSn)_{12}O_{14}(OH)_6]^{2+}$ ($R = {}^iPr$, ^nBu) have been analysed by ESI MS in dichloromethane-methanol solution, and gave the parent dication as the sole species. Over time, two of the hydroxide groups were replaced by methoxide (from the solvent).[22] The reaction of $[(BuSn)_{12}O_{14}(OH)_6]^{2+}$ with diphenylphosphinic acid (Ph_2POOH) gave $[\{BuSn(OH)O_2PPh_2\}_3O]^+$ which showed the parent ion together with species which have the OH groups replaced by OMe groups from the solvent.

Several studies have focused on the interactions of organotin species with biologically relevant molecules. The interactions between Et_2SnCl_2 and various nucleotides have been studied by NMR spectroscopy and ESI MS.[23] ESI MS has also been employed in a study of the interactions of Bu_2SnBr_2 with ceroid lipofuscin protein (subunit C).[24]

As a result of industrial applications of organotin compounds, and hence their occurrence in the environment, there have been a number of quantitative studies on organotin compounds by ESI MS and the related ionspray MS technique, including coupling with extraction and chromatographic techniques.[25] Mass spectrometric analysis allows the speciation to be identified, which is of importance since the chemical and toxicological properties of organotins depend highly on the type of species. Tributyltin species are perhaps the most widely used, and the most important in the marine environment, and have been analysed by ionspray-MS/MS.[26] In this work it was possible to estimate the quantities of tributyltin species at low levels (about five picogrammes) by monitoring the daughter-parent ion pair at m/z 179 and 291, these ions being $BuSnH_2^+$ and Bu_3Sn^+. LC-ESI MS methods for the analysis of organotin compounds in aqueous media have been reported.[27]

6.2.4   Organolead Compounds

In comparison with tin, relatively little work appears to have been carried out with organolead compounds. Trialkyllead halides R_3PbX, like their tin counterparts, behave as neutral metal halide complexes in ESI MS analysis (Section 5.3), giving solvated $[R_3Pb(solvent)]^+$ cations at low cone voltages, and bare $[R_3Pb]^+$ at higher cone voltages. This is illustrated for Me_3PbOAc in Figure 6.2. At a low cone voltage (Figure 6.2a) various ions are observed, corresponding to the solvated $[Me_3Pb(MeOH)]^+$ ion as expected, together with dinuclear species $[(Me_3Pb)_2OH]^+$, $[(Me_3Pb)_2OMe]^+$ and $[(Me_3Pb)_2OAc]^+$, formed from the various anions present in the solution. The very low intensity of any simple mono-lead parent ions of the type $[Me_3PbOAc + cation]^+$ (cation = e.g. hydrogen, sodium etc) is worth noting. A reported study of aryl lead(IV)

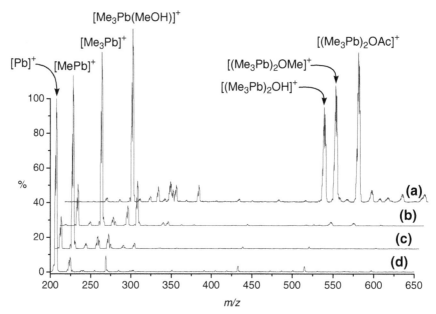

Figure 6.2
Positive-ion ESI mass spectra of Me_3PbOAc in methanol solution at a range of cone voltages: (a) 5 V, (b) 50 V, (c) 80 V and (d) 140 V. The spectra show the $[Me_3Pb]^+$ cation and derivatives thereof at the lower cone voltages, while at the higher cone voltages, fragmentation to $[MePb]^+$ and then $[Pb]^+$ is observed

carboxylates did however give a reliable $[M + Na]^+$ signal in the presence of excess sodium carboxylate.[28]

When the cone voltage is increased, to 50 V (Figure 6.2b), the base peak becomes the unsolvated $[Me_3Pb]^+$ ion. On further increasing the cone voltage, to 80 V (Figure 6.2c), fragmentation of the $[Me_3Pb]^+$ ion occurs giving $[MePb]^+$, and increasing the cone voltage still further (to 140 V, Figure 6.2d) results in the formation of the bare $[Pb]^+$ ion. This is another example of an electrospray mass spectrometer operating in *bare metal ion* or *elemental* mode (Section 4.2.2) – the parent ion, $[Me_3Pb]^+$ and solvated analogues thereof, are fragmented to give bare $[Pb]^+$ ions. Of course, since these $[Pb]^+$ ions are freed from all attendant ligands, the isotope pattern of this ion gives a direct measure of the isotopic composition of the element. In the case of lead, the isotopic composition can be somewhat variable, depending on the geochemical source of the element.* Figure 6.3

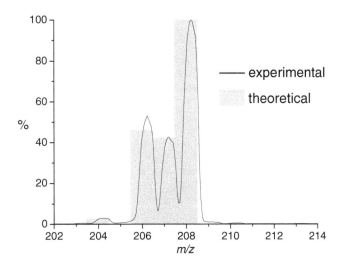

Figure 6.3
Experimental and theoretical isotope distribution patterns for the $[Pb]^+$ ion, demonstrating the isotopic composition of atomic lead

shows a comparison of the experimental and calculated isotope distribution patterns for the $[Pb]^+$ ion, obtained by CID in the Me_3PbOAc system. The use of bare metal ion mode can also be used in the quantitative analysis of organolead compounds. In this way, a series of organolead compounds of the type R_3PbX (R = e.g. propyl, butyl, phenyl; X = Cl, O_2CCH_3) have also been analysed by quantitative ESI MS using the bare metal-ion mode, which gave Pb^+ as the dominant ion.[29]

When a strong donor ligand such as pyridine is added to a solution of an organolead halide, the loss of the halide ion is promoted. This is illustrated by ESI mass spectra in

*Because lead can have various origins including from radioactive decay processes, its isotopic composition can vary depending on the source's geochemical history. The atomic mass of lead is therefore often quoted to a relatively small number of significant figures because of this inherent inter-sample variation.

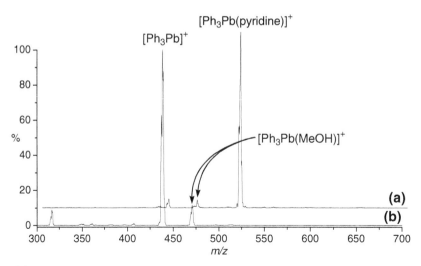

Figure 6.4
Positive-ion ESI mass spectra of Ph_3PbCl in methanol solvent (a) at a cone voltage of 20 V with added pyridine, showing formation of the $[Ph_3Pb(pyridine)]^+$ cation and (b) at a cone voltage of 60 V, showing the bare $[Ph_3Pb]^+$ cation

Figure 6.4a for Ph_3PbCl. At 20 V, the base peak in the spectrum is $[Ph_3Pb(pyridine)]^+$ (*m/z* 518). The spectrum at a higher cone voltage (60 V, Figure 6.4b) without added pyridine shows the $[Ph_3Pb]^+$ ion (*m/z* 438) and a low intensity methanol solvated ion $[Ph_3Pb(CH_3OH)]^+$ (*m/z* 470).

Organometallic Derivatives of Group 15 Elements

6.3.1   Organophosphorus Compounds

Organophosphorus compounds are not typically classified as true organometallic compounds, however their use is intimately linked with coordination and organometallic chemistry, and so they are discussed in this chapter.

Analysis of Phosphine Ligands by In Situ Formation of Coordination
Complexes, Oxide or Alkylated Derivatives

Phosphine (phosphane) ligands, R_3P, are ubiquitous in coordination and organometallic chemistry. Any issue of a Journal covering coordination chemistry will have multiple papers describing some aspect of the chemistry of phosphine complexes. Paralleling studies on their coordination chemistry, there is a related field of research concerned with the synthesis and characterisation of new phosphines with tailored characteristics, such as solubility in water or fluorinated solvents, chirality, presence of additional functional groups etc. Mass spectrometry plays an important role in the characterisation of these ligands. EI and FAB MS techniques have been used for some time, due to the volatility of many phosphines. However, given the gentle nature of ES ionisation, it is desirable to have developed methods for the ESI MS analysis of phosphines (and their metal complexes).

Some ligands, such as water-soluble sulfonated triphenylphosphine derivatives and metal complexes thereof can be analysed directly using negative-ion ESI MS, because they are negatively charged.[30,21] However, many other phosphines cannot be directly analysed so easily, so for ESI MS analysis, it is desirable to convert the phosphine into a charged derivative. There are several reasons for doing this:

1. Phosphines have relatively poor basicity, so that formation of $[M + H]^+$ ions does not lead to good signal intensity.
2. Phosphines have a strong affinity for soft metal centres such as silver(I) and mercury(II). Residual traces of these metals left in an instrument can be stripped by a phosphine analyte, giving ions such as $[Ag(PPh_3)_n]^+$ (n = 0 to 2), and hence a potentially confusing spectrum.
3. Phosphines are often contaminated with traces of phosphine oxide R_3PO (formed by atmospheric oxidation), where the oxygen atom has considerable basicity and readily form protonated or metallated ions. This can give a distorted picture of the composition (and hence the apparent purity) of a phosphine sample.

By derivatising the phosphine of interest prior to analysis, these complications can generally be avoided. A number of simple *in situ* derivatisation methods can be used:

(a) *Formation of metal complexes in situ.* Phosphine ligands have the ability to form stable coordination complexes with a wide range of metal ions. By addition of a soft monocation such as Ag^+ or $Cu^{+,\dagger}$ charged coordination complexes are formed of the type [n(ligand) + cation]$^+$, where the value of n is dependent on the metal centre, the ligand, the metal to ligand mole ratio, and the cone voltage.[32] These cations are then readily analysed by positive-ion ESI MS. For example, Figure 6.5 shows positive ion ESI mass spectra of triphenylphosphine with added silver nitrate, at cone voltages of 10 V and 80 V. A range of easily assigned silver-PPh$_3$ species are observed with high signal intensity, and the dominance of the $[Ag(PPh_3)_2]^+$ ion (*m/z* 633, 635) at high cone voltages confirms the stability of this species. Species containing a nitrate anion, such as $[Ag_2(PPh_3)_2NO_3]^+$ (*m/z* 802) are also seen. The number of silver ions in the species can easily be determined by examination of the isotope pattern of the ion. Silver has two isotopes, ^{107}Ag and ^{109}Ag, of relative abundances 100 % and 92.9 % respectively. The calculated isotope patterns for the ions $[Ag(PPh_3)_2]^+$ and $[Ag_2(PPh_3)_2(NO_3)]^+$ in Figure 6.6 clearly illustrate this.

(b) *Oxidation to the phosphine oxide.* Treatment of tertiary phosphines with hydrogen peroxide (or another oxidant, which may be carried out *in situ*) gives the corresponding phosphine oxide, $R_3P{=}O$, which is more basic than the phosphine, and ionises by protonation or metallation with Na^+ etc. As an example, Figure 6.7 shows ESI MS spectra for triphenylphosphine oxide (Ph$_3$PO). In spectrum (a) at a cone voltage of 5 V with no added ionisation aid, a range of ions are observed, but on addition of sodium chloride, the spectrum (b) is simplified, giving predominantly the range of ions

---

†For silver, any soluble silver salt can be used, for example AgNO$_3$ or AgClO$_4$. For copper(I), the ion $[Cu(MeCN)_4]^+$, as either its BF_4^- or PF_6^- salt, provides a convenient, soluble source of copper(I) ions.

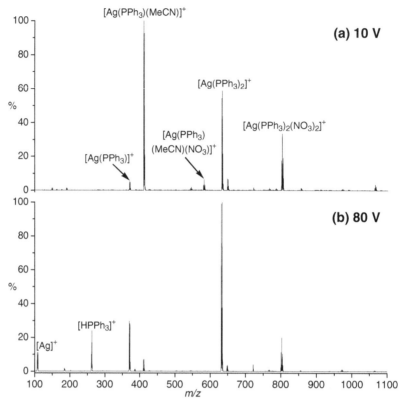

Figure 6.5
Positive-ion ESI mass spectra of a mixture of triphenylphosphine and $AgNO_3$, in MeCN solvent, at cone voltages of (a) 10 V and (b) 80 V. The high stability of the $[Ag(PPh_3)_2]^+$ ion is shown by its occurrence as the base peak even at the high cone voltage of 80 V

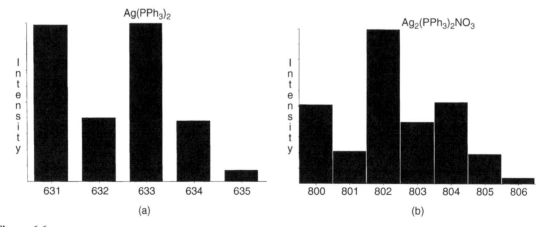

Figure 6.6
Theoretical isotope patterns for the ions (a) $[Ag(PPh_3)_2]^+$ and (b) $[Ag_2(PPh_3)_2(NO_3)]^+$, showing the dominant effect of the silver isotopes

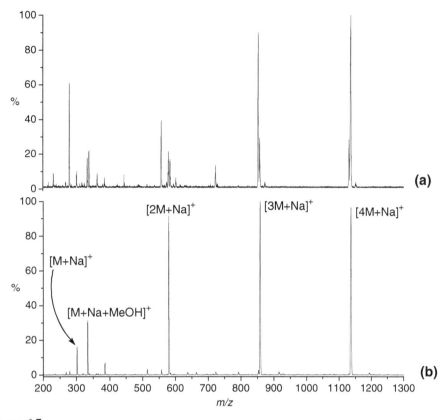

Figure 6.7
(a) Positive-ion ESI mass spectrum of triphenylphosphine oxide, Ph_3PO in methanol solution at a cone voltage of 5 V, showing a wide range of ions, including adducts with adventitious H^+, NH_4^+ and Na^+ ions. The spectrum is considerably simplified upon addition of sodium chloride, (b), resulting in an easily interpreted spectrum comprising sodium adduct ions

$[nPh_3PO + Na]^+$, where n is 0 to 4. This spectrum is now very easily interpreted, and the observed ions are easily related to the Ph_3PO analyte. It is worth commenting on the high intensity of the $[4Ph_3PO + Na]^+$ ion (m/z 1135) and absence of the $[5Ph_3PO + Na]^+$ ion; due to the steric bulk of the Ph_3PO ligand it is not possible to fit five ligands around the small sodium cation.

(c) *Conversion to a charged phosphonium salt.* This can easily be achieved by addition of an alkyl halide, such as methyl iodide (which is widely used in the literature, but is rather volatile and toxic), or a less volatile (solid) alkyl halide such as *p*-nitrobenzyl bromide. The general reaction is:

$$R_3P + R'X \longrightarrow R_3R'P^+X^-$$

Using this method, a range of mono-, di- and tri-tertiary phosphines have been converted to their methylphosphonium salts and analysed by positive-ion ESI MS.[33] Quaternary phosphonium cations have long been known to give simple parent cations in

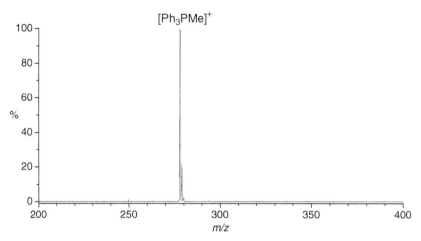

Figure 6.8
The analysis of a tertiary phosphine such as PPh_3 can readily be carried out as in this example, by addition of excess of methyl iodide, forming exclusively the $[Ph_3PMe]^+$ cation, which gives an intense positive-ion ESI MS signal

ESI MS analysis, for example, $[Ph_3PCH_2Ph]^+$.[34] The related bis(triphenylphosphine) iminium ($Ph_3PNPPh_3^+$ = PPN^+) cation is widely used as a countercation, particularly in the chemistry of large metal carbonyl clusters, and gives an intense m/z 538 ion.

As an example, the ESI MS spectrum of triphenylphosphine, derivatised *in situ* by addition of excess methyl iodide, is shown in Figure 6.8. The $[Ph_3PMe]^+$ ion (m/z 277) is the sole ion observed in the spectrum, making this a simple derivatisation method for analysis of phosphines. Generally, parent mono- or dications are obtained, but some polyphosphines show some oxidation of non-methylated phosphine groups. The same methodology can be applied to tertiary arsines and to tertiary stibines.[35] Quaternary phosphonium (and arsonium etc.) salts of this type show high ESI responses, and a tendency to 'linger' in the capillary systems of ESI instruments; Figure 4.2 illustrates the high ESI response (ionisation efficiency) for the $[Ph_4As]^+$ ion.

Alternatively, addition of an excess of aqueous formaldehyde and hydrochloric acid to a tertiary phosphine R_3P gives the phosphonium cation $[R_3PCH_2OH]^+$ which can be analysed by ESI MS. In this way, air-sensitive primary (RPH_2) and secondary (R_2PH) phosphines give air-stable products where each P-H bond is converted to a $-CH_2OH$ group, i.e. $RP(CH_2OH)_3^+$ and $R_2P(CH_2OH)_2^+$ respectively, so this is a useful derivatisation method for these reactive phosphines. A number of phosphines, including ferrocene-derived phosphines have been analysed using this derivatisation method.[31,36]

Electrospray-Friendly Phosphine Ligands

As discussed in the preceding section, tertiary phosphine ligands such as PPh_3 are often poorly basic, and give weak $[M + H]^+$ signals in ESI MS. Upon coordination to a metal centre, and in the absence of any other ionisation pathways, the metal complex can be 'invisible' to the ES ionisation process. Examples of such compounds are $Fe(CO)_4(PPh_3)$

and $Ru_3(CO)_{11}(PPh_3)$. This problem can be overcome by replacing the triphenylphosphine ligand with an analogue containing one or more groups that undergo ready protonation (or other adduct formation, deprotonation, oxidation etc). *Para*-substituted phosphines of the type $PPh_n(C_6H_4R)_{3-n}$ (R = OMe or NMe$_2$) have been studied for this purpose.[37] They are easily synthesised (are commercially available in some cases), and because they are *para*-substituted they are sterically and electronically similar to the parent PPh$_3$, so that the overall ligand properties are perturbed as little as possible, while still adding an electrospray friendly substituent.

An 'electrospray-friendly' phosphine analogue of widely-used PPh$_3$.

The application of such ligands in metal carbonyl chemistry is described in Chapter 7.

Miscellaneous Organophosphorus Compounds

Organophosphorus compounds find numerous applications, particularly in the agricultural industry, but also in chemical weapons. Several studies have investigated the ESI MS behaviour of neutral organophosphorus compounds, including pesticides.[38,39] Simple adduct ions $[M + H]^+$, $[M + NH_4]^+$ etc. are typically observed for this class of compound.

Phosphonic acids (RPO_3H_2) and phosphinic acids (R_2PO_2H) are relatively strong acids that give simple $[M - H]^-$ ions in the negative ion ESI mass spectra. An ESI MS study on methylphosphonic acid, $CH_3PO_3H_2$ gave aggregate ions $[(MePO_3H) + n(MePO_3H_2)]^-$ (n = 0 to 9) formed by hydrogen bonding interactions.[40] This behaviour contrasts with arsonic acids ($RAsO_3H_2$), which undergo alkylation in alcohol solution, discussed in Section 6.3.2.

6.3.2 Organoarsenic Compounds

Tertiary Arsines

Tertiary arsines, AsR_3, have many common characteristics with phosphine ligands (discussed in Section 6.3.1) but are far less widely used as ligands in coordination chemistry. One reason for this is that arsines are generally conceived to be highly toxic (though it should be noted that phosphines are themselves toxic), but probably the major reason is that arsenic does not have a nucleus readily accessible by nuclear magnetic resonance (NMR) spectroscopy. This contrasts with phosphorus chemistry, which has flourished in part due to the ease with which the ^{31}P nucleus can be observed. The development of tertiary arsine chemistry would be significantly enhanced by the application of techniques such as ESI MS to the field. Using the technique of silver derivatisation described in Section 6.3.1, a wide range of tertiary arsine ligands have been characterised by positive-ion ESI MS. Using this derivatisation method it is possible to obtain information on the purity of the ligands.[41]

Organoarsenic Acids

Arsonic acids, $RAsO_3H_2$, give alkylated monoanions $[RAsO_3alkyl]^-$ when analysed by negative-ion ESI MS in alcohol solvent; this is in contrast to their phosphonic acid counterparts that give simple deprotonated $[RPO_3H]^-$ ions and hydrogen bonded aggregates thereof (Section 6.3.1).[42] For example, negative-ion ESI mass spectra of phenylarsonic acid, $PhAsO_3H_2$, in methanol solution at cone voltages of 20 V and 40 V are shown in Figure 6.9. At a cone voltage of 20 V, the base peak is the monoester

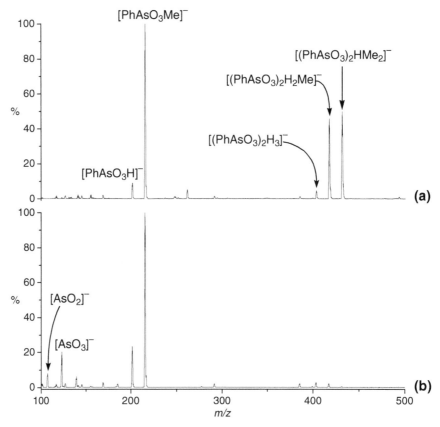

Figure 6.9
Negative-ion ESI mass spectra of phenylarsonic acid $(PhAsO_3H_2)$ in methanol solvent, at cone voltages of (a) 20 V and (b) 40 V. At both of these cone voltages the ion $[PhAsO_3Me]^-$, formed by reaction with the solvent, is the base peak

$[PhAsO_3Me]^-$ at *m/z* 215, together with a trace of the deprotonated parent acid, $[PhAsO_3H]^-$ (*m/z* 201). Additionally, aggregate ions $[(PhAsO_3)_2H_3]^-$ (*m/z* 403), $[(PhAsO_3)_2H_2Me]^-$ (*m/z* 417) and $[(PhAsO_3)_2HMe_2)]^-$ (*m/z* 431) are also seen, but not $[(Ph\ AsO_3)_2Me_3]^-$. For these aggregate ions, at least one proton is necessary in order for the hydrogen bond to form between the two arsonate species. At 40 V, the monoester $[PhAsO_3Me]^-$ remains the base peak, the aggregate ions are of low intensity, and fragment ions $[AsO_2]^-$ (*m/z* 107) and $[AsO_3]^-$ (*m/z* 123) are observed, as a result of cleavage of the carbon-arsenic bond.

Miscellaneous Organoarsenic Compounds

A considerable number of studies have concerned the ESI MS analysis of organoarsenic compounds having biochemical and environmental importance. Advances in this field have been summarised in several reviews.[43] Fragmentation pathways of various organoarsenic compounds have been studied by CID,[44] including an FTICR MS study.[45] The $[As]^+$ ion at *m/z* 75 is a useful fragment ion for confirmation of an arsenic species. However, the formation of this ion has found to be very dependent on the purity of the nitrogen drying gas used.[46] Various compounds detected in natural systems include arsenobetaine $Me_3As^+CH_2CO_2^-$, arsenocholine $Me_3As^+CH_2CH_2OH$, tetramethylarsonium iodide $Me_4As^+I^-$, methylarsonic acid $MeAs(O)(OH)_2$, dimethylarsinic acid $Me_2As(O)OH$, trimethylarsine oxide, arylarsonic acids, and a range of arsenosugars. The majority of the organoarsenic studies have concerned quantitative analysis, for example using a coupled separation technique such as HPLC,[47] ion chromatography[48] or capillary electrophoresis.[49] These methods have been applied to analysis of organoarsenic compounds in seaweeds,[50] algae,[51] shellfish,[52] urine[53] and water samples.[54]

Arsenic compounds related to the chemical weapon Lewisite ($ClCH=CH-AsCl_2$) including chlorovinylarsonous acid have been derivatised by reaction with 2-mercaptopyridine and analysed using APCI MS or ESI MS, giving improved sensitivity relative to the underivatised compounds.[39] The $Me_2AsSe_2^-$ ion, synthesised in solution by reaction of sodium selenite, sodium dimethylarsinate and glutathione, has also been characterized using ESI MS.[15]

6.3.3 Organoantimony and -Bismuth Compounds

Relatively little work has been done on the mass spectrometric analysis of organoantimony and organobismuth compounds. Like phosphines and arsines, tertiary stibines are poorly basic molecules, but triphenylantimony ($SbPh_3$) can be analysed by addition of silver(I) ions, which predominantly gives $[Ag(SbPh_3)_4]^+$.[32] This contrasts with the situation for triphenylphosphine, which predominantly forms $[Ag(PPh_3)_2]^+$. The difference is presumably related to both the better donor ligand properties of the phosphine, and the shorter silver-phosphorus bond distance compared to silver-antimony, with the consequence that four stibine donors are able to coordinate around an Ag^+ ion, satisfying its electronic requirements.

As for phosphines, tertiary stibines such as $1,3\text{-}C_6H_4(CH_2SbMe_2)_2$ and $1,3\text{-}C_6H_4(SbMe_2)_2$ can be converted into quaternary stibonium salts (by reaction with excess MeI) e.g. $1,3\text{-}C_6H_4(CH_2SbMe_3)_2^{2+}$, which can be analysed by ESI MS.[35] Neutral metal carbonyl complexes of these two ligands, such as $[\{Fe(CO)_4\}_2\{1,3\text{-}C_6H_4(SbMe_2)_2\}]$ – which are not likely to give ions in the ESI MS spectra without prior derivatisation – do give $[M]^+$ ions using the harsher ionisation technique of API MS.

Triphenylantimony oxide, Ph_3SbO is considerably more basic than Ph_3Sb, and gives solely the protonated $[M + H]^+$ ion in both ESI MS and APCI MS analyses.[56] This study also investigated, using these two ionisation techniques with similar results, the air oxidation of trimethylstibine (Me_3Sb), which is produced naturally by fungi, for example. A wide range of species with between one and four antimony atoms were observed, with

$[Me_3SbOH]^+$, $[(Me_3SbO)_2H]^+$, and $[(Me_4Sb)_2OH]^+$ having the greatest intensity. The ESI MS analysis of Me_3SbCl_2 has also been carried out using two different solvents.[57] In methanol-water various Me_3Sb species were observed, such as $[Me_3SbOH]^+$, $[Me_3SbOH + H_2O]^+$, $[Me_3SbCl + MeOH]^+$, and $[Me_3SbOH + MeOH]^+$. These observations are as expected, given that the antimony-carbon bonds are relatively strong compared to the antimony-chloride bonds, the latter undergoing solvolysis reactions in solution. In methanol-water-acetic acid (HOAc) solution, $[Me_3Sb(OAc)]^+$ was the sole species observed.

Coordination complexes of organoantimony ligands have also been investigated using ESI MS; cationic rhodium(III) complexes of ditertiary stibines $[Rh(L-L)_2X_2]^+$ (L-L = $Ph_2Sb(CH_2)_3SbPh_2$ or o-$C_6H_4(SbMe_2)_2$; X = Cl, Br, I)[58] giving the expected ions. The organoantimony nickel carbonyl cluster $[Ni_{10}(SbPr^i)_2(CO)_{18}]^{2-}$ has also been characterised by ESI MS.[59]

Organometallic Derivatives of Group 16 Elements; Organosulfur, -selenium and -tellurium Compounds

A number of organothiolate anions (RS^-) have been investigated by negative-ion ESI MS.[60] Air oxidation of the thiolate solutions was readily observed, giving RSO_2^- and RSO_3^- ions. For dithiols with an excess of added NaOMe base, no $[M - 2H]^{2-}$ ions were observed, and instead the monoanions $[M - H]^-$ and $[M - 2H + Na]^-$ were seen.

Simple organoselenium compounds, such as selenocystamine and selenomethionine, have been analysed by ESI MS, giving simple $[M + H]^+$ ions; on-line CE-ESI MS was also successfully employed for the analysis and detection of these selenium species together with selenocysteine.[61]

A number of phenylselenoethers of the type p-$XC_6H_4SeCH_2CH(R)NH_2$ (X = H, Cl, F, Cl; R = H, Me) are promising agents for the treatment of hypertension, and have also been studied by ESI MS. Extensive fragmentation occurred, even under the mildest CID conditions. This was overcome by addition of 18-crown-6, which formed a complex enabling enhanced sensitivity and reduced fragmentation.[62] Thioethers, selenoethers and telluroethers are well-known to act as ligands towards transition metal centres, and ESI MS is being increasingly used in their characterisation. A range of cationic manganese(I) and rhenium(I) tricarbonyl complexes of multidentate thioether, selenoether and telluroether ligands gave the expected $[M]^+$ cations in ESI MS analysis.[63] Cationic macrocyclic selenoether complexes of ruthenium(II)[64] and chromium(III)[65] behave in the same way, as do chromium(III) complexes of macrocyclic thioethers[66] and ditelluroether complexes of ruthenium(II).[67]

Organomercury Compounds

While mercury is a d-block element, its compounds show characteristics of main group metal compounds, and so are included in this chapter for comparison. Organomercury(II) compounds have had widespread industrial use in the past. Certain microorganisms have the ability to biomethylate Hg^{2+} ions into considerably more toxic CH_3Hg^+, and this species has a widespread occurrence in the environment. Accordingly, there have been several studies concerned with the analysis of organomercury species, and the interactions of organomercury compounds with biologically relevant molecules such as amino acids, peptides and proteins.

Not surprisingly, the greatest amount of work has been carried out with methylmercury derivatives. Interactions of CH_3Hg^+ with amino acids (AA) have been investigated by ESI MS; species $[MeHg(AA)]^+$ with a 1:1 stoichiometry, together with 2:1 species $[(MeHg)_2(AA-H)]^+$ were observed.[68] Methylmercury derivatives of nitrogen bases (such as N-methylpyrazoles, bis(pyridin-2-yl)methane, creatinine), of the type $[CH_3HgL]^+$ give the expected parent ion. This study was also one of the first to demonstrate the utility of ESI MS as a technique for monitoring ligand exchange reactions in solution.[69]

The interactions of the $[MeHg]^+$ and $[PhHg]^+$ cations with peptides containing cysteinyl (cys) residues have also been investigated by ESI MS.[68,70] The amino acid cysteine contains an -SH group, and thus shows a strong affinity for complexation to mercury(II) by formation of very stable Hg-S-cys groups. The ESI MS study showed that mercury addition occurred primarily at the SH group, with evidence for isomeric species in which the Hg is bound to an amino or carboxylate group. $[MeHg]^+$ showed almost exclusive affinity for the SH group. The tripeptide glutathione (Glu) gave evidence for formation of a 3:1 species $[(MeHg)_3(Glu-2H)]^+$.[68]

A simple organomercury halide such as PhHgCl gives only weak ions when analysed by ESI MS in methanol. However, upon addition of pyridine, the ion $[PhHg(pyridine)]^+$ (*m/z* 358) is the only species observed up to cone voltage of 20 V. At higher cone voltages, e.g. 40 V, the pyridine was lost, giving additional ions $[PhHg]^+$ (*m/z* 279) and its methanol-solvated analogue $[PhHg(MeOH)]^+$ (*m/z* 311).

A study of the behaviour of the hydroxide PhHgOH in aqueous solutions showed that $[PhHgOH_2]^+$, $[(PhHg)_2OH]^+$ and $[(PhHg)_3O]^+$ co-exist in a pH-dependent equilibrium in solution.[71] The oxonium ion $[(PhHg)_3O]^+$ was found to be particularly stable towards cone voltage-induced fragmentation, while $[PhHgOH_2]^+$ (which can be considered to be protonated PhHgOH or solvated $PhHg^+$) readily lost H_2O, giving the bare $PhHg^+$ cation.

The aryl-mercury compounds $(C_6F_4Hg)_3$ and o-$C_6F_4(HgCl)_2$ have been studied by negative-ion nanoelectrospray MS.[72] These mercury compounds act as polydentate Lewis acids, and bind halide ions. $(C_6F_4Hg)_3$ gives adducts of the type $[(C_6F_4Hg)_3 + halide]^-$ and $[2(C_6F_4Hg)_3 + halide]^-$, with preferred binding in the order $I^- > Br^- > Cl^- > F^-$. The compound o-$C_6F_4(HgCl)_2$ with excess halide gives different results, depending on the halide; chloride gives $[o$-$C_6F_4(HgCl)_2 + Cl]^-$, bromide gives $[o$-$C_6F_4(HgBr)_2 + Br]^-$, and iodide gives the condensation species $[(C_6F_4Hg)_3 + I]^-$. The organomercury derivative of the nucleobase 1,3-dimethyluracil (dmura) **6.3** has been analysed by ESI MS in DMSO and aqueous solutions, and gave hydrolysed ions of the type $[Hg_3(dmura)_3O]^+$, $[Hg_6(dmura)_6O_2]^{2+}$, $[Hg_6(dmura)_6O_2(NO_3)]^+$ and $[Hg(dmura)(DMSO)]^+$.[73] The compound forms the $[(RHg)_3O]^+$ oxonium species, which then dimerises via Hg$\cdots$Hg contacts; these Hg$\cdots$Hg interactions are strong enough to survive the ES ionisation process.

6.3

In the area of environmental analysis, MeHgCl, EtHgCl and PhHgCl have been analysed using ESI and APCI ionisation techniques. To minimise adsorption of the organomercury species onto components of the injection system (notably the injection loop) 2-mercaptoethanol was added to the analyte solutions to minimise this effect through the formation of stable thiolate complexes. It was concluded that APCI ionisation was the preferred technique, and adduct ions $[RHgSCH_2CH_2OH + H]^+$ were formed between the analytes and 2-mercaptoethanol. A qualitative analysis of methylmercury in a spiked fish extract was also carried out.[74]

Other Organometallic Derivatives

For labile organometallic compounds, for example those of the group 1 and group 2 metals, the **coldspray** ionisation technique[75] (whereby the sample is introduced at a low temperature, to preserve labile species) may prove to have uses in the analysis of solution species. An example of this is the analysis of the Grignard reagent MeMgCl in tetrahydrofuran (THF) solution. Grignard reagents are widely used in organic and organometallic chemistry, and have been known since 1900. However, the nature of the actual species present in solution has been the subject of some debate over the years. Using coldspray ionisation MS at a spray temperature of $-20\,°C$, a solution of MeMgCl in THF showed four major ions assigned as the solvated dinuclear species $[MeMg_2Cl_3(THF)_n - H]^+$ (where n = 4 to 6), proposed to have the structure **6.4**. Upon CID of the $[MeMg_2Cl_3(THF)_6 - H]^+$ ion, sequential loss of THF molecules was observed, confirming the highly solvated nature of these ions.

$$(THF)_n-Mg \underset{Cl}{\overset{Cl}{<}} \overset{Cl}{\underset{}{>}} Mg \underset{(THF)_n}{\overset{Me}{<}}$$

6.4

Finally, in early ESI MS studies of main group organometallic compounds, methyl-indium and -thallium compounds yielded ions formed by loss of coordinated water molecules; for example $[Me_2Tl(phen)(H_2O)]NO_3$ yielded the ion $[Me_2Tl(phen)]^+$.[76]

Summary

- Main group element halides tend to give ions formed by halide loss and solvolysis (see also Section 5.3).
- For ligands such as phosphines, arsines, stibines, thioethers and selenoethers, cationisation with Ag^+ or quaternisation with alkyl halides should give a charged derivative with a high ionisation efficiency.
- Reduction of organo-derivatives may occur at elevated cone voltages, e.g. $Me_3Pb^+ \longrightarrow MePb^+ \longrightarrow Pb^+$.

References

1. W. Henderson, B. K. Nicholson and L. J. McCaffrey, *Polyhedron*, 1998, **17**, 4291; J. C. Traeger, *Int. J. Mass Spectrom.*, 2000, **200**, 387.

2. A. Chapeaurouge, L. Biglerr, A. Schäfer and S. Bienz, *J. Am. Soc. Mass Spectrom.*, 1995, **6**, 207.

3. M. Yoshida, S. Tsuzuki, M. Goto and F. Nakanishi, *J. Chem. Soc., Dalton Trans.*, 2001, 1498.

4. L. K. Frevel, W.-L. Lee and R. E. Tecklenburg, *J. Am. Soc. Mass Spectrom.*, 1999, **10**, 231.

5. T. Fukuo, N. Kubota, M. Tamura, and R. Arakawa, *J. Mass Spec. Soc. Japan*, 2000, **48**, 353.

6. T. Løver, W. Henderson, G. A. Bowmaker, J. M. Seakins and R. P. Cooney, *J. Mater. Chem.*, 1997, 1553.

7. J. H. Lamb and G. M. A. Sweetman, *Rapid Commun. Mass Spectrom.*, 1996, **10**, 594.

8. J. Wei, J. Chen and J. M. Miller, *Rapid Commun. Mass Spectrom.*, 2001, **15**, 169.

9. J. Beckmann, D. Horn, K. Jurkschat, F. Rosche, M. Schürmann, U. Zachwieja, D. Dakternieks, A. Duthie and A. E. K. Lim, *Eur. J. Inorg. Chem.*, 2003, 164.

10. W. Henderson and M. J. Taylor, *Polyhedron*, 1996, **15**, 1957.

11. G. Lawson, R. H. Dahm, N. Ostah and E. D. Woodland, *Appl. Organomet. Chem.*, 1996, **10**, 125.

12. D. Dakternieks, A. E. K. Lim, and K. F. Lim, *Phosphorus, Sulfur Silicon Related Elem.*, 1999, **150 – 151**, 339.

13. M. Kemmer, H. Dalil, M. Biesemans, J. C. Martins, B. Mahieu, E. Horn, D. de Vos, E. R. T. Tiekink, R. Willem and M. Gielen, *J. Organomet. Chem.*, 2000, **608**, 63.

14. M. Kemmer, M. Gielen, M. Biesemans, D. de Vos and R. Willem, *Metal-Based Drugs*, 1998, **5**, 189.

15. M. Kemmer, L. Ghys, M. Gielen, M. Biesemans, E. R. T. Tiekink and R. Willem, *J. Organomet. Chem.*, 1999, **582**, 195.

16. R. Altmann, K. Jurkschat, M. Schürmann, D. Dakternieks and A. Duthie, *Organometallics*, 1997, **16**, 5716.

17. M. Mehring, I. Vrasidas, D. Horn, M. Schürmann and K. Jurkschat, *Organometallics*, 2001, **20**, 4647.

18. M. Pellei, G. G. Lobbia, M. Ricciutelli and C. Santini, *Polyhedron*, 2003, **22**, 499.

19. D. Dakternieks, K. Jurkschat, S. van Dreumel and E. R. T. Tiekink, *Inorg. Chem.*, 1997, **36**, 2023.

20. H. Dalil, M. Biesemans, R. Willem and M. Gielen, *Main Group Metal Chem.*, 1998, **21**, 741.

21. M. Mehring, I. Paulus, B. Zobel, M. Schürmann, K. Jurkschat, A. Duthie and D. Dakternieks, *Eur. J. Inorg. Chem.*, 2001, 153.

22. D. Dakternieks, H. Zhu, E. R. T. Tiekink and R. Colton, *J. Organomet. Chem.*, 1994, **476**, 33.

23. L. Ghys, M. Biesemans, M. Gielen, A. Garoufis, N. Hadjiliadis, R. Willem and J. C. Martins, *Eur. J. Inorg. Chem.*, 2000, **3**, 513.

24. E. M. Ryan, A. Buzy, D. E. Griffiths, K. R. Jennings and D. N. Palmer, *Biochem. Soc. Trans.*, 1996, **24**, 290S.

25. J. M. Bayona and Y. Cai, *Trends Anal. Chem.*, 1994, **13**, 327.

26. K. W. M. Siu, G. J. Gardner and S. S. Berman, *Anal. Chem.*, 1989, **61**, 2320.

27. T. L. Jones-Lepp, K. E. Varner and B. E. Hilton, *Appl. Organomet. Chem.*, 2001, **15**, 933; T. L. Jones-Lepp, K. E. Varner, M. McDaniel and L. Riddick, *Appl. Organomet. Chem.*, 1999, **13**, 881.

28. R. T. Aplin, J. E. H. Buston and M. G. Moloney, *J. Organomet. Chem.*, 2002, **645**, 176.

29. G. Zoorob, F. B. Brown and J. Caruso, *J. Analyt. Atom. Spectrom.*, 1997, **12**, 517.

30. D. J. F. Bryce, P. J. Dyson, B. K. Nicholson and D. G. Parker, *Polyhedron*, 1998, **17**, 2899.

31. W. Henderson and G. M. Olsen, *Polyhedron*, 1998, **17**, 577.

32. L. S. Bonnington, R. K. Coll, E. J. Gray, J. I. Flett and W. Henderson, *Inorg. Chim. Acta*, 1999, **290**, 213.

33. R. Colton, J. C. Traeger and J. Harvey, *Org. Mass Spectrom.*, 1992, **27**, 1030.

34. R. Colton and J. C. Traeger, *Inorg. Chim. Acta*, 1992, **201**, 153.

35. W. Levason, M. L. Matthews, G. Reid and M. Webster, *J. Chem. Soc., Dalton Trans.*, 2004, 51.

36. W. Henderson and G. M. Olsen, *Polyhedron*, 1996, **15**, 2105.

37. C. Decker, W. Henderson and B. K. Nicholson, *J. Chem. Soc., Dalton Trans.*, 1999, 3507.

38. A. J. Bell, D. Despeyroux, J. Murrell and P. Watts, *Int. J. Mass Spectrom. Ion Proc.*, 1997, **165/166**, 533; J. Banoub, E. Gentil, and J. Kiceniuk, *Int. J. Environ. Anal. Chem.*, 1995, **61**, 143; C. Molina, P. Grasso, E. Benfenati and D. Barceló, *J. Chromatogr. A*, 1996, **737**, 47; C. Molina, P. Grasso, E. Benfenati and D. Barceló, *Int. J. Environ. Anal. Chem.*, 1996, **65**, 69; D. Barceló, *Biomed. Environ. Mass Spectrom.*, 1988, **17**, 363.

39. W. R. Creasy, *J. Am. Soc. Mass Spectrom.*, 1999, **10**, 440.

40. V. T. Borrett, R. Colton and J. C. Traeger, *Eur. Mass Spectrom.*, 1995, **1**, 131.

41. M. G. Fitzpatrick, L. R. Hanton, W. Henderson, P. E. Kneebone, E. G. Levy, L. J. McCaffrey and D. A. McMorran, *Inorg. Chim. Acta*, 1998, **281**, 101.

42. S. R. Alley and W. Henderson, *J. Organomet. Chem.*, 2001, **637–639**, 216.

43. Z. Gong, X. Lu, M. Ma, C. Watt and X. C. Le, *Talanta*, 2002, **58**, 77; L. Ebdon, S. Fitzpatrick and M. E. Foulkes, *Chemia Analityczna*, 2002, **47**, 179; M. Careri, A. Mangia and M. Musci, *J. Chromatogr. A*, 1996, **727**, 153; J. W. McLaren, K. W. M. Siu and S. S. Berman in *Instrum. Trace Org. Monit.*, R. E. Clement, K. W. M. Siu and H. H. Hill Jr (Eds), 1992, 195.

44. B. R. Larsen, C. Astorga-Llorens, M. H. Florencio and A. M. Bettencourt, *J. Chromatog. A*, 2001, **926**, 167; S. N. Pedersen and K. A. Francesconi, *Rapid Commun. Mass Spectrom.*, 2000, **14**, 641; M. H. Florencio, M. F. Duarte, A. M. M. de Bettencourt, M. L. Gomes and L. F. V. Boas, *Rapid Commun. Mass Spectrom.*, 1997, **11**, 469.

45. R. Pickford, M. Miguens-Rodriguez, S. Afzaal, P. Speir, S. A. Pergantis and J. E. Thomas-Oates, *J. Analyt. Atom. Spectrom.*, 2002, **17**, 173.

46. D. Kuehnelt, W. Goessler and K. A. Francesconi, *Rapid Commun. Mass Spectrom.*, 2003, **17**, 654.

47. See for example S. A. Pergantis, W. Winnik and D. Betowski, *J. Analyt. Atom. Spectrom.*, 1997, **12**, 531.

48. P. A. Gallagher, J. A. Shoemaker, X. Wei, C. A. Brockhoff-Schwegel and J. T. Creed, *Fresenius' J. Anal. Chem.*, 2001, **369**, 71; K. W. M. Siu, R. Guevremont, J. C. Y. Le Blanc, G. J. Gardner and S. S. Berman, *J. Chromatog.*, 1991, **554**, 27.

49. O. Schramel, B. Michalke and A. Kettrup, *J. Analayt. Atom. Spectrom.*, 1999, **14**, 1339.

50. M. Van Hulle, C. Zhang, X. Zhang and R. Cornelis, *Analyst*, 2002, **127**, 634; M. Miguens-Rodriguez, R. Pickford, J. E. Thomas-Oates and S. A. Pergantis, *Rapid Commun. Mass Spectrom.*, 2002, **16**, 323; S. McSheehy, M. Marcinek, H. Chassaigne and J. Szpunar, *Anal. Chim. Acta*, 2000, **410**, 71; P. A. Gallagher, X. Wei, J. A. Shoemaker, C. A. Brockhoff and J. T. Creed, *J. Analyt. Atom. Spectrom.*, 1999, **14**, 1829.

51. S. McSheehy, P. Pohl, D. Velez and J. Szpunar, *Analyt. Bioanalyt. Chem.*, 2002, **372**, 457; A. Geiszinger, W. Goessler, S. N. Pedersen and K. A. Francesconi, *Env. Toxicol. and Chem.*, 2001, **20**, 2255; S. McSheehy, P. Pohl, R. Lobinski and J. Szpunar, *Anal. Chim. Acta*, 2001, **440**, 3; A. D. Madsen, W. Goessler, S. N. Pedersen and K. A. Francesconi, *J. Analyt. Atom. Spectrom.*, 2000, **15**, 657; S. McSheehy and J. Szpunar, *J. Analyt. Atom. Spectrom.*, 2000, **15**, 79; S. A. Pergantis, S. Wangkam, K. A. Francesconi and J. E. Thomas-Oates, *Anal. Chem.*, 2000, **72**, 357.

52. S. McSheehy, J. Szpunar, R. Lobinski, V. Haldys, J. Tortajada and J. S. Edmonds, *Anal. Chem.*, 2002, **74**, 2370; D. Sanchez-Rodas, A. Geiszinger, J. L. Gomez-Ariza and K. A. Francesconi, *Analyst*, 2002, **127**, 60; S. McSheehy, P. Pohl, R. Lobinski and J. Szpunar, *Analyst*, 2001, **126**, 1055; K. A. Francesconi and J. S. Edmonds, *Rapid Commun. Mass Spectrom.*, 2001, **15**, 1641.

53. H. R. Hansen, R. Pickford, J. Thomas-Oates, M. Jaspars and J. Feldmann, *Angew. Chem., Int. Ed. Engl.*, 2004, **43**, 337; K. A. Francesconi, R. Tanggaard, C. J. McKenzie and W. Goessle, *Clinical Chem.*, 2002, **48**, 92; Z. Gong, G. Jiang, W. R. Cullen, H. V. Aposhian and X. C. Le, *Chem. Res. Toxicol.*, 2002, **15**, 1318; Y. Inoue, Y. Date, T. Sakai, N. Shimizu, K. Yoshida, H. Chen, K. Kuroda and G. Endo, *Appl. Organomet. Chem.*, 1999, **13**, 81.

54. J. Wu, Z. Mester and J. Pawliszyn, *Anal. Chim. Acta*, 2000, **424**, 211; M. H. Florencio, M. F. Duarte, S. Facchetti, M. L. Gomes, W. Goessler, K. J. Irgolic, H. A. van't Klooster, L. Montanarella, R. Ritsema, L. F. V. Boas and A. M. M. de Bettencourt, *Analusis*, 1997, **25**, 226.

55. J. Gailer, G. N. George, H. H. Harris, I. J. Pickering, R. C. Prince, A. Somogyi, G. Buttigieg, R. S. Glass and M. B. Denton, *Inorg. Chem.*, 2002, **41**, 5426.

56. P. J. Craig, S. A. Forster, R. O. Jenkins, G. Lawson, D. Miller and N. Ostah, *Appl. Organomet. Chem.*, 2001, **15**, 527.

57. O. Schramel, B. Michalke and A. Kettrup, *J. Chromatogr. A*, 1998, **819**, 231; J. Lintschinger, O. Schramel and A. Kettrup, *Fresenius' J. Anal. Chem.*, 1998, **361**, 96.

58. A. M. Hill, W. Levason and M. Webster, *Inorg. Chim. Acta*, 1998, **271**, 203.

59. P. D. Mlynek and L. F. Dahl, *Organometallics*, 1997, **16**, 1641.

60. T. D. McCarley and R L. McCarley, *Anal. Chem.*, 1997, **69**, 130.

61. O. Schramel, B. Michalke and A. Kettrup, *J. Chromatogr. A*, 1998, **819**, 231.

62. W. Z. Shou, M. M. Woznichak, S. W. May and R. F. Browner, *Anal. Chem.*, 2000, **72**, 3266.

63. J. Connolly, A. R. J. Genge, W. Levason, S. D. Orchard, S. J. A. Pope and G. Reid, *J. Chem. Soc., Dalton Trans.*, 1999, 2343.

64. W. Levason, J. J. Quirk, G. Reid and S. M. Smith, *J. Chem. Soc., Dalton Trans.*, 1997, 3719.

65. W. Levason, G. Reid and S. M. Smith, *Polyhedron*, 1997, **16**, 4253.

66. S. J. A. Pope, N. R. Champness and G. Reid, *J. Chem. Soc., Dalton Trans.*, 1997, 1639.

67. W. Levason, S. D. Orchard, G. Reid and V.-A. Tolhurst, *J. Chem. Soc., Dalton Trans.*, 1999, 2071.

68. A. J. Canty, R. Colton, A. D'Agostino and J. C. Traeger, *Inorg. Chim. Acta*, 1994, **223**, 103.

69. A. J. Canty and R. Colton, *Inorg. Chim. Acta*, 1994, **215**, 179.

70. A. D'Agostino, R. Colton, J. C. Traeger and A. J. Canty, *Eur. Mass Spectrom.*, 1996, **2**, 273.

71. W. J. Jackson, A. Moen, B. K. Nicholson, D. G. Nicholson and K. A. Porter, *J. Chem. Soc., Dalton Trans.*, 2000, 491.

72. J. M. Koomen, J. E. Lucas, M. R. Haneline, J. D. Beckwith King, F. Gabbai and D. H. Russell, *Int. J. Mass Spectrom.*, 2003, **225**, 225.

73. F. Zamora, M. Sabat, M. Janik, C. Siethoff and B. Lippert, *Chem. Commun.*, 1997, 485.

74. C. F. Harrington, J. Romeril and T. Catterick, *Rapid Commun. Mass Spectrom.*, 1998, **12**, 911.

75. K. Yamaguchi and S. Sakamoto, *JEOL News*, 2002, **28A**, 2; K. Yamaguchi, *J. Mass Spectrom.*, 2003, **38**, 473; S. Sakamoto, T. Imamoto and K. Yamaguchi, *Org. Lett.*, 2001, **3**, 1793.

76. A. J. Canty, R. Colton and I. M. Thomas, *J. Organomet. Chem.*, 1993, **455**, 283.

7 The ESI MS Behaviour of Transition Metal and Lanthanide Organometallic Compounds

Introduction

The transition and lanthanide metals have an extensive and diverse organometallic chemistry. A wide range of complexes are formed, containing ligands such as CO, isonitriles (RNC), alkyl and aryl groups, as well as π-bonded ligands such as cyclopentadienyl, arene, alkene, alkyne and allyl ligands, and many more. In this chapter, the ESI MS behaviour of transition metal and lanthanide organometallic compounds is discussed according to the type(s) of organometallic ligand in the complex. Naturally, there is some overlap with other sections and no attempt to be fully comprehensive has been made.

The application of electrospray mass spectrometry to characterisation of transition metal organometallic species has been covered in several reviews up to the year 2000.[1] The use of spectroscopic and mass spectrometric techniques for the characterisation of metal carbonyl clusters has also been recently reviewed.[2]

Metal Carbonyl Complexes

Metal carbonyl complexes are an extremely important class of organometallic compound; they are of interest in their own right for structure and bonding considerations, as precursors to a wide range of other carbonyl-containing metal complexes and clusters and as precursors to catalysts. The ability to analyse such compounds by modern mass spectrometry methods, particularly ESI and MALDI techniques, is therefore highly desirable.

Metal carbonyl compounds are here divided into several classes, according to charge and nuclearity. The analysis of *ionic* metal carbonyl complexes and clusters by ESI MS is typically straightforward, but neutral binary metal carbonyls of the type $M_x(CO)_y$ usually present more difficulties due to the difficulty in ionising these materials. Derivatisation strategies are therefore necessary for these compounds and these are discussed in Section 7.2.3.

Mass Spectrometry of Inorganic, Coordination and Organometallic Compounds   W. Henderson and J. S. McIndoe
© 2005 John Wiley & Sons, Ltd   ISBNs: 0-470-85015-9 (HB); 0-470-85016-7 (PB)

7.2.1 Ionic Mononuclear Metal Carbonyl Compounds

The analysis of charged metal complexes using electrospray ionisation involves simple transfer of the solution ions into the gas phase, with the dominant theme being observation of intact $[M]^+$ or $[M]^-$ ions when low fragmentation conditions are used. When the skimmer (cone) voltage is raised, parent ions often lose CO ligands, and the ability to vary this voltage allows the fragmentation to be easily investigated. Electrospray appears to be the technique of choice for these compounds and a comparative study of a range of the mononuclear carbonyl complexes $[Fe(CO)_3(\eta^5\text{-dienyl})]BF_4$, $[Fe(CO)_2Cp(\eta^2\text{-alkene})]BF_4$ and $[M(CO)_3(\eta\text{-}C_7H_7)]BF_4$ (M = Cr, Mo, W) by ESI, FAB and field desorption ionisation methods concluded that ESI was the preferred technique.[3] Cationic η^4-dienyl-phosphonium complexes $[(\eta^4\text{-diene-PPh}_3)Fe(CO)_3]^+$ gave the expected $[M]^+$ cation in ESI MS, but on replacement of a CO by a more electron-donating PPh_3 to give $[(\eta^4\text{-diene-PPh}_3)Fe(CO)_2(PPh_3)]^+$ the $[M]^+$ ion was weak, with the major ion formed by loss of PPh_3 from the diene ligand.[4]

In addition to those described above, a wide range of other cationic metal carbonyl complexes have been analysed by ESI mass spectrometry, including *trans*-$[Cr(CO)_2$ $(dppe)_2]^+$ (dppe = $Ph_2PCH_2CH_2PPh_2$),[5,6] *trans*-$[Mn(CO)_2(dppm)_2]^+$,[7] $[M(CO)_2$ (diphosphine)X]$^+$ (diphosphine = dppe or $Ph_2PCH_2PPh_2$; M = Mo, W; X = F, Cl, Br, I),[8] *cis/mer*-$[Re(CO)_2Br(Ph_2PCH_2PPh_2Me)$ $(Ph_2PCH_2PPh_2)]^+$,[5] $[Re(CO)_2(Ph_2PCH_2CH_2$ $PPhCH_2CH_2PPh_2)X]^+$ (X = Cl, Br),[9] and manganese(I) and rhenium(I) carbonyl complexes of the phosphines $(Ph_2PCH_2)_3CCH_3$ and $(Ph_2PCH_2CH_2)_3P$.[10] The amount of deuteration in $[Mo(CO)_2(dppe)_2D]^+$ has also been determined using ESI MS.[11] Manganese(I) and rhenium(I) carbonyl complexes of a range of tridentate thio-, seleno- and telluro-ethers, arsines and phosphines, and monodentate primary and secondary phosphines have also been characterised using ESI MS; in general the parent cations were observed, in some cases with CO loss fragment ions.[12]

ESI MS has also been used to detect a labile rhenium carbonyl reaction intermediate in a CO_2-fixation reaction. Photolysis of $[Re(CO)_3(bipy)(PPh_3)]^+$ in $HC(O)NMe_2$-triethanolamine (L) under CO_2 gave the formate complex $[Re(CO)_3(bipy)(OCOH)]^+$ and two intermediate complexes, one of which was found by ESI MS to be the triethanolamine (L) complex $[Re(CO)_3(bipy)(L)]^+$.[13]

Anionic complexes are equally easily analysed, for example $[M(CO)_5X]^-$, (M = Cr, Mo, W; X = Cl, Br, I) and the manganese and rhenium complexes *cis*-$[M(CO)_4X_2]^-$ (X = Cl, Br, I).[8] However, there are fewer reported examples of the application of ESI MS to anionic mononuclear systems, possibly due to the air sensitivity of many complexes of this type.

7.2.2 Ionic Metal Carbonyl Clusters

The application of ESI MS to analysis of charged metal clusters has been carried out in a number of systems and the number of applications is rapidly growing. As examples, the clusters $[Ru_3Co(CO)_{13}]^-$,[14,15] $[Ru_6(CO)_{18}]^{2-}$,[14] $[Ru_6C(CO)_{16}]^{2-}$ [14,16] $[Os_{10}C$ $(CO)_{24}]^{2-}$,[16,17] $[HOs_5(CO)_{15}]^-$,[17] $[PtRu_5C(CO)_{15}]^-$,[17] $[Os_{17}(CO)_{36}]^{2-}$,[17] $[Pt_3Ru_{10}C_2$ $(CO)_{32}]^{2-}$,[17] and $[Pd_6Ru_6(CO)_{24}]^{2-}$,[17] all give solely the parent anions by ESI MS at low cone voltages. Positively-charged metal carbonyl clusters are less common, but one example is the radical cation $[(CF_3CCCF_3)Co_2(CO)_2\{P(OMe)_3\}_4]^+$ which gives the parent cation together with CO-loss fragment ions.[18]

Clusters containing other main group heteroatoms are just as easily characterised using ESI MS. The manganese carbonyl clusters $[E_2Mn_4(CO)_{12}]^{2-}$ give the expected dianions at m/z 310 (E = S) and m/z 358 (E = Se), and the related $[Mn(CO)_3(CSe_4)]_2^{2-}$ also gave the dianion species.[19] Similarly, the monoanions $[EFe_2Co(CO)_9]^-$ (E = selenium, tellurium) and $[E_2Fe_2Mn(CO)_9]^-$ (E = selenium, tellurium) gave the parent ions in ESI MS.[20]

In some cases, ESI MS does not give exclusively parent ions, but useful information is still obtained. As a good example of this type of system, the air-sensitive and labile platinum-carbonyl anion $[Pt_{12}(CO)_{24}]^{2-}$ has been analysed by ESI MS as its Ph_4P^+ salt.[21] The parent dianion (m/z 1324) together with the ion paired monoanion $[Pt_{12}(CO)_{24}(PPh_4)]^-$ (m/z 3351) are the two most intense ions in the spectrum. Other ions $[Pt_9(CO)_{18}(PPh_4)_2]^-$, $[Pt_9(CO)_{13}(PPh_4)_2]^-$ and $[Pt_9(CO)_8]^{2-}$ are formed by loss of one of the Pt_3 triangles (a common feature in the chemistry of this platinum-carbonyl anion system). ESI MS analysis of the phosphido-bridged dinuclear nickel complex $[Ni_2(CO)_4(\mu\text{-}PPh_2)_2]^{2-}$ showed the one-electron oxidation product $[Ni_2(CO)_4(\mu\text{-}PPh_2)_2]^-$ and the ions $[Ni_2(CO)_n(\mu\text{-}PPh_2)_2]^-$ (n = 1 to 3) formed by loss of CO ligands; the peak for the unoxidised dianion was very weak or absent.[22] The air-sensitive $[Ni_{10}(BiMe)_2(CO)_{18}]^{2-}$ cluster gave a series of eight ions corresponding to $[M-2Me-nCO]^{2-}$ formed by CO and methyl group loss.[23] By comparison, the related antimony cluster $[Ni_{10}(Sb^iPr)_2(CO)_{18}]^{2-}$ gave the parent ion, together with ions formed by loss of CO and isopropyl groups.[24]

The study of the metal cluster anions $[Fe_3(CO)_{11}HgM]^-$ [M = $CpMo(CO)_3$, $Co(CO)_4$, $CpFe(CO)_2$, $CpW(CO)_3$ and $Mn(CO)_5$] were studied using ESI MS with a cone voltage of 50 V.[25] At this cone voltage the parent ions were observed, but the base peak was formed by loss of a CO ligand; the use of lower cone voltages was not reported, but would be expected to result in decreased fragmentation. FAB MS was also used for these clusters, but it was concluded that ESI MS was the preferred technique. Related heterometal clusters $[Fe_6C(CO)_{16}(AuPPh_3)]^-$ [25] and $[Fe_4C(CO)_{12}HgM]^-$ [M = $CpMo(CO)_3$, $Co(CO)_4$, $CpFe(CO)_2$, $CpW(CO)_3$ and $Mn(CO)_5$][26] also gave the parent anions as the base peak in the negative-ion ESI mass spectra.

Clusters that can become ionic as a result of ionisable ancillary ligands (e.g. carboxylic or surfonic acids) require no additional derivatisation step and typically give good ESI MS spectra. An example of such a ligand is the sulfonated triphenylphosphine ligands **7.1**. Clusters containing this ligand have been analysed by ESI MS; $[Ru_3(CO)_{12-n}L_n]^{3n-}$ (n = 1 to 3), $[H_4Ru_4(CO)_{11}L]^{3-}$ and $[Ru_6C(CO)_{16}L]^{3-}$ where L is $P(m\text{-}C_6H_4SO_3)_3]^{3-}$ give aggregate peaks formed by association with the Na^+ counterions, due to their high negative charge.[27] For example, $Na_6[Ru_3(CO)_{10}L_2]$ gave ions $[Ru_3(CO)_{10}L_2Na]^{5-}$, $[Ru_3(CO)_{10}L_2Na_2]^{4-}$, $[Ru_3(CO)_{10}L_2Na_3]^{3-}$ and $[Ru_3(CO)_{10}L_2Na_4]^{2-}$. In principle, many other clusters bearing this type of ligand should be able to be directly analysed using ESI MS.

7.1

One 'unsuccessful' application of ESI MS in metal cluster chemistry is known, when attempts were made to characterise the high nuclearity copper-nickel cluster

$[Cu_xNi_{35-x}(CO)_{40}]^{5-}$, where x is either three or five. Despite the use of a high resolution FTICR instrument, it was not possible to distinguish the two possible stoichiometries, due to the similar masses of copper and nickel, and the minimal effect on the overall isotope distribution envelope on metal atom substitution.[28]

7.2.3   Neutral Metal Carbonyl Compounds

Neutral metal carbonyl compounds that bear an ancillary ligand that is protonatable usually present few difficulties in ESI mass spectrometric analysis. As an example, compounds such as $[Fe(CO)_4(\eta^4\text{-diene})]$ and $[Cr(CO)_3(\eta^6\text{-arene})]$ where the organic ligand has carbonyl (>C=O), amine ($-NH_2$) or imine (>C=N) groups give $[M+H]^+$ ions in their ESI mass spectra.[29] A range of clusters containing the $\{CCo_3(CO)_9\}$ core functionalised with carbohydrate groups also give $[M+NH_4]^+$ or $[M+K]^+$ ions in positive-ion ESI MS.[30]

However, simple ESI MS applied to neutral binary metal carbonyl species generally yields few or no ions. This is because the carbonyl ligand cannot easily be protonated in solution, despite the coordinated CO ligand formally containing a lone pair of electrons on the oxygen atom. In the solid-state however, interactions of coordinated CO ligands with electrophiles and hydrogen-bond donors have been observed[31] but generally these are not strong enough to permit ionisation by the electrospray method. As a result, a derivatisation step must usually be carried out on this class of compound prior to analysis, although some metal carbonyl compounds do give ions without prior derivatisation. This can also be extended to derivatives where a CO ligand has been replaced with a non-ionisable ligand, such as triphenylphosphine (PPh_3) in $[Fe(CO)_4(PPh_3)]$.

Various derivatisation methods can be applied to the analysis of neutral metal carbonyl clusters; these are detailed in the subsequent sections.

Alkoxide Derivatisation

The most widely applicable method for metal carbonyl derivatisation reported to date involves reaction with alkoxide ions.[16] The alkoxide ion undergoes nucleophilic attack at a coordinated carbonyl ligand:

$$M-C{\equiv}O \longrightarrow \left[M-C\underset{OR}{\overset{O}{\diagup}} \right]^-$$
$$\underset{OR^-}{}$$

The resulting negative ions, formed in the majority of cases by addition of the OR^- ion, are then readily analysed in negative-ion mode. The derivatisation is easily achieved experimentally by dissolving the compound in the alcohol solvent (in which most compounds have some solubility), and adding a small amount of the corresponding sodium alkoxide.* As a derivatisation technique, reaction with alkoxide ions has the advantage of being rapid, and favours the formation of charged products. The technique has been found to be successful for a wide range of metal carbonyl compounds, from

---

*Dilute solutions of sodium alkoxide are easily prepared by addition of a small piece of sodium metal to the dry alcohol. For best results, the alkoxide solution should be freshly prepared.

Table 7.1 A selection of metal carbonyl compounds which give clean $[M + OMe]^-$ ions in their negative-ion ES mass spectra, after derivatisation with NaOMe in methanol

Mononuclear compounds

$[Cr(CO)_6]$	$[Mo(CO)_6]$	$[W(CO)_6]$	$[Fe(CO)_5]$
$[Mo(CO)_3(Cp)(C_4Ph)]$			

Dinuclear compounds

$[Mn_2(CO)_{10}]$	$[Re_2(CO)_{10}]$	
$[Fe(CO)_2Cp]_2$	$[Ru(CO)_2Cp]_2$	$[Mo(CO)_3Cp]_2$

Trinuclear compounds

$[Ru_3(CO)_{12}]$	$[Os_3(CO)_{12}]$	$[Fe_3(CO)_{11}(CN^tBu)]$
$[Ru_3(CO)_{11}(PPh_3)]$	$[Ru_3H(C_2^tBu)(CO)_9]$	

Tetranuclear and higher nuclearity clusters

$[Rh_4(CO)_{12}]$	$[Ir_4(CO)_{12}]$	$[Ru_4H_4(CO)_{12}]$
$[NiRu_3H_3(CO)_9Cp]$	$[Ru_5(CO)_9(C_6H_5Me)]$	
$[Ru_6C(CO)_{17}]$	$[Ru_6C(CO)_{14}(C_6H_5Me)]$	
$[Rh_6(CO)_{16}]$		

mononuclear complexes such as $[M(CO)_6]$ (M = Cr, Mo, W) to higher nuclearity clusters [e.g. $M_3(CO)_{12}$ (M = Ru, Os), $Rh_4(CO)_{12}$, $Rh_6(CO)_{16}$] to clusters with a range of other ancillary ligands; a selection of representative species that give alkoxide-derivatised ions are listed in Table 7.1.[32,33] The technique has also been used in the analysis of a mixture of high nuclearity osmium clusters which demonstrates their colloid-like properties.[34]

Although most work has been done using methanol, in principle, it is possible to vary the alcohol (and hence the alkoxide ion), for example in cases of peak overlap, or for species confirmation. A complication of this technique may arise if metal-hydride groups are present, since these may undergo competing deprotonation, giving $[M - H]^-$ ions.

Figure 7.1 shows an example of a metal carbonyl $[Mn_2(CO)_{10}]$ analysed by the alkoxide addition method. The methoxide-adduct ion, $[Mn_2(CO)_{10} + OMe]^-$ is the base peak at *m/z* 421 and the ion $[Mn(CO)_5]^-$ (*m/z* 195) is formed as a fragment ion in this case.

Azide Derivatisation

In the same way that the derivatisation of a metal carbonyl compound with alkoxide ions involves nucleophilic attack at a coordinated CO ligand, the analogous reaction with azide ions, N_3^-,[35] provides an alternative ionisation technique for metal carbonyl compounds. The resulting ions are formed by elimination of dinitrogen from the intermediate addition product:

$$[L_nM\text{-}CO] + N_3^- \longrightarrow [L_nM\text{-}NCO]^- + N_2$$

Using this method, a wide range of metal carbonyls has been investigated, and most yield the isocyanate species $[L_nM\text{-}NCO]^-$ by the above process. Evidence for an

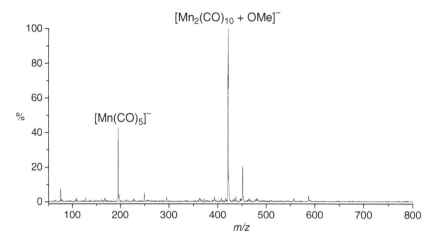

Figure 7.1
Negative-ion ESI mass spectrum of $Mn_2(CO)_{10}$ in methanol solution with added NaOMe, showing the formation of the $[Mn_2(CO)_{10} + OMe]^-$ adduct ion

intermediate species in the azide addition reaction has also come from these ESI MS studies.[36]

There may be some ambiguity associated with ion identification using this method, since the masses of CO and N_2 are identical (under low resolution conditions), but studies have shown that the N_2 loss is much more facile. The reaction with azide also appears to be less specific than the alkoxide method, since other ions which can be formed are $[M + N_3]^-$, $[M - CO + NCO]^-$, $[M - H]^-$, and in one case, $[M - CO + CN]^-$. As an example of deprotonation, the azide ion can abstract a proton giving from the hydride-containing cluster $[Ru_4H_4(CO)_{12}]$ giving $[M - H]^-$ in competition with formation of isocyanate product ions. Hence the azide addition method appears to be less useful than the alkoxide method.

Metal Ion Adduction

In a limited number of cases, metal carbonyl species can be analysed by ESI mass spectrometry by adduction of metal ions, either adventitious or introduced, giving positive ions. Metal carbonyl complexes that are electron–rich (by substitution of CO ligands with phosphine or related ligands) can give $[M + Na]^+$ ions, for example when treated with NaOMe to provide $[M + OMe]^-$ ions. Thus, $[Ru_3(CO)_9(PPh_3)_3]$ gives a $[M + Na]^+$ ion, but the corresponding unsubstituted cluster $[Ru_3(CO)_{12}]$ is less electron-rich, and does not.[16] It is probable that the sodium ion interacts with the carbonyl oxygen (which will be more electron rich in a more highly substituted cluster), giving the arrangement $L_nM\text{-}CO\cdots Na^+$.

As another example, a number of rhenium(I) carbonyl complexes which contain coordinated 2,2′-bipyridine-type ligands (R_2L) of the type $[ReX(CO)_3(R_2L)]$ (X = CN, Cl, Br, HCO_2; R = H, CH_3, CF_3) were analysed in methanol solution, but yielded no ions of the expected type $[Re(CO)_3(R_2L)]^+$ formed by halide loss (Section 4.3). However, addition of sodium nitrate resulted in detection of $[ReX(CO)_3(R_2L) + Na]^+$ ions.[37]

The nitrosyl (NO) ligand is closely related to the carbonyl ligand, and although complexes are not strictly classified as organometallic, they might be expected to show similar characteristics in mass spectrometric analysis. Some ruthenium-nitrosyl complexes of the type $[Ru(NO)Cl_3(EPh_3)_2]$ (E = phosphorus, arsenic, antimony) have been found to give $[M + cation]^+$ ions when analysed by ESI mass spectrometry with added alkali metal cations.[38] It is possible that the cations interact with the oxygen atom of the nitrosyl ligand in this case, although interaction with the halide ligands is also possible.

Addition of silver ions can also provide ionisation of certain neutral metal carbonyls, though the method is less widely applicable than the other methods described above. The derivatisation is easily carried out by addition of a dilute solution of a soluble silver salt (e.g. $AgNO_3$, $AgBF_4$, $AgClO_4$) in the solvent (methanol and acetonitrile have been used) to the metal carbonyl solution. The resulting ions are of the type $[M + Ag(solvent)_x]^+$ (x = 0 or 1, depending on the degree of fragmentation employed). Metal carbonyls that have been successfully analysed by this ionisation technique using ES ionisation are $[Re_2(CO)_{10}]$, $[Ru_3(CO)_{12}]$, $[Os_3(CO)_{12}]$ and $[SiFe_4(CO)_{16}]$.[16,39] This ionisation technique probably occurs by attachment of the Ag^+ ion to a metal-metal bond, rather than attachment in an isocarbonyl-type mode $L_nM-CO\cdots M^+$, as for the other metal derivatives.

Hydride Addition

There is one example in the literature involving addition of hydride ion to a metal carbonyl cluster, in order to provide a negatively-charged derivative suitable for ESI MS analysis. Reaction of the neutral di- and triclusters $Me[PFe_3(CO)_9P]_nMe$ (n = two or three), containing P-P bridges, with one equivalent of $NaBH_4$ resulted in hydride addition to the terminal μ_3-PMe ligand to produce a μ_2-PHMe group. The resulting hydride addition products were stable in solution.[40] The addition of hydride ions to clusters is potentially a more widely utilisable chemical ionisation technique for neutral clusters.

7.2.4 Oxidation and Reduction Processes Involving Metal Carbonyls

The binary carbonyl $[Fe_3(CO)_{12}]$ gives a negative ion $[Fe_3(CO)_{12}]^-$ with no prior derivatisation, in addition to CO-loss fragment ions, as shown in Figure 7.2.[41] The ES ionisation source can act as an electrochemical cell and reduction of $[Fe_3(CO)_{12}]$ could occur in this manner, or alternatively by the reaction with impurities which are known to effect such reduction reactions.[42]

The neutral rhenium(I) carbonyl complexes $[CpRe(CO)_3]$ and $[(Ind)Re(CO)_3]$ (Ind = η^5-indenyl) give the ion $[Re(CO)_3(MeCN)_3]^+$ in acetonitrile solution, even though the parent species are stable in bulk acetonitrile solution. It was suggested that labile seventeen-electron cations were formed, which then undergo ligand substitution reactions with loss of the Cp or Ind ligand. The related alkyne complexes $[LRe(CO)_2(butyne)]$ (L = Cp or Ind) however remain more intact, and give the ions $[LRe(CO)(MeCN)(butyne)]^+$.[43]

7.2.5 Characterisation of Reaction Mixtures Involving Metal Carbonyl Clusters

The methods described above for the ionisation of metal carbonyl compounds have typically been developed by analysis of (generally pure) single compounds. In many

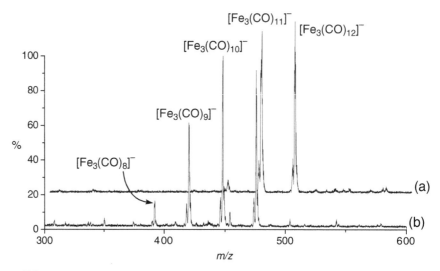

Figure 7.2
(a) Negative-ion ESI mass spectrum of $Fe_3(CO)_{12}$ in methanol solution at a cone voltage of 10 V, showing the formation of $[Fe_3(CO)_{12}]^-$ and the fragment ion $[Fe_3(CO)_{11}]^-$. At a higher cone voltage (20 V, b), further loss of CO ligands occurs

cases, mixtures of products are commonly formed in reactions involving metal carbonyl compounds. This is particularly true in metal carbonyl cluster chemistry, where the use of thermolysis methods for higher nuclearity carbonyl cluster synthesis often results in metal core and ligand transformations. The classical approach has generally been to separate the resulting mixture by chromatography and to characterise each component individually. This relatively laborious approach may lead to minor and/or unstable product components being missed and the methodology is not very conducive to quantitation or to screening of reaction mixtures. Charged clusters are also often not amenable to chromatographic recovery, so undoubtedly products have been previously overlooked. Additionally, characterisation by X-ray crystallography is typically carried out on a single crystal, which might not be representative of the bulk sample. The application of mass spectrometry to the direct analysis of reaction mixtures has obvious advantages over traditional techniques that can be used for the same purpose, such as infrared spectroscopy, since the latter can be plagued by superposition of bands from different species, which becomes particularly problematical for large clusters, where differences occur only for weak bands.[44] Charged clusters can be analysed directly, whereas mixtures of neutral clusters may need to be derivatised prior to analysis, for example by the methoxide addition method. Care would need to be taken with reaction mixtures that contained both charged and neutral clusters, since the high ionisation efficiency of charged clusters would tend to dominate the spectra.

As an example of this approach, the pyrolysis of $[Os_3(CO)_{12}]$ has been often used in the synthesis of higher nuclearity osmium carbonyl clusters. Pyrolysis of $[Os_3(CO)_{12}]$ at 210 °C gives a mixture of higher nuclearity clusters containing between four and seven osmium atoms, which can be directely analysed by extraction with ethyl acetate and derivatisation with NaOMe.[6]

The analysis of reaction mixtures which produce charged products is particularly easily accomplished using ESI MS. As an example, the reaction of $[\mu_4\text{-Ge}\{Co_2(CO)_7\}_2]$ with $[Co(CO)_4]^-$ monitored by ESI MS identified a series of known germanium-cobalt clusters, together with the new species $[Ge_2Co_{10}(CO)_{24}]^{2-}$, which was subsequently isolated and characterised by an X-ray diffraction study.[45] The reactions between $[Ru_6C(CO)_{17}]$ and $[M(CO)_4]^-$ ions (M = cobalt, iridium) have similarly been monitored using the energy-dependent electrospray ionisation (EDESI) technique (Section 1.3.2).[46] In this work, the anions $[Ru_5CoC(CO)_{16}]^-$, $[Ru_3Co(CO)_{13}]^-$, $[RuCo_3(CO)_{12}]^-$, $[HRu_4Co_2C(CO)_{15}]^-$, $[Ru_5IrC(CO)_{16}]^-$, $[Ru_3Ir(CO)_{13}]^-$ and $[RuIr_3(CO)_{12}]^-$ were identified from the reaction mixtures. A cluster synthesis reaction has also been studied;[47] $RhCl_3$ and Na_2PtCl_6 supported on MgO were reduced by CO to give (after extraction from the support) the product cluster ion $[PtRh_5(CO)_{15}]^-$ and fragment ions $[PtRh_5(CO)_{14}]^-$ and $[PtRh_5(CO)_{13}]^-$.

High nuclearity metal carbonyl clusters are a class of compound that are starting to benefit from application of modern mass spectrometric methods. These clusters contain many metal atoms (and CO ligands) and it is recognised that large metal clusters and colloidal metals have many similar characteristics, there being a size continuum between the two formal classifications. The soft ionisation provided by electrospray has been used to determine the exact composition of samples of the cluster $[Os_{20}(CO)_{40}]^{2-}$.[34] A sample of this material separated by chromatography showed primarily $[Os_{20}(CO)_{40}]^{2-}$, together with lesser amounts of the trianion $[Os_{20}(CO)_{40}]^{3-}$ and the nineteen-metal cluster $[Os_{19}(CO)_{39}]^{2-}$. Imprecision in the number of metal atoms in a particle is the signature of a colloidal material, and thus the ESI MS results clearly suggest that the Os_{20} cluster system is at the early stages of showing characteristics of a colloidal system.

7.2.6 Fragmentation of Transition Metal Carbonyl Clusters; Electrospray as a Source of Bare Metal Clusters

When a metal carbonyl cluster anion is subjected to CID, for example by use of an applied cone voltage, loss of CO ligands typically occurs. As an example, ESI MS of $[CoRu_3(CO)_{13}]^-$ generates the parent $[M]^-$ ion, and the CO ligands can be sequentially stripped away by CID, giving the series of ions $[CoRu_3(CO)_n]^-$ (n = 0 to 13).[14,48] Carbide clusters, containing a strongly-bound (typically interstitial) carbide carbon, behave similarly. For example, $[CoRu_5C(CO)_{16}]^-$ undergoes sequential loss of all CO ligands down to the bare $[CoRu_5C]^-$ core.[49] These processes are best visualised using the EDESI technique (Section 1.3.2), which visually presents data from spectra acquired under a range of CID conditions.[50] The behaviour of a multiply-charged anionic cluster, such as $[PtRu_5C(CO)_{15}]^{2-}$ is a little different; initial CO loss gives the series of ions $[PtRu_5C(CO)_n]^{2-}$ (n = 8 to 15), which is then followed by the loss of an electron (*electron autodetachment*) giving $[PtRu_5C(CO)_n]^-$ (n = 0 to 9). This can be seen in an EDESI MS map, where the monoanionic fragments have a higher *m/z* than the dianionic clusters.[17]

The loss of CO ligands from a parent carbonyl cluster anion by CID thus affords a method for obtaining a wide range of bare metal cluster species in the gas phase. Traditionally, gas phase metal clusters have been prepared by vaporising a metal target (e.g. using a laser), condensing the metal atoms into clusters, which are then investigated using mass spectrometry. The metal carbonyl cluster could be either homo- or polymetallic, or even a cluster mass-selected from reaction mixtures that generate mixtures of clusters. The gas-phase reactions of these bare-metal clusters can then be investigated using MS techniques. For example, reactions of gas-phase metal clusters with small molecules have been extensively studied as models for catalytic reactions on metal surfaces. By using an ESI source coupled with an FTICR MS, individual cluster ions were selected and their gas phase reactions with small molecules such as methane and hydrogen investigated.[15] This method offers much potential for the simple generation of gas-phase metal clusters from simple precursors, for investigation of their gas-phase chemistry with a wide range of molecular substrates.

Using the EDESI MS technique (Section 1.3.2) it has been possible to observe the different fragmentation pathways of the clusters $[Ru_6C(CO)_{16}(COOMe)]^-$ and $[Rh_6(CO)_{15}(COOMe)]^-$, generated from the parent, neutral clusters by nucleophilic addiction of methoxide ions.[51] Both clusters undergo loss of formaldehyde, though at different stages. $[Ru_6C(CO)_{16}(COOMe)]^-$ undergoes loss of two CO ligands, followed then by loss of HCHO giving the hydride cluster $[HRu_6C(CO)_{15}]^-$, but in contrast, $[Rh_6(CO)_{15}(COOMe)]^-$ only loses HCHO after loss of seven CO ligands. These observations were correlated with solution properties, since $[Ru_6C(CO)_{17}]$ undergoes ready reduction with methanolic potassium hydroxide giving $[Ru_6C(CO)_{16}]^{2-}$, while $[Rh_6(CO)_{16}]$ requires more forcing conditions (e.g. sodium amalgam).

7.2.7   The Use of 'Electrospray-Friendly' Ligands in Organometallic Chemistry

Triphenylphosphine derivatives of metal carbonyls, such as $Fe(CO)_4(PPh_3)$ and $Mo(CO)_4(PPh_3)_2$ do not give any ions in their positive-ion ESI mass spectra. However, by using a phosphine analogue which contains protonatable groups–*i.e.* an **electrospray-friendly ligand** such as $P(C_6H_4OMe)_3$–this problem can be overcome. The concept of electrospray friendly ligands is introduced in Section 6.3.1.[52] A range of metal carbonyl derivatives of these phosphines gave good $[M + H]^+$ ions in their ESI spectra. While a single OMe group was found to be sufficient to give ions, the ion intensity increased with increasing numbers of OMe groups, or when the more basic NMe_2 substituent was employed. Other electrospray derivatives can also be used, including the tertiary arsine $As(C_6H_4OMe)_3$, stibine $Sb(C_6H_4OMe)_3$, and thiolate $MeOC_6H_4S^-$.

The use of electrospray-friendly ligands also offers some possibilities in the direct analysis of reaction mixtures involving neutral metal clusters. In particular, they should find application where a chemical derivatisation step, such as with alkoxide ions, is not possible. As an example of the deliberate incorporation of basic groups into a cluster to facilitate mass spectrometric reaction monitoring, the reactions of $[Co_4(\mu_4-SiR)_2(CO)_{11}]$ (R = H, OMe, NMe_2) clusters with isocyanides has been monitored by ESI MS. Evidence for substitution of up to nine CO ligands by isocyanide was observed, with the substituted products giving $[M + H]^+$ ions when R = NMe_2, but oxidised $[M]^+$ ions when R = H or OMe.[53]

This general approach can, in principle, be extended to many other classes of compound, by incorporation of suitable groups that are ionic, or can be charged by protonation, deprotonation, oxidation, or reduction.

Metal Isocyanide Complexes

The isocyanide ligand, RNC, is isoelectronic to the CO ligand, and shows many similarities when coordinated to metal centres.[54] Due to the presence of the substituent

R, which is typically an alkyl or aryl group, the steric and electronic properties of RNC ligands can be systematically tuned in a way that CO cannot.

A range of isocyanide derivatives of $[Fe_3(CO)_{12}]$ and $[Ru_3(CO)_{12}]$, formed by replacement of one or more CO ligands by RNC ligands all give $[M + H]^+$ ions in positive-ion ESI MS, even when the isocyanide was PhNC.[55] The ionisation efficiency increases with increasing isocyanide substitution (with spectra from the monosubstituted products only being obtainable from concentrated solutions of the purified substances), suggesting that the protonation occurs on either a M-M bond or on a bridging carbonyl ligand. ESI MS also gave evidence for previously-unknown highly substituted products such as $[Fe_3(CO)_6(CNPh)_6]$.

Metal Cyclopentadienyl and Related Complexes

The cyclopentadienyl (C_5H_5, Cp) ligand occurs in many organometallic complexes, in which it typically binds as an η^5 ligand; the metal is typically bonded equidistantly to all five carbon atoms, as in **7.2**, though other binding modes are known. The coordinated Cp ligand cannot easily be protonated (with the exception of the decaphenylferrocene system) and so the analysis of metal-Cp complexes is dependent on other ionisation pathways. Compounds are classified into metallocenes, and other compounds containing a coordinated Cp ligand.

7.2

7.4.1   Ferrocene-Based Compounds

The classical metallocene is **ferrocene**, $Fe(\eta^5\text{-}C_5H_5)_2$ (Cp_2Fe, FcH) and analogues formed by many other transition metals are also well known. Ferrocene can be protonated but only by strongly acidic media; no $[M + H]^+$ ion has been observed in mass spectrometric studies of ferrocene itself. Ferrocene does undergo a fully reversible, one-electron oxidation-reduction process to the ferrocenium radical cation $[Cp_2Fe]^{\bullet+}$; thus ferrocene is widely used as a reference material in electrochemical studies. Because of the relative ease of oxidation of ferrocene and the ability of the metal capillary of an electrospray ion source to act as an electrochemical cell (Section 3.9.1), ferrocene gives the Cp_2Fe^+ cation in ESI analysis; phenylferrocene behaves the same way.[56,57,58] The high-resolution positive-ion ESI MS spectrum of ferrocene in methanol solution is shown in Figure 7.3. Alternatively, the ferrocene can be preoxidised prior to analysis, for example with the chemical oxidant $NO^+BF_4^-$.[59] The advantage of direct chemical oxidation is that all ferrocene molecules can be converted into ionic ferrocenium cations, thus improving signal intensity for weak samples.

A study of biferrocenes produced $[M]^{2+}$ and $[M]^+$ ions by oxidation in the electrospray capillary; the intensity ratio of these ions was found to be dependent on the redox potentials of the ferrocene units, the distance between them, the concentration, and the flow rate of the analyte solution. For octa(ferrocenyldimethylsilane), the $[M]^{4+}$ cation was observed by the electrolytic oxidation process.[60]

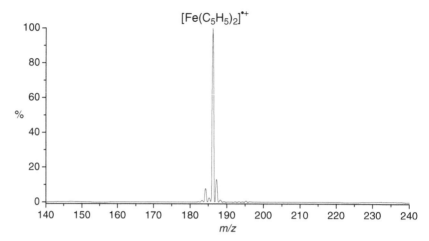

Figure 7.3
High-resolution positive-ion ESI mass spectrum of ferrocene (Cp$_2$Fe) dissolved in methanol at a cone voltage of 20 V, showing the oxidised [Cp$_2$Fe]$^+$ ion

ESI MS of the blue form of decaphenylferrocene, which is the isomer [Fe(η^6-C$_6$H$_5$-C$_5$Ph$_4$)(η^5-C$_5$Ph$_5$)], in methanol-water solution gave the protonated ion [Fe(η^6-C$_6$H$_5$-C$_5$HPh$_4$)(η^5-C$_5$Ph$_5$)]$^+$, known from other techniques to be protonated at the ligand. When a basic solution was used, a mixture of protonated and oxidised [Fe(η^6-C$_6$H$_5$-C$_5$Ph$_4$)(η^5-C$_5$Ph$_5$)]$^+$ ions were observed, the latter formed by electrochemical oxidation in the capillary. Oxidation of [Fe(η^6-C$_6$H$_5$-C$_5$Ph$_4$)(η^5-C$_5$Ph$_5$)] with NO$^+$BF$_4^-$ gave a solution which contained only protonated [Fe(η^6-C$_6$H$_5$-C$_5$HPh$_4$)(η^5-C$_5$Ph$_5$)]$^+$ rather than the oxidised species; the protonated species was independently analysed by addition of triflic acid (CF$_3$SO$_3$H) to the neutral [Fe(η^6-C$_6$H$_5$-C$_5$Ph$_4$)(η^5-C$_5$Ph$_5$)]. Analysis of the isomeric species [HFe(η^5-C$_5$Ph$_2$)$_2$]$^+$, which contains a protonated iron atom, gave the expected [M]$^+$ ion in ESI MS.[61]

Other substituted ferrocenes, FcR, behave according to the nature of the functional group R. Ionic ferrocene-based compounds give the expected parent ions, such as the ferrocene-thiolate anion [FcCO$_2$(CH$_2$)$_{16}$S]$^-$,[62] and the ferrocene-ammonium cation [FcCH$_2$NMe$_3$]$^+$.[63] Host-guest complexes between a range of related ferrocene-ammonium salts and cyclodextrins have been observed.[64] Ferrocene-derived phosphonic acids Fc(CH$_2$)$_n$PO$_3$H$_2$ (n = 0, 1, 2) give the expected [M − H]$^-$ ions in negative-ion ES analysis.[65] However, the corresponding arsonic acid FcCH$_2$CH$_2$AsO$_3$H$_2$ and phenylarsonic acid (PhAsO$_3$H$_2$) undergo facile alkylation in alcohol (ROH) solvent, giving [FcCH$_2$CH$_2$As(O)(OR)O]$^-$ and [PhAs(O)(OR)O]$^-$ ions respectively (R = Me, Et), Section 6.3.2. [M − H]$^-$ and [2M − H]$^-$ ions were observed for 1,1'-ferrocenedicarboxylic acid Fe(η^5-C$_5$H$_4$CO$_2$H)$_2$ in negative-ion mode.[66] In positive-ion mode, the compound was found to exist as a cyclic tetramer in methanol or acetonitrile solutions, complexed more strongly with an Na$^+$ ion than any other alkali metal ion.

For neutrally-charged, substituted ferrocenes, oxidation of the ferrocene and protonation of the substituent can occur competitively, with varying amounts of [M]$^+$ and [M + H]$^+$ ions observed. The detection of both of these ions requires close monitoring of the high-resolution spectrum (isotope pattern), since the two ions differ by only 1 *m/z* and the presence of one of the ions may easily be missed in a low-resolution spectrum. The relative intensities of ions formed by protonation versus oxidation is dependent on the redox potential of the metallocene (a more electron-rich metallocene will be oxidised more readily) and on the basicity of the protonatable substituent. The solvent flow rate can also influence the relative amounts of protonated and oxidised ions formed, since a greater residence time in the metal capillary will tend to increase the amount of oxidation. This is illustrated by acetyl ferrocene, $FcC(O)CH_3$, which gives a mixture of oxidised $[FcC(O)CH_3]^+$ and protonated $[FcC(O)CH_3 + H]^+$ ions at m/z 228 and 229 respectively, when analysed in methanol solution. Figure 7.4 shows the isotope pattern in the m/z

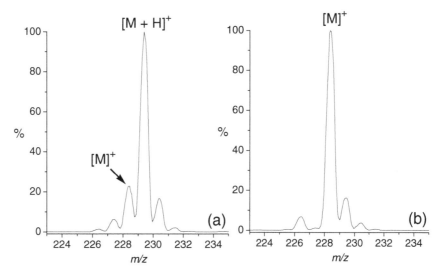

Figure 7.4
Positive-ion ESI mass spectra of acetyl ferrocene, $FcC(O)CH_3$ in methanol solution at a cone voltage of 20 V: (a) initial spectrum, showing a mixture of oxidised (minor, m/z 228) and protonated (major, m/z 229) ions and (b) spectrum obtained after addition of pyridine base, showing exclusive formation of oxidised ions

region 223–235. The left spectrum (a) is the original spectrum, containing the protonated species (m/z 229) as the major contributor, and the oxidised ion (m/z 228) as a minor contributor. Upon addition of pyridine base, spectrum (b), the availability of protons is drastically reduced, and the ESI mass spectrum shows essentially solely the oxidised species at m/z 228. This indicates that the pH of the solution (and hence the solvent quality) can bear a major influence on the ionisation pathways for certain ferrocene compounds. Other neutral ferrocene compounds behave according to the basicity of the substituent. As expected, the highly basic ferrocenyl amine $FcCH_2NMe_2$ gives only $[M + H]^+$, while the lower basicity phosphonate $FcCH_2P(O)(OPh)_2$ gave both oxidised and protonated ions.[63]

The behaviour of substituted ferrocenes $FcCH_2R$ at elevated cone voltages is also of interest. Because of resonance, the $[FcCH_2]^+$ cation is stabilised and is therefore a common fragment ion for compounds of this type. Thus, $[FcCH_2NMe_3]^+$ readily yields $[FcCH_2]^+$ at higher cone voltages as a result of the neutral NMe_3 being a good leaving

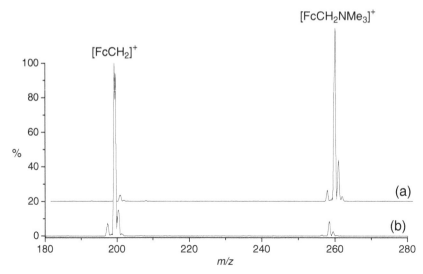

Figure 7.5
Positive-ion ESI mass spectra of $[FcCH_2NMe_3]^+ I^-$ at cone voltages of (a) 10 V, showing the parent $[FcCH_2NMe_3]^+$ cation, and (b) 30 V, showing the facile loss of NMe_3, giving the stabilised carbocation $[FcCH_2]^+$ (*m/z* 199)

group. Figure 7.5 shows the positive-ion ESI mass spectra of $FcCH_2NMe_3^+ I^-$ at cone voltages of 10 and 30 V. At 10 V, the $[FcCH_2NMe_3]^+$ parent ion dominates the spectrum, but $[FcCH_2]^+$ is the base peak at 30 V, reflecting the ease of loss of NMe_3.[63]

FcCH$_2$OH and other derivatives form the $[FcCH_2]^+$ cation less readily, reflecting the ease with which the different substituents can act as leaving groups on protonation. For $FcCH_2P(O)(OPh)_2$ the unsubstituted C_5H_5 ring was selectively lost at elevated cone voltages, and the resulting cation is probably stabilised by an $Fe\cdots O{=}P$ interaction, as in **7.3**. Likewise, at elevated cone voltages, $FcPO_3H_2$ fragments to $[C_5H_4PO_2H]^-$ whereas $FcCH_2PO_3H_2$ and $FcCH_2CH_2PO_3H_2$ fragment to $[Fe\{C_5H_4(CH_2)_nPO_3\}]^-$ ions, presumably due to an intramolecular interaction between the iron and the phosphonate group. In contrast, the arsonic acids $FcCH_2CH_2AsO_3H_2$ and $PhAsO_3H_2$ (which alkylate in the alcohol solvent used) undergo arsenic-carbon bond cleavage.[65]

7.3

Metallocene-based phosphines, typified by 1,1′-bis(diphenylphosphino)ferrocene (dppf) **7.4** (and to a far lesser extent the ruthenocene analogue **7.5**) are important ligands, widely used in coordination chemistry.[67] A range of ferrocene-based phosphines have been analysed by ESI MS, including application of ionisation reagents such as Ag^+ ions and the use of a formaldehyde-HCl mixture for ESI MS analysis of the ligands themselves.[68] Various metal complexes containing dppf ligands have been characterised by ESI MS.[69,70] A number of dppr complexes of ruthenium, for example

$[CpRu(dppr)(MeCN)]^+$, $[CpRu(dppr)(CO)]^+$ and $[CpRu(dppr)(C=CHPh)]^+$ have also been investigated by ESI MS, and give the parent cations.[71] In these dppf and dppr examples, the metallocene-based ligand is a spectator and the complexes are either charged or readily protonatable on other functional groups.

7.4 M = Fe; dppf
7.5 M = Ru, dppr

Other ferrocene-containing systems that have been studied by ESI MS are ferrocene-containing calixarene ligands and their lanthanide complexes.[72]

7.4.2 Use of Ferrocene Derivatives as Electroactive Derivatisation Agents for Electrospray Ionisation

As a result of their facile electrochemical (or chemical) ionisation to the charged ferrocenium ion, several ferrocene-based reagents have been developed as ionisation aids for mass spectrometric analysis of organic substrates.

Ferrocenoyl azide, $FcC(O)N_3$[†] has been applied to the derivatisation of alcohols, sterols and phenols for ESI MS analysis.[73] The reaction proceeds by formation of a ferrocenyl carbamate, *via* FcNCO as a reactive intermediate species, for reaction with an alcohol ROH.

$$FcC(O)N_3 \longrightarrow FcNCO \xrightarrow{ROH} FcNHC(O)OR$$

Fragmentation patterns for 27 such derivatives of saturated primary, secondary and tertiary alcohols using tandem MS, allow some distinction of structural isomers, especially primary from tertiary alcohols.[74] The screening of alcohols and phenols in natural product mixtures such as oil of cloves, lemon oil and oil of peppermint has been carried out by derivatisation with $FcC(O)N_3$.[75] While a total analysis of volatile components was still considered to be best carried out by the traditional GC MS technique, the modified ESI method was considered complementary as it does not require chromatographic separation, and has improved speed and selectivity for detection of a targeted alcohol.

Boronic acids, $RB(OH)_2$ have a strong affinity for 1,2-diols and carbohydrates, by the formation of boronate esters, $RB(OR)_2$, hence by incorporation of a ferrocene moiety into a boronic acid, an electrochemically-active derivatisation agent is obtained. Thus, ferrocene-boronic acid $FcB(OH)_2$ has been used in the analysis of various carbohydrates. Using this reagent it is possible to distinguish between the diastereomers of several neutral mono- and di-saccharides.[76] The estrogens 2- and 4-hydroxyestradiol, after derivatisation with $FcB(OH)_2$, can also be distinguished by CID studies,[77] and the diol, pinacol[73] has also been derivatised for analysis using $FcB(OH)_2$. A method for the

---

[†]Ferrocenoyl azide is easily synthesised by reaction of commercially-available $FcCO_2H$ with PCl_5, giving FcCOCl, followed by reaction with NaN_3.

analysis of alkenes (which cannot easily be ionised in ESI MS) involves conversion to a 1,2-diol (using $KMnO_4$ or OsO_4) followed by conversion to the boronate ester using $FcB(OH)_2$. This ester can then be analysed by electrochemical ionisation in ESI MS, and the methodology has been demonstrated using 2,3-dimethylbutene, $Me_2C=CMe_2$.[78]

Finally, the thiol derivative $Fc(CH_2)_6SH$ undergoes a facile addition reaction with the α,β-unsaturated carbonyl group of the hepatotoxin derivative microcystin-LR, allowing analysis by ESI MS.[79]

7.4.3 Other Metallocene Systems

In contrast to the considerable body of work concerning ferrocene derivatives, little has been reported with other metallocene systems. For ruthenocene (Cp_2Ru) and osmocene (Cp_2Os), ESMS analysis gave in each case the $[Cp_2M]^+$ cation as a result of capillary-based electrochemical oxidation, but for osmocene there was competing protonation giving $[Cp_2OsH]^+$.[57] Osmocene and ruthenocene are more basic than ferrocene, so that metal-based protonation is more facile for the heavier metallocenes. In dichloromethane solvent, chloride-containing metal(IV) complexes $[Cp_2RuCl]^+$ and $[Cp_2OsCl]^+$ were observed. The related substituted osmocene cation $[(C_5Me_5)_2OsH]^+$ has also been characterised by ESI MS as the salt with the $[Os_2Br_8]^{2-}$ anion, formed by reaction of H_2OsBr_6 with C_5Me_5H and ethanol.[80]

ESI MS has been used to characterise a series of titanium(IV) β-diketonate complexes $[Cp_2Ti(\beta\text{-diketonate})]^+$ which gave the expected $[M]^+$ cations.[81] Several studies have concerned zirconium-cyclopentadienyl compounds. The three complexes Cp_2ZrMe_2, $(MeCp)_2ZrMe_2$ ($MeCp = \eta^5\text{-}C_5H_4Me$) and $(MeCp)_2Zr(BH_4)_2$ have been studied by EI and ESI MS techniques.[82] The fragmentation of the resulting ions was investigated in detail and occurs primarily through loss of the Me or BH_4 groups. Ziegler-Natta like alkene polymerisation activity has been demonstrated using ESI MS with a zirconium metallocene complex. The zirconocene cation $[Cp_2ZrMe]^+$ [as its $B(C_6F_5)_4^-$ salt] has been analysed by ESI MS in acetonitrile solution, giving the $[Cp_2ZrMe(MeCN)]^+$ cation. When the analysis was carried out using a number of alkenes, the corresponding alkyl derivatives were observed, formed by alkene insertion.[83]

A study on molybdocene compounds has also appeared. The reaction products of the antitumour-active metallocene molybdocene dichloride (Cp_2MoCl_2) with biological thiols (cysteine and glutathione), of the type $Cp_2Mo(SR)_2$, have been characterised by positive-ion ESI MS.[84]

7.4.4 Monocyclopentadienyl Complexes

This class of compound covers a wide range of structural types. The tripodal oxygen donor ligands **7.6** have many properties in common with cyclopentadienyl ligands themselves.

7.6

Using negative-ion electrospray, the sodium salts of these ligands (L^-) gave ions of the type $[NaL_2]^-$, $[Na_2L_3]^-$ and $[Na_3L_4]^-$ which are related to the trimeric and polymeric structures adopted by the sodium salts[85] and this behaviour is as observed for many other sodium salts (Chapter 4). The behaviour with added alkali metal ions was also studied. For example, with added sodium chloride, salt cluster ions of the type $[(NaCl)_n + L]^-$ ($n = 1$ to 9) and $[(NaCl)_n + NaL_2]^-$ ($n = 1$ to 8) were observed. A study of the complexes of this ligand with monovalent, divalent and trivalent metal ions has also been carried out using ESI MS.[86] Group 2 metals react with a deficiency of L^- giving $[HML(OAc)]^+$ (acetate ions came from the mobile phase used) and $[HML_2]^+$ when sufficient L^- is present. Metals from Groups 3 and 13 and the lanthanides give $[ML(OAc)]^+$ and $[HML(OAc)_2]^+$ with a deficiency of L^-, and $[ML_2]^+$ with excess L^-.[86]

Reduction of $[Cp_2^*Mo_2O_5]$ and $[Cp_2^*Mo_2O_4]$ in methanol-water-trifluoroacetate solutions have been investigated using electrochemistry coupled on line with ESI MS, revealing the existence of a wide range of previously unknown complexes.[87]

A number of pentamethylcyclopentadienyl rhodium(III) and iridium(III) complexes have also been studied by ESI MS, including rhodium complexes of dioxolene ligands.[88] Gas-phase molecular recognition of aromatic acids, by non-covalent π-π interactions, has been carried out using the complex $[Cp^*Rh(2'\text{-deoxyadenosine})]_3^{3+}$.[89]

The ruthenium-pentamethylcyclopentadienyl η^6-arene complexes $[Cp^*Ru(C_6H_6)]$ (CF_3SO_3) and $[Cp^*Ru(indole)](CF_3SO_3)$ give the expected parent cation together with the $[Cp^*Ru]^+$ ion in ESI MS. No $[Ru(arene)]^+$ ions were observed, and it was concluded that the Ru-Cp* bond is stronger than the ruthenium-arene bond.[57]

Metal η^3-Allyl Complexes

The η^3-allyl ligand has some similarities, from a mass spectrometry perspective, to the η^5-cyclopentadienyl ligand, in that it behaves as a spectator ligand. Ionisation is therefore dependent on other processes. As an illustrative example, the dinuclear allyl complex $[Pd(\eta^3\text{-}C_3H_5)Cl]_2$ behaves as a metal-halide complex in ESI MS analysis. Thus, in methanol solution with added pyridine (py), the ion $[Pd(allyl)(py)_2]^+$ (m/z 305) is the base peak at a cone voltage of 5V (Figure 7.6a) together with the ion $[\{Pd(allyl)\}_2Cl]^+$, which retains a chloride bridge from the parent complex. When the cone voltage is raised to 40 V (Figure 7.6b) the mono pyridine ion $[Pd(allyl)(py)]^+$ (m/z 226) dominates, and at 70 V (Figure 7.6c), the bare cation $[Pd(allyl)]^+$ (m/z 147) is the base peak. At these higher cone voltages a range of other minor ions with coordinated methanol and pyridine solvent molecules are also seen. The complexes $[Pt(\eta^3\text{-}C_3H_5)(PPh_3)X]$ (X = Cl, Br, I) also give the ion $[Pt(C_3H_5)(PPh_3)]^+$ by halide loss, together with various solvated analogues.[90]

In contrast, the platinum allyl cyclopentadienyl complex $[Pt(C_5H_4Me)(\eta^3\text{-}C_3H_5)]$ gives a relatively complex ESI MS spectrum, with $[M]^+$ ions formed by oxidation, with a contribution from the $[M + H]^+$ ion; other aggregate ions were observed, and the principal fragment ion was $[Pt(C_5H_4Me)]^+$.[91]

The in situ formation of palladium π-allyl complexes has been demonstrated by Bayer *et al.* as an example of Coordination Ionspray Mass Spectrometry, Section 5.14.[92] The terpene alcohol isopulegol was complexed by addition of palladium(II) acetate, and the CIS MS spectrum, with no electric field applied, showed principal ions $[M + Pd-H]^+$ and $[2M + Pd-H]^+$.

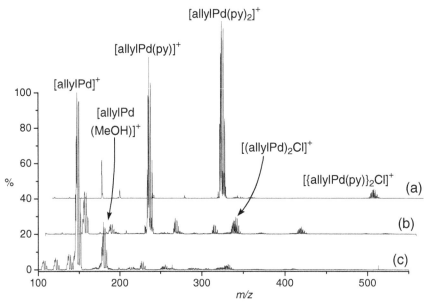

Figure 7.6
The dinuclear palladium allyl complex [Pd(η^3-C$_3$H$_5$)Cl]$_2$ behaves as a metal halide complex with a spectator allyl ligand, as shown by positive-ion ESI mass spectra in methanol solution with added pyridine. At a cone voltage of 5 V (a) the [Pd(allyl)(pyridine)$_2$]$^+$ cation (*m/z* 305) dominates, while at higher cone voltages (b) 40 V and (c) 70 V, the pyridine ligands are lost, ultimately giving the 'bare' [Pd(allyl)]$^+$ cation at *m/z* 147

Metal Arene Complexes

When coordinated to metal centers in a π fashion, and in the absence of substituents that promote ionisation, the arene group is electrospray invisible and thus ionisation is dependent on other features in the complex. The ruthenium complex [Ru$_4$(η^6-C$_6$H$_6$)$_4$(OH)$_4$]$^{4+}$, which has a cubane structure, has been analysed by ESI MS in aqueous solution, which yielded high intensity monocationic peaks due to tetrameric and dimeric species and low intensity peaks due to tri- and mono-ruthenium species. This provided evidence for the existence of the cubane unit in solution, which had previously been questioned.[93] The ruthenium complex [Ru(η^6-*p*-cymene)(η^2-triphos)Cl]PF$_6$ has been analysed in an ionic liquid solution of [bmim][PF$_6$] (bmim = 1-butyl-3-methylimidazolium).[94] This is noteworthy in that ionic liquids are involatile solvents, and the identification of an analyte in the presence of a high salt concentration can be difficult.

Formation of π-Hydrocarbon Complexes and their Use as an Ionisation Aid

Silver(I) ions have an affinity for π-electron density; arene rings, alkenes and alkynes are all included in this category. Because the silver ion is charged, it can be used as an ionisation aid for a wide range of neutral analyte molecules that do not ionise well by other mechanisms, such as by oxidation, protonation or deprotonation. Likewise, silver ions form strong complexes with soft donor ligands such as phosphines and arsines and have been used to form charged derivatives of these ligands for ESI MS analysis (Section 6.3.1).

An early ESI MS study of complexes formed by silver(I) ions and arenes (benzene, toluene, *o*-xylene and mesitylene) gave ions $[Ag(arene)_2]^+$ and $[Ag(arene)]^+$, the latter becoming more prominent at elevated cone voltages. For benzene, only the ion $[Ag(benzene)]^+$ was observed.[95] By using mixtures of polycyclic aromatic hydrocarbons (PAHs), the relative silver ion affinities of a range of aromatic hydrocarbons was found to be: benzene < naphthalene < phenanthrene < anthracene < fluoranthene < pyrene < triphenylene < chrysene < benz[a]anthracene.[96] Isomeric PAHs can also be distinguished from each other using silver cationisation. Thus, in a study of 13 PAHs, the ions $[PAH + Ag]^+$ and $[2PAH + Ag]^+$ were readily generated and CID produced the radical cation $[PAH]^{\bullet +}$. By measuring the relative intensities of the $[PAH + Ag]^+$, $[2PAH + Ag]^+$ and $[PAH]^{\bullet +}$ ions, the PAHs can be distinguished, with the $[PAH]^{\bullet +}$ to $[PAH + Ag]^+$ ratio increasing with decreasing ionisation potential of the PAH, as would be expected.[97]

Silver ions have been used as an ionisation aid for various neutral hydrocarbons, which otherwise do not give ions using ES ionisation. Aromatic hydrocarbons containing long alkyl substituents, such as phenylnonane, were used as model compounds and gave adduct ions $[M + Ag]^+$ and lesser amounts of $[2M + Ag]^+$ with added Ag^+ ions; the abundance of the latter ion can be reduced by increasing the cone voltage. The silver ionisation method has been applied to the analysis of a heavy aromatic petroleum fraction, and an involatile residue collected after vacuum distillation of a crude oil at 565 °C.[98] For polyalkenes such as carotenoids, $[M]^{\bullet +}$ radicals were also observed with the $[M + Ag]^+$ ion.[99] Peroxidation products of cholesterol linoleate and cholesterol arachidonate have also been characterised after derivatisation with Ag^+ ions.[100]

Silver(I) ions also form complexes with calixarenes (analysed by SIMS and ESI MS),[101] and a range of hydrogen-bonded supramolecular assemblies, for MALDI-TOF analysis. Analysis by ESI MS, or by MALDI-TOF without added Ag^+ ions, was unsuccessful.[102] Again, in both of these cases, the Ag^+ ions were proposed to form a π complex with benzene rings in the assemblies.

The in situ derivatisation of metal acetylide complexes has also been carried out and is included in Section 7.8.

Metal-Acetylene/Acetylide Complexes and Complexes of Metal-Acetylides

Acetylenes, R–C≡C–R′, are soft donor ligands and can coordinate in a π-mode to many metal centres. The association of various dicyanophenylacetylenes of the type $NCC_6H_4C≡CC_6H_4CN$ with silver ions have been studied by positive-ion ESI MS.[103] However, crystal structures of five derivatives showed no interactions between the Ag^+ and the acetylene group, with silver coordination involving the cyano groups and in some cases the counteranion.

Most binary metal acetylides $(M-C≡CR)_n$ are insoluble polymeric materials, and thus have low solubilities in common organic solvents. Accordingly, it might be anticipated that ESI MS might not provide very useful data on the structures of these materials. However, the more forcing conditions offered by MALDI ionisation has allowed the analysis of silver phenylacetylide $(AgC≡CPh)_n$. Ions of the type $[n(AgC≡CPh) + Ag]^+$

were observed by silver ion attachment to oligomeric acetylide units, with values of n up to 20 observed using the high sensitivity of the linear mode of the MALDI-TOF instrument.[104]

A transition metal acetylide complex can also act as a metallo-ligand towards other metal centres such as Cu^+ and Ag^+. Preformed cationic complexes of acetylide metalloligands, such as the diplatinum complexes $[\{(dppf)Pt(C{\equiv}CPh)_2\}_2M]^+$ (M = Cu, Ag) [69] and the rhenium(I) complexes $[\{Re(CO)_3(bipy)(C{\equiv}CPh)\}_2M]^+$ (M = Cu, Ag),[105] give the intact cations in their ESI mass spectra. The binding of sodium ions by a copper(I) acetylide complex bearing a pendant crown ether has been studied by ESI MS, though in this case the acetylide was inactive as far as the sodium binding was concerned.[106] The in situ derivatisation of a metal acetylide complex with Ag^+ ions has also been reported to be successful for the poly-yne complex $[Cp(CO)_3W\text{-}(C{\equiv}C)_4\text{-}W(CO)_2Cp]$ in acetonitrile, which gave $[M + Ag(MeCN)]^+$ and $[2M + Ag]^+$ ions by addition of Ag^+ ions in situ.[33]

Transition Metal σ-Alkyl and Aryl Complexes

As for many other organometallic compounds, unfunctionalised alkyl or aryl substituents on an organometallic complex are typically invisible to the ES ionisation process. The metal-carbon bonds are often strong–especially so for many late transition metal derivatives–so the metal carbon bond is not cleaved except under forcing CID conditions. A good example of this has been given in Section 5.3, where the platinum(II)-methyl complex *trans*-$[PtMeI(PPh_3)_2]$ only undergoes loss of the methyl group at a high cone voltage.

In a rare study on early transition metal alkyls by ESI MS the zirconium(IV) alkyl complex $[Cp_2ZrMe]^+[B(C_6F_5)_4]^-$ gave the solvated cation $[Cp_2ZrMe(MeCN)]^+$ in acetonitrile solution.[83]

Early studies concerned cationic methyl derivatives of gold(III) [and also indium(III) and thallium(III)] containing ancillary nitrogen-donor ligands.[107] The gold complexes, of the type $[Me_2AuL]^+$ gave the expected cations as the principal ion, though loss of both methyl groups occurred upon CID.

In a novel application, ESI MS has been applied to the kinetics of the base-catalysed hydrolysis of CH_3ReO_3 to ReO_4^- and CH_4; weak peaks due to an intermediate species $[ReO_3(OH)(H_2O)_4]^-$ were observed when the cone voltage was reduced to the very low value of 3 V.[108] This study emphasised the importance of examining the effect of different collision (fragmentation) conditions in order to gain the most accurate picture of solution speciation.

Organocopper complexes are important synthetic reagents in organic chemistry, and ESI MS has been applied to an investigation of the aggregation states of organocuprates R_2CuLi (R = thienyl, $MeOCMe_2C{\equiv}C$, and Me_3SiCH_2) generated in solution from RLi and CuX (X = CN, I, Br).[109] Spectra were generally quite complex, although most reveal the major species to be dimeric in nature. Anionic species ranging from monomers to trimers were observed, while tetramers and pentamers could occasionally be seen for R = Me_3SiCH_2. However, attempts to obtain ESI MS spectra on cuprates such as Me_2CuLi and $^nBu_2CuLi$ were unsuccessful.

Using a quadrupole ion trap and CID studies the hitherto unknown silver(I) methyl species $[Ag(CH_3)_2]^-$ has been generated in the gas phase. ESI MS of silver(I) acetate gave predominantly the ion $[Ag(O_2CCH_3)_2]^-$. CID on this ion results in loss of CO_2 to give $[CH_3Ag(O_2CCH_3)]^-$, which in turn yields $[Ag(CH_3)_2]^-$ plus CO_2, and $[Ag(CH_3)]$ plus $CH_3CO_2^-$ by two competing pathways.[110]

The platinum(II) ylide complex $[Pt(CH_3)(Ph_3PCHCO_2Et)(dppe)]^+$ (dppe = Ph_2PCH_2 CH_2PPh_2) has been investigated by ESI MS and gave the parent $[M]^+$ cation, together with $[Pt(CH_3)(dppe)]^+$, formed by loss of the neutral ylide Ph_3PCHCO_2Et; the protonated ylide (*i.e.* phosphonium cation) $[Ph_3PCH_2CO_2Et]^+$ was observed. These observations indicate that the ylide ligand is bonded relatively weakly to the platinum, and is relatively easily dissociated. These observations follow the usual fragmentation pattern for such a cation, involving loss of the most weakly bonded neutral ligand.[111]

Mass Spectrometry of Lanthanide Organometallic Complexes

In marked contrast to the burgeoning field of organometallic ESI MS of transition metal and main group metal systems, few studies have been carried out with organometallic lanthanide complexes. In part, this is likely to be due to the high air and/or moisture sensitivity of many of these compounds. A study of some highly reactive compounds of this type has been carried out recently using glove-box techniques,[112] including both divalent and trivalent compounds, and neutral and cationic species. The trivalent bis(pentamethylcyclopentadienyl) complexes $[LnCp_2^*]^+BPh_4^-$ (Ln = Nd, Sm, Y, Tm) give solvated parent ions $[LnCp_2^*(MeCN)_x]^+$ (x = 0 to 2). However, divalent complexes $[LnCp_2^*(THF)_x]$ give observed ions that are dependent on the lanthanide. With europium, the divalent ions $[EuCp^*(MeCN)_x]^+$ (x = 0 to 3) were observed, together with poly-metallic ions $[Eu_2Cp_3^*(MeCN)_x]^+$ (x = 0 to 2). For the more reducing complexes $[SmCp_2^*(THF)_2]$ and $[YbCp_2^*(THF)]$ the trivalent cations $[LnCp_2^*(MeCN)_x]^+$ (x = 0 to 2) were seen, due to oxidation during ESI MS analysis. Monocyclopentadienyl complexes $[LnCp^*(\mu\text{-}I)(THF)_2]_2$ (Ln = Sm, Yb, Eu) gave different sets of ions depending on the metal, e.g. the samarium complex gave $[SmCp_2^*(MeCN)_x]^+$ (x = 0 to 2) while the europium complex gave divalent $[EuCp^*(MeCN)_x]^+$ (x = 0 to 3).

Summary

- As for other inorganic compounds, singly-charged species give intense parent ions.
- Derivatisation may be necessary for neutral metal carbonyl complexes and clusters; addition of methoxide or azide are good options to try. Protonation may work in some cases, if there is an isonitrile or protonatable ligand present. Expect sequential loss of CO ligands at elevated cone voltages.
- Ligands such as cyclopentadienyl, allyl, alkyl, aryl etc. – in the absence of any ionisable sites – are invisible to the ESI ionisation process, and behave largely as spectator ligands. Ionisation in such species is dependent on other pathways.

References

1. J. C. Traeger, *Int. J. Mass Spectrom.*, 2000, **200**, 387; W. Henderson, B. K. Nicholson and L. J. McCaffrey, *Polyhedron*, 1998, **17**, 4291; R. Colton, A. D'Agostino and J. C. Traeger, *Mass Spectrom. Rev.*, 1995, **14**, 79.

2. B. F. G. Johnson and J. S. McIndoe, *Coord. Chem. Rev.*, 2000, **200 – 202**, 901.

3. L. A. P. Kane-Maguire, R. Kanitz and M. M. Sheil, *J. Organomet. Chem.*, 1995, **486**, 243.

4. A. F. H. Siu, L. A. P. Kane-Maguire, S. G. Pyne and R. H. Lambrecht, *J. Chem. Soc., Dalton Trans.*, 1996, 3747.

5. R. Colton and J. C. Traeger, *Inorg. Chim. Acta*, 1992, **201**, 153.

6. A. M. Bond, R. Colton, J. B. Cooper, J. C. Traeger, J. N. Walter and D. M. Way, *Organometallics*, 1994, **13**, 3434.

7. J. C. Eklund, A. M. Bond, R. Colton, D. G. Humphrey, P. J. Mahon and J. N. Walter, *Inorg. Chem.*, 1999, **38**, 2005.

8. I. Ahmed, A. M. Bond, R. Colton, M. Jurcevic, J. C. Traeger and J. N. Walter, *J. Organomet. Chem.*, 1993, **447**, 59.

9. A. M. Bond, R. Colton, R. W. Gable, M. F. Mackay and J. N. Walter, *Inorg. Chem.*, 1997, **36**, 1181.

10. A. M. Bond, R. Colton, A. van den Bergen and J. N. Walter, *Inorg. Chem.*, 2000, **39**, 4696.

11. F. Marken, A. M. Bond and R. Colton, *Inorg. Chem.*, 1995, **34**, 1705.

12. J. Connolly, A. R. J. Genge, W. Levason, S. D. Orchard, S. J. A. Pope and G. Reid, *J. Chem. Soc., Dalton Trans.*, 1999, 2343; M. K. Davies, M. C. Durrant, W. Levason, G. Reid and R. L. Richards, *J. Chem. Soc., Dalton Trans.*, 1999, 1077.

13. H. Hori, F. P. A. Johnson, K. Koike, K. Takeuchi, T. Ibusuki and O. Ishitani, *J. Chem. Soc., Dalton Trans.*, 1997, 1019; H. Hori, O. Ishitani, K. Koike, K. Takeuchi and T. Ibusuki, *Anal. Sci.*, 1996, **12**, 587.

14. C. P. G. Butcher, B. F. G. Johnson, J. S. McIndoe, X. Yang, X.-B. Wang and L.-S. Wang, *J. Chem. Phys.*, 2002, **116**, 6560.

15. C. P. G. Butcher, A. Dinca, P. J. Dyson, B. F. G. Johnson, P. R. R. Langridge-Smith and J. S. McIndoe, *Angew. Chem., Int. Ed. Engl.*, 2003, **42**, 5752.

16. W. Henderson, J. S. McIndoe, B. K. Nicholson and P. J. Dyson, *J. Chem. Soc., Dalton Trans.*, 1998, 519; W. Henderson, J. S. McIndoe, B. K. Nicholson and P. J. Dyson, *J. Chem. Soc., Chem. Commun.*, 1996, 1183.

17. C. P. G. Butcher, P. J. Dyson, B. F. G. Johnson, T. Khimyak and J. S. McIndoe, *Chem. Eur. J.*, 2003, 944.

18. N. W. Duffy, C. J. McAdam, B. H. Robinson and J. Simpson, *J. Organomet. Chem.*, 1998, **565**, 19.

19. K. C. Huang, Y.-C. Tsai, G.-H. Lee, S.-M. Peng and M. Shieh, *Inorg. Chem.*, 1997, **36**, 4421.

20. M. Shieh, T.-F. Tang, S.-M. Peng, *Inorg. Chem.*, 1995, **34**, 2797.

21. C. E. C. A. Hop and R. Bakhtiar, *J. Chem. Educ.*, 1996, **73**, A162.

22. B. Keşanli, D. R. Gardner, B. Scott and B. W. Eichhorn, *J. Chem. Soc., Dalton Trans.*, 2000, 1291.

23. P. D. Mlynek and L. F. Dahl, *Organometallics*, 1997, **16**, 1655.

24. P. D. Mlynek and L. F. Dahl, *Organometallics*, 1997, **16**, 1641.

25. M. Ferrer, R. Reina, O. Rossell, M. Seco and G. Segalés, *J. Organomet. Chem.*, 1996, **515**, 205.

26. R. Reina, O. Riba, O. Rossell, M. Seco, P. Gómez-Sal and A. Martin, *Organometallics*, 1997, **16**, 5113.

27. D. J. F. Bryce, P. J. Dyson, B. K. Nicholson and D. G. Parker, *Polyhedron*, 1998, **17**, 2899.

28. P. D. Mlynek, M. Kawano, M. A. Kozee and L. F. Dahl, *J. Cluster. Sci.*, 2001, **12**, 313.

29. L. A. P. Kane-Maguire, R. Kanitz and M. M. Sheil, *Inorg. Chim. Acta*, 1996, **245**, 209.

30. J. L. Kerr, B. H. Robinson and J. Simpson, *J. Chem. Soc., Dalton Trans.*, 1999, 4165.

31. D. Braga and F. Grepioni, *Acc. Chem. Res.*, 1997, **30**, 81; D. Braga, F. Grepioni, E. Tedesco, K. Biradha and G. R. Desiraju, *Organometallics*, 1996, **15**, 2692; D. Braga, F. Grepioni and G. R. Desiraju, *J. Organomet. Chem.*, 1997, **548**, 33; D. Braga and F. Grepioni, *Chem. Commun.*, 1996, 571.

32. M. I. Bruce, B. W. Skelton, A. H. White and N. N. Zaitseva, *Aust. J. Chem.*, 1997, **50**, 163; M. I. Bruce, B. W. Skelton, A. H. White and N. N. Zaitseva, *Inorg. Chem. Commun.*, 1999, **2**, 17.

33. M. I. Bruce, M. Ke and P. J. Low, *Chem. Commun.*, 1996, 2405.

34. P. J. Dyson, B. F. G. Johnson, J. S. McIndoe and P. R. R. Langridge-Smith, *Inorg. Chem.*, 2000, **39**, 2430.

35. H. Werner, W. Beck and H. Engelmann, *Inorg. Chim. Acta*, 1969, **3**, 331.

36. J. S. McIndoe and B. K. Nicholson, *J. Organomet. Chem.*, 1999, **573**, 232.

37. H. Hori, J. Ishihara, K. Koike, K. Takeuchi, T. Ibusuki and O. Ishitani, *Chem. Lett.*, 1997, 273; H. Hori, J. Ishihara, K. Koike, K. Takeuchi, T. Ibusuki and O. Ishitani, *Chem. Lett.*, 1997, 1249.

38. S. Chand, R. K. Coll and J. S. McIndoe, *Polyhedron*, 1998, **17**, 507.

39. W. Henderson and B. K. Nicholson, *J. Chem. Soc., Chem. Commun.*, 1995, 2531.

40. M. R. Jordan, P. S. White, C. K. Schauer and M. A. Mosley III, *J. Am. Chem. Soc.*, 1995, **117**, 5403.

41. C. Decker, PhD thesis, University of Waikato, 2002.

42. F.-H. Luo, S.-R. Yang, C.-S. Li, J.-P. Duan and C.-H. Cheng, *J. Chem. Soc., Dalton Trans.*, 1991, 2435.

43. C. E. C. A. Hop, J. T. Brady and R. Bakhtiar, *J. Am. Soc. Mass Spectrom.*, 1997, **8**, 191.

44. S. F. A. Kettle, E. Diana, R. Rossetti and P. L. Stanghellini, *J. Am. Chem. Soc.*, 1997, **119**, 8228.

45. C. Evans, K. M. Mackay and B. K. Nicholson, *J. Chem. Soc., Dalton Trans.*, 2001, 1645.

46. P. J. Dyson, A. K. Hearley, B. F. G. Johnson, T. Khimyak, J. S. McIndoe and P. R. R. Langridge-Smith, *Organometallics*, 2001, **20**, 3970.

47. Z. Xu, S. Kawi, A. L. Rheingold and B. C. Gates, *Inorg. Chem.*, 1994, **33**, 4415.

48. C. P. G. Butcher, A. Dinca, P. J. Dyson, B. F. G. Johnson, P. R. R. Langridge-Smith and J. S. McIndoe, *Angew. Chem., Int. Ed. Engl.*, 2003, **42**, 5752.

49. P. J. Dyson, A. K. Hearley, B. F. G. Johnson, J. S. McIndoe, P. R. R. Langridge-Smith and C. Whyte, *Rapid Commun. Mass Spectrom.*, 2001, **15**, 895.

50. C. P. G. Butcher, P. J. Dyson, B. F. G. Johnson, P. R. R. Langridge-Smith, J. S. McIndoe and C. Whyte, *Rapid Commun. Mass Spectrom.*, 2002, **16**, 1595.

51. P. J. Dyson, N. Feeder, B. F. G. Johnson, J. S. McIndoe and P. R. R. Langridge-Smith, *J. Chem. Soc., Dalton Trans.*, 2000, 1813.

52. C. Decker, W. Henderson and B. K. Nicholson, *J. Chem. Soc., Dalton Trans.*, 1999, 3507.

53. C. Evans and B. K. Nicholson, *J. Organomet. Chem.*, 2003, **665**, 95.

54. E. Singleton and H. E. Oosthuizen, *Adv. Organomet. Chem.*, 1983, **22**, 209; F. Bonati and G. Minghetti, *Inorg. Chim. Acta*, 1974, **9**, 95.

55. C. Decker, W. Henderson and B. K. Nicholson, *J. Organomet. Chem.*, 2004, **689**, 1691.

56. G. J. Van Berkel and F. Zhou, *Anal. Chem.*, 1995, **67**, 3958.

57. X. Xu, S. P. Nolan and R. B. Cole, *Anal. Chem.*, 1994, **66**, 119.

58. L. A. P. Kane-Maguire, R. Kanitz and M. M. Sheil, *Inorg. Chim. Acta*, 1996, **245**, 209.

59. R. Colton and J. C. Traeger, *Inorg. Chim. Acta*, 1992, **201**, 153.

60. T. D. McCarley, M. W. Lufaso, L. S. Curtin and R. L. McCarley, *J. Phys. Chem. B*, 1998, **102**, 10078.

61. A. M. Bond, R. Colton, D. A. Fiedler, L. D. Field, T. He, P. A. Humphrey, C. M. Lindall, F. Marken, A. F. Masters, H. Schumann, K. Sühring and V. Tedesco, *Organometallics*, 1997, **16**, 2787.

62. T. D. McCarley and R. L. McCarley, *Anal. Chem.*, 1997, **69**, 130.

63. W. Henderson, A. G. Oliver and A. L. Downard, *Polyhedron*, 1996, **15**, 1165.

64. R. Bakhtiar and A. E. Kaifer, *Rapid Commun. Mass Spectrom.*, 1998, **12**, 111.

65. S. R. Alley and W. Henderson, *J. Organomet. Chem.*, 2001, **637 – 639**, 216.

66. N. Kubota, T. Fukuo and R. Akawa, *J. Am. Soc. Mass Spectrom.*, 1999, **10**, 557.

67. K.-S. Gan and T. S. A. Hor in *Ferrocenes: Homogeneous catalysis, organic synthesis and materials science,* A. Togni and T. Hayashi (Eds), VCH, 1995.

68. W. Henderson and G. M. Olsen, *Polyhedron*, 1998, **17**, 577.

69. W.-Y. Wong, G.-L. Lu and K.-H. Choi, *J. Organomet. Chem.*, 2002, **659**, 107.

70. C. Jiang, T. S. A. Hor, Y. K. Yan, W. Henderson and L. J. McCaffrey, *J. Chem. Soc., Dalton Trans*, 2000, 3197; J. S. L. Yeo, G. Li, W.-H. Yip, W. Henderson, T. C. W. Mak and T. S. A. Hor, *J. Chem. Soc., Dalton Trans.*, 1999, 435; W. Henderson, L. J. McCaffrey, M. B. Dinger and B. K. Nicholson, *Polyhedron*, 1998, **17**, 3137; C. C. H. Chin, J. S. L. Yeo, Z. H. Loh, J. J. Vittal and T. S. A. Hor, *J. Chem. Soc., Dalton Trans.*, 1998, 3777; J. Fawcett, W. Henderson, R. D. W. Kemmitt, D. R. Russell and A. Upreti, *J. Chem. Soc., Dalton Trans.*, 1996, 1897.

71. S. P. Yeo, W. Henderson, T. C. W. Mak and T. S. A. Hor, *J. Chem. Soc., Dalton Trans.*, 1999, 171.

72. G. D. Brindley, O. D. Fox and P. D. Beer, *J. Chem. Soc., Dalton Trans.*, 2000, 4354.

73. G. J. Van Berkel, J. M. E. Quirke, R. A. Tigani, A. S. Dilley and T. R. Covey, *Anal. Chem.*, 1998, **70**, 1544.

74. J. M. E. Quirke and G. J. Van Berkel, *J. Mass Spectrom.*, 2001, **36**, 179.

75. J. M. E. Quirke, Y.-L. Hsu and G. J. Van Berkel, *J. Nat. Prod.,* 2000, **63**, 230.

76. D. Williams and M. K. Young, *Rapid Commun. Mass Spectrom.,* 2000, **14**, 2083; M. K. Young, N. Dinh and D. Williams, *Rapid Commun. Mass Spectrom.*, 2000, **14**, 1462.

77. D. Williams, S. Chen and M. K. Young, *Rapid Commun. Mass Spectrom.*, 2001, **15**, 182.

78. G. J. Van Berkel, J. M. E. Quirke and C. L. Adams, *Rapid Commun. Mass Spectrom.*, 2000, **14**, 849.

79. K. K.-W. Lo, D. C.-M. Ng, J. S.-L. Lau, R. S.-S. Wu and P. K.-S. Lam, *New J. Chem.*, 2003, **27**, 274.

80. C. L. Gross, S. R. Wilson and G. S. Girolami, *Inorg. Chem.*, 1995, **34**, 2582.

81. A. M. Bond, R. Colton, U. Englert, H. Hügel and F. Marken, *Inorg. Chim. Acta*, 1995, **235**, 117.

82. S. Codato, G. Carta, G. Rossetto, P. Zanella, A. M. Gioacchini and P. Traldi, *Rapid Commun. Mass Spectrom.*, 1998, **12**, 1981.

83. D. Feichtinger, D. A. Plattner and P. Chen, *J. Am. Chem. Soc.*, 1998, **120**, 7125.

84. J. B. Waern and M. M. Harding, *Inorg. Chem.*, 2004, **43**, 206.

85. R. Colton, A. D'Agostino, J. C. Traeger and W. Kläui, *Inorg. Chim. Acta*, 1995, **233**, 51.

86. R. Colton and W. Kläui, *Inorg. Chim. Acta*, 1993, **211**, 235.

87. J. Gun, A. Modestov, O. Lev and R. Poli, *Eur. J. Inorg. Chem.*, 2003, 2264.

88. W. Henderson, J. Fawcett, R. D. W. Kemmitt and D. R. Russell, *J. Chem. Soc., Dalton Trans.*, 1995, 3007.

89. R. Bakhtiar, H. Chen, S. Ogo and R. H. Fish, *Chem. Commun.*, 1997, 2135.

90. S. Favaro, L. Pandolfo and P. Traldi, *Rapid Commun. Mass Spectrom.*, 1997, **11**, 1859.

91. G. Carta, G. Rossetto, L. Borella, D. Favretto and P. Traldi, *Rapid Commun. Mass Spectrom.*, 1997, **11**, 1315.

92. E. Bayer, P. Gfrörer and C. Rentel, *Angew. Chem., Int. Ed. Engl.,* 1999, **38**, 992.

93. P. J. Dyson, K. Russell and T. Welton, *Inorg. Chem. Comm.*, 2001, **4**, 571.

94. P. J. Dyson, J. S. McIndoe and D. Zhao, *Chem Comm.*, 2003, 508.

95. A. J. Canty and R. Colton, *Inorg. Chim. Acta*, 1994, **220**, 99.

96. K. M. Ng, N. L. Ma and C. W. Tsang, *Rapid Commun. Mass Spectrom.*, 1998, **12**, 1679.

97. K. M. Ng, N. L. Ma and C. W. Tsang, *Rapid Commun. Mass Spectrom.*, 2003, **17**, 2082.

98. S. G. Roussis and R. Proulx, *Anal. Chem.*, 2002, **74**, 1408.

99. E. Bayer, P. Gfrörer and C. Rentel, *Angew. Chem., Int. Ed. Engl.*, 1999, **38**, 992.

100. C. M. Havrilla, D. L. Hachey and N. A. Porter, *J. Am. Chem. Soc.*, 2000, **122**, 8042.

101. F. Inokuchi, Y. Miyahara, T. Inazu and S. Shinkai, *Angew. Chem., Int. Ed. Engl.*, 1995, **34**, 1364.

102. K. A. Jolliffe, M. C. Calama, R. Fokkens, N. M. M. Nibbering, P. Timmerman and D. N. Reinhoudt, *Angew. Chem., Int. Ed. Engl.*, 1998, **37**, 1247.

103. K. A. Hirsch, S. R. Wilson and J. S. Moore, *J. Am. Chem. Soc.*, 1997, **119**, 10401.

104. Y. H. Xu, X. R. He, C. F. Hu, B. K. Teo and H. Y. Chen, *Rapid Commun. Mass Spectrom.*, 2000, **14**, 298.

105. V. W.-W. Yam, S. H.-F. Chong, K. M.-C. Wong and K.-K. Cheung, *Chem. Comm.*, 1999, 1013.

106. V. W.-W. Yam, C.-H. Lam and K.-K. Cheung, *Inorg. Chim. Acta*, 2001, **316**, 19.

107. A. J. Canty, R. Colton and I. M. Thomas, *J. Organomet. Chem.,* 1993, **455**, 283.

108. J. H. Espenson, H. Tan, S. Mollah, R. S. Houk and M. D. Eager, *Inorg. Chem.*, 1998, **37**, 4621.

109. B. H. Lipshutz, J. Keith and D. J. Buzard, *Organometallics*, 1999, **18**, 1571.

110. R. A. J. O'Hair, *Chem. Commun.*, 2002, 20.

111. F. Benetollo, R. Bertani, P. Ganis, L. Pandolfo and L. Zanotto, *J. Organomet. Chem.*, 2001, **629**, 201.

112. W. J. Evans, M. A. Johnston, C. H. Fujimoto and J. Greaves, *Organometallics*, 2000, **19**, 4258.

8   A Selection of Special Topics

Introduction

In this short chapter some specific applications of ESI and MALDI MS techniques are reviewed. These topics have been selected from ones where there is considerable current interest (e.g. detection of reaction intermediates), where the topic is interdisciplinary in nature (e.g. dendrimers, supramolecular chemistry), or in developing areas from our own research (reaction screening using ESI MS).

Characterisation of Dendrimers Using ESI and MALDI-TOF MS Techniques

Dendrimers are a unique class of materials that possess features of both small molecules and polymers. They are highly branched molecules that are synthesised in successive 'generations', and they therefore have a very distinct size, shape and architecture. Thus a dendrimer of any particular generation has a precisely defined molecular weight, which contrasts with more classical chain-type polymers, where a molecular weight distribution is typical. Dendrimers have been derived from a diverse range of building blocks, ranging from organic molecules, to organosilanes and -germanes. Likewise, the surface of dendrimers have been functionalised with many different groups, including phosphine ligands and ferrocene groups, to name but two.

Because of the ability of electrospray ionisation to generate multiply-charged gas-phase ions, this MS technique has been widely employed in the characterisation of a wide range of dendrimer materials and in this section some 'inorganic' examples are briefly discussed. Selected examples include dendrimers based on a polyhedral silsesquioxane core,[1] highly-charged polysilane dendrimers with Cp^*Ru^+ functionalities,[2] dendrimers based on the silatrane group,[3] organopalladium dendrimers,[4] dendrimers with capping ferrocene and cobaltocenium groups[5] and dendrimers bearing metal-binding ligands.[6]

The behaviour of large dendrimer molecules in ESI MS analysis is as follows: a large, multiply-charged dendrimer will give a series of multiply-charged ions formed by association with a variable number of counterions. For example, the silane dendrimer G2-Ru24 **8.1** yielded the series of multiply-charged ions $[M - 4OTf]^{4+}$, $[M - 5OTf]^{5+}$, $[M - 6OTf]^{6+}$ and $[M - 7OTf]^{7+}$; using an FTICR MS, the individual isotopic envelopes, and hence the charge, could be resolved for these multiply-charged ions.[2] In contrast, a neutrally-charged dendrimer will tend to undergo multiple charging from protons or other solution cations such as NH_4^+.[3] This behaviour is reminiscent of that

Mass Spectrometry of Inorganic, Coordination and Organometallic Compounds   W. Henderson and J. S. McIndoe
© 2005 John Wiley & Sons, Ltd   ISBNs: 0-470-85015-9 (HB); 0-470-85016-7 (PB)

shown by other large molecules such as polyethers, peptides and proteins, which also undergo multiple charging in ESI MS analysis.

8.1   G2-Ru24

MALDI-TOF has also been extensively used for dendrimer characterisation and produces primarily singly-charged ions. As an example, dendrimeric molecules based on the tetrahedral $Co_2(CO)_6(RC{\equiv}CR)$ core gave monocations with loss of some CO ligands.[7] Organogermanium dendrimers have also been characterised by MALDI MS.[8] A study of metal-binding dendrimers by both MALDI and ESI MS techniques found ESI MS to be the preferred technique, since MALDI data could not be obtained for some dendrimers.[6] Other dendrimers that have been studied by MALDI-TOF MS are discussed in Section 3.7.3.

Investigating the Formation of Supramolecular Coordination Assemblies Using ESI MS

Supramolecular chemistry is a field that has expanded rapidly in recent years and involves the spontaneous self-assembly of complementary components, held together by intermolecular forces. The resulting assemblies are typically large, ordered, and may be highly charged. The building blocks may be simple organic molecules or more complex coordination compounds.

Techniques such as X-ray crystallography can be used to identify these materials once they have been crystallised but in order to provide characterisation data on the materials in solution, or to identify components of reaction mixtures, ESI MS is the preferred technique. Furthermore, because ESI is able to transfer multiply-charged ions from solution into the gas phase, the original charge information of the supramolecular system may be retained in the mass spectrum, providing the charge density on the resulting gas phase ion is not too high. A technique such as MALDI would lose this charge information, because ions produced by MALDI ionisation are invariably singly-charged (Section 3.7).

To illustrate the general usefulness of ESI MS in this field of chemistry, the specific example of a metal helicate, formed by spontaneous self-assembly of two ligand 'strands' around several metal ions, will be examined. For example, two molecules of the polydentate ligand **La** spontaneously self-assemble with five Cu^+ ions to give the helicate complex **Ha** (Figure 8.1). This species has a 5+ charge but because of the large size the charge density remains low, and the parent pentacation is observed as the base peak in the positive ion ESI mass spectrum, Figure 8.2, along with ion pair aggregates formed with the BF_4^- counterion, $[\mathbf{Ha} + BF_4]^{4+}$ and $[\mathbf{Ha} + 2BF_4]^{3+}$.[9]

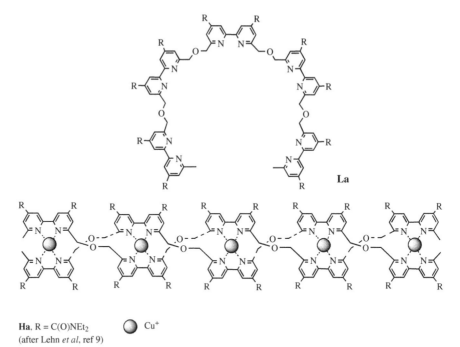

Figure 8.1
Self-assembly
of polydentate
ligand (La) to
helicate
complex (Ha)

Ha, R = C(O)NEt₂

Ha, R = C(O)NEt$_2$ ⬤ Cu⁺
(after Lehn *et al*, ref 9)

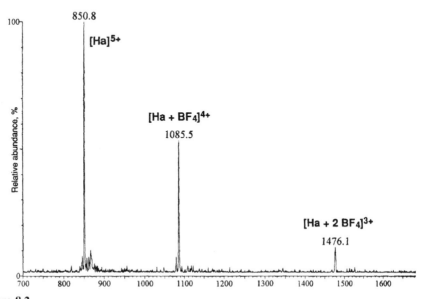

Figure 8.2
The positive-ion ESI mass spectrum of the self-assembled helical complex **Ha** (Reproduced with
permission from Lehn *et al.*,)[9]

Many of the studies in this area have employed polypyridine ligands, for their excellent coordinating properties to metal centres, with helicates being a particularly common motif in studies by ESI MS.[10] Other ESI MS studies have investigated molecular squares[11] and other metallomacrocycles,[12] cages,[13] and catenanes comprising two interlocking rings.[14]

Because the attractive forces that assemble supramolecular systems are often relatively weak, for example hydrogen bonds, or π-π interactions, they may not survive even the fairly gentle ES ionisation process. In such cases, the coldspray ionisation variant has been shown to be a very effective method, operating at low temperatures, which helps to preserve even the weakest supramolecular interactions. A number of examples of this application have been given in recent reviews.[15]

ESI MS as a Tool for Directing Chemical Synthesis: a Case Study Involving the Platinum Metalloligands [Pt$_2$(μ-E)$_2$(PPh$_3$)$_4$]

8.4.1 Background

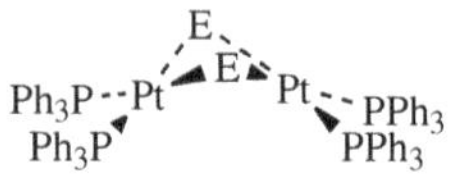

8.2

The complexes [Pt$_2$(μ-E)$_2$(PPh$_3$)$_4$] (E = S, Se) **8.2** contain highly nucleophilic sulfide or selenide atoms in a {Pt$_2$E$_2$} core, and the complexes show a rich chemistry, particularly for the more extensively studied sulfide system, which has been summarised in recent reviews.[16,17] Derivatives containing other phosphine ligands have also been studied, but to a far lesser extent. The main modes of reactivity are towards organic electrophiles, RX (resulting in alkylation of a sulfide ligand producing a bridging thiolate ligand, SR), Scheme 8.1, and towards coordination complexes (resulting in the formation of multi-metallic sulfide-bridged aggregates), Scheme 8.2. Many different structural types are

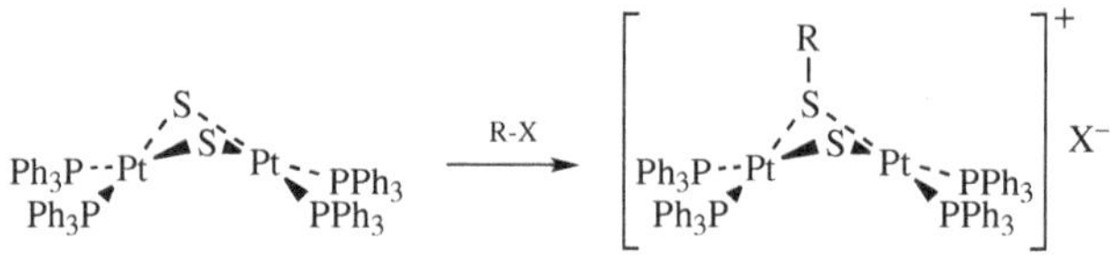

Scheme 8.1
Reaction of [Pt$_2$(μ-S)$_2$(PPh$_3$)$_4$] with an organic electrophile

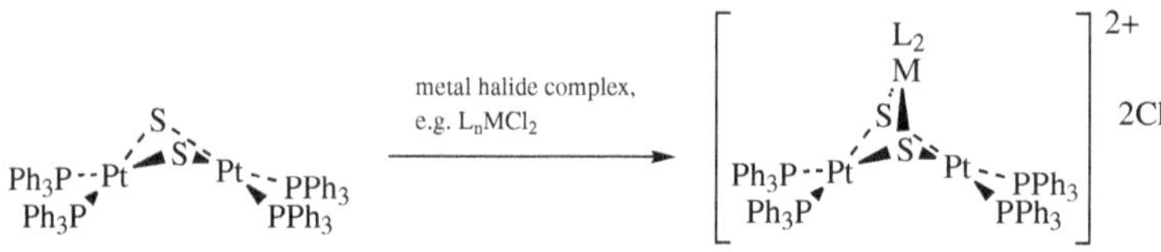

Scheme 8.2
Typical reaction of [Pt$_2$(μ-S)$_2$(PPh$_3$)$_4$] with a metal-chloride complex L$_n$MCl$_2$, where L is a neutral ancillary ligand such as a phosphine etc.

known, so the reactions depicted in Schemes 8.1 and 8.2 show only the more common behaviour. The $\{Pt_2S_2\}$ unit has been found to coordinate to a substantial number of metals in the Periodic Table, including main group and transition metals, and even the actinide metal uranium.

In this section research carried out towards probing the chemistry of $[Pt_2(\mu\text{-}E)_2(PPh_3)_4]$ using an ESI MS based approach is summarised. There are a number of advantages offered by this mass spectrometry-directed synthesis strategy, as listed below:

1. ESI MS is a solution based technique, thus it can be used to directly analyse *reaction* solutions. The experiment simply involves dissolving a small quantity of $[Pt_2(\mu\text{-}E)_2(PPh_3)_4]$ and the reagent of choice in a suitable solvent (usually methanol), allowing the reaction to proceed, and analysing the products by ESI MS. The results of the mass spectrometry screening process are then used to target subsequent syntheses on the macroscopic scale. The isolated products are then characterised by macroscopic techniques, such as NMR spectroscopy, elemental analysis, X-ray crystallography etc.
2. As a result of the sensitivity of MS techniques, miniscule amounts of material can be used in these micro-syntheses, thus conserving expensive platinum starting materials.
3. The methodology allows large numbers of substrates to be screened, in a combinatorial-type approach. Promising reaction systems for further study – including the formation of novel species – are thus identified by their characteristic mass spectral signatures. The use of theoretical isotope patterns (Section 1.7) is an important tool in species assignment.
4. Because the reactivity of $[Pt_2(\mu\text{-}E)_2(PPh_3)_4]$ invariably involves the formation of charged products, ESI MS is an ideal technique for investigating the chemistry in this system.

Using such an ESI MS based approach, a wide range of processes can easily be studied and some of these are briefly discussed below.

8.4.2 Analysis of the Metalloligands; Formation of Protonated Species

Although the methylated derivative $[Pt_2(\mu\text{-}S)(\mu\text{-}SMe)(PPh_3)_4]^+$ was one of the earliest characterised derivatives of the $\{Pt_2S_2\}$ core,[18] the related protonated species was not isolated until relatively recently, following ESI MS studies.[19,20] The sulfide atoms of $[Pt_2(\mu\text{-}S)_2(PPh_3)_4]$ are sufficiently basic to permit the ready detection of the neutral complex in ESI MS, as its $[M + H]^+$ ion at *m/z* 1504. Even though the complex is only sparingly soluble in common organic solvents such as methanol, there is sufficient solubility to obtain the low concentrations required for MS analysis. A typical spectrum is shown in Figure 8.3, where the base peak is the monoprotonated $[Pt_2(\mu\text{-}S)_2(PPh_3)_4 + H]^+$; in some spectra, an additional low intensity ion can be identified as the diprotonated $[Pt_2(\mu\text{-}S)_2(PPh_3)_4 + 2H]^{2+}$ (*m/z* 752). This observation subsequently led to the macroscopic synthesis and isolation of the protonated complex $[Pt_2(\mu\text{-}S)(\mu\text{-}SH)(PPh_3)_4]^+$, formed by addition of one equivalent of hydrochloric acid to the parent complex in methanol, followed by precipitation with a large anion, *viz.* PF_6^-.[19]

For the related selenide system, addition of acid to $[Pt_2(\mu\text{-}Se)_2(PPh_3)_4]$ results in isolation of the diprotonated species $[Pt_2(\mu\text{-}SeH)_2(PPh_3)_4]^{2+}$, and with one equivalent of acid a mixture of mono and diprotonated species is observed in the ESI MS spectrum.[21]

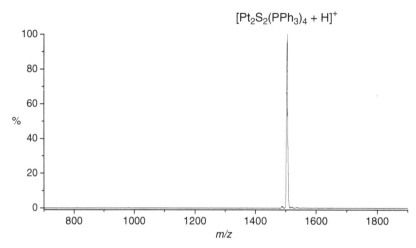

Figure 8.3
Positive-ion ESI mass spectrum of a methanolic solution of $[Pt_2(\mu\text{-}S)_2(PPh_3)_4]$ at a cone voltage of 20 V, showing the monoprotonated ion $[Pt_2(\mu\text{-}S)_2(PPh_3)_4 + H]^+$ (*m/z* 1503)

8.4.3   Reactivity of $[Pt_2(\mu\text{-}E)_2(PPh_3)_4]$ Towards Metal-Halide Complexes

The metalloligands have been known for quite some time to display a high reactivity towards metal-halide complexes.[16] The reactions proceed (e.g. Scheme 8.2) by displacement of one or more halide ligands, and formation of a cationic sulfide- or selenide-bridged polymetallic complex, containing the new heterometal fragment (usually with ancillary ligands attached, depending on the metal). In recent years, ESI MS has been used to investigate reactions of this type and adducts of $[Pt_2(\mu\text{-}S)_2(PPh_3)_4]$ or $[Pt_2(\mu\text{-}Se)_2(PPh_3)_4]$ with various metal substrates have been probed using an ESI MS based approach; examples include mercury(II)[22], nickel(II)[23], palladium(II) and platinum(II)[24], and organometallic derivatives of rhodium(III), iridium(III), ruthenium(II) and osmium(II).[25]

Metal complexation studies leads to formulation of the following general rules:

1. The absence of complexation is indicated by observation of peaks due to the starting materials $[Pt_2(\mu\text{-}E)_2(PPh_3)_4]$, as described above. An example of the simplicity of the ESI MS-monitored reaction screening is given by the reaction of $[Pt_2(\mu\text{-}S)_2(PPh_3)_4]$ with $[Rh(cod)Cl]_2$ (cod = cyclo-octa-1,5-diene). As can be seen from the ESI MS spectrum (Figure 8.4) at a low cone voltage (20 V), a single monocationic adduct $[Pt_2(\mu\text{-}S)_2(PPh_3)_4Rh(cod)]^+$ is seen. The observed isotope distribution pattern shows an envelope of peaks, with adjacent peaks separated by 1 *m/z*, which is indicative of a monocation, and the pattern agrees well with the theoretical pattern.
2. The labile ligands are typically but not exclusively halide ligands. The resulting adducts are then positively charged. The use of low cone voltages (typically 20 V) permits the detection of parent ions, with minimal or no fragmentation.
3. It may be possible to add either one or two metal centres, depending on the metal complex and the mole ratio used, resulting in the formation of either monocationic

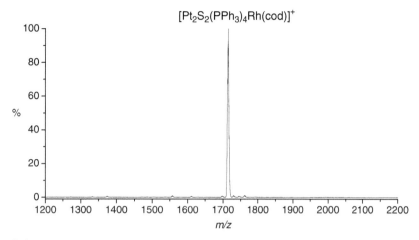

Figure 8.4
Positive-ion ESI mass spectrum of a mixture of $[Pt_2(\mu\text{-}S)_2(PPh_3)_4]$ and $[Rh(cod)Cl]_2$ (cod = cyclo-octadiene) at a cone voltage of 20 V, showing exclusive formation of monocationic $[Pt_2(\mu\text{-}S)_2(PPh_3)_4Rh(cod)]^+$

and/or dicationic products. For example, Figure 8.5 shows the ESI MS spectrum of a reaction mixture of $[Pt_2(\mu\text{-}S)_2(PPh_3)_4]$ with Ph_3PAuCl, giving $[Pt_2(\mu\text{-}S)_2(PPh_3)_4AuPPh_3]^+$ and $[Pt_2(\mu\text{-}S)_2(PPh_3)_4(AuPPh_3)_2]^{2+}$.

4. The use of elevated cone voltages (e.g. 60 V and above) results in the loss of a PPh_3-ligand as the first stage of fragmentation.

In some rare cases, notably involving $[Pt_2(\mu\text{-}Se)_2(PPh_3)_4]$, the use of ESI MS as the primary screening technique allows *metal scrambling* to be observed. Instead of the

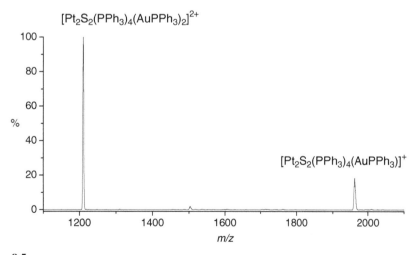

Figure 8.5
Positive-ion ESI mass spectrum of mixture of $[Pt_2(\mu\text{-}S)_2(PPh_3)_4]$ and Ph_3PAuCl at a cone voltage of 5 V, showing formation of monocationic $[Pt_2(\mu\text{-}S)_2(PPh_3)_4AuPPh_3]^+$ and dicationic $[Pt_2(\mu\text{-}S)_2(PPh_3)_4(AuPPh_3)_2]^{2+}$

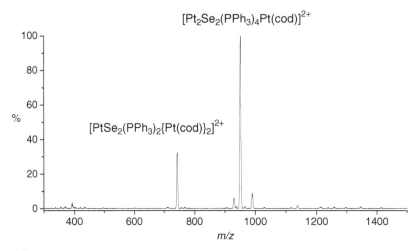

Figure 8.6
The phenomenon of metal-scrambling occurs readily for $[Pt(\mu\text{-}Se)_2(PPh_3)_4]$ as shown by the positive-ion ESI MS spectrum of a mixture of $[Pt_2(\mu\text{-}Se)_2(PPh_3)_4]$ with $[PtCl_2(cod)]$ in methanol

reaction of $[Pt_2(\mu\text{-}Se)_2(PPh_3)_4]$ with a metal halide substrate $LMCl_2$ giving solely the simple adduct $[Pt_2(\mu\text{-}Se)_2(PPh_3)_4ML]^{2+}$, the metal-scrambled species $[Pt(\mu\text{-}Se)_2(PPh_3)_2(ML)_2]^{2+}$ is also observed. The metal scrambled species still has the same $\{M_3Se_2\}$ core. This is illustrated in Figure 8.6 for the reaction of $[Pt_2(\mu\text{-}Se)_2(PPh_3)_4]$ with $[PtCl_2(cod)]$ (cod = cyclo-octa-1,5-diene); the base peak at m/z 950 is due to the simple adduct $[Pt_2(\mu\text{-}Se)_2(PPh_3)_4Pt(cod)]^{2+}$, while the lower intensity peak at m/z 742 is due to the metal-scrambled species $[Pt(\mu\text{-}Se)_2(PPh_3)_2\{Pt(cod)\}_2]^{2+}$. Various platinum(II), gold(III) and rhodium(III) substrates behave in this way.[26] Such metal scrambling has scarcely been seen for the related $[Pt_2(\mu\text{-}S)_2(PPh_3)_4]$.

Applications of ESI MS in the Detection of Reactive Intermediates and Catalyst Screening

8.5.1   Detection of Intermediates in Reactions of Organic Compounds

Because ESI MS relies on the introduction of the analyte sample as a *solution*, it is a very simple matter to introduce reaction mixtures directly, appropriately diluted if necessary. This can be used to either investigate the end products of a reaction or, alternatively, to detect **reaction intermediates**. The gentle nature of the ESI technique means that species observed in the gas phase are often very closely related to the actual species in solution. Of course, the usual provisos will apply, namely that the species in question is charged, or easily able to be converted into a charge species by oxidation, protonation etc., and that the intermediate has a sufficiently long lifetime to be detected. An advantage of ESI MS is that it can analyse paramagnetic species as easily as diamagnetic ones.

In an early application of this kind, various organophosphorus reactions (Wittig, Staudinger and Mitsunobu) were studied by ESI MS; ionic intermediates were detected directly and $[M + H]^+$ ions were observed for many of the neutrally-charged intermediates.[27] Since then, a range of reaction intermediates have been detected by ESI MS and the following examples hopefully serve to give a flavour of what it possible. It is certain that the technique shows huge potential for further development. In contrast, techniques such as MALDI and FAB show much less utility, since they are unable to *directly* analyse reaction solutions.

One of the early applications involving detection of transition metal intermediates by ESI MS was the detection of nickel(II) intermediates in the Raney-nickel catalysed coupling of 2-bromo-6-methylpyridine (dmbp).[28] This reaction gave a complex of composition [NiBr$_2$(dmbp)], which was found to be dimeric in solution, since it gave a strong [Ni$_2$Br$_3$(dmbp)$_2$]$^+$ ion, in addition to [NiBr(dmbp)]$^+$. When equimolar amounts of two different bromopyridines (2-bromopyridine and dmbp) were coupled using nickel(0), eight different intermediates corresponding to all possible complexes of 2,2′-bipyridine, 6-methyl-2,2′-bipyridine and 6,6′-dimethyl-2,2′-bipyridine were detected. Upon hydrolysis, the free bipyridines were liberated.

The Suzuki reaction is widely used in synthetic organic chemistry for the coupling of aryl halides with aryl boronic acids, and is catalysed by palladium(0) complexes. ESI MS has been used to detect intermediates in this reaction. Catalytic intermediates have also been observed in the palladium(0)-catalysed Suzuki coupling reaction between aryl halides and arylboronic acids.[29] The proposed mechanism of this reaction involves oxidative addition of the aryl halide to palladium(0), transmetallation of boron for palladium (giving a *trans*-biaryl palladium complex), *cis-trans* isomerisation, followed by reductive elimination of the biaryl. For studying this reaction by ESI MS, the aryl halide was replaced by a pyridyl bromide, so that the basic pyridine nitrogen is available for protonation of any species which might be otherwise invisible to electrospray ionisation, because of lack of an ionisation site. Pyridyl-palladium(II) species were observed as intermediates, as were pyridyl aryl palladium(II) species, confirming the presence of the proposed intermediates in the Suzuki reaction. In related chemistry, reaction intermediates in the palladium-catalysed oxidative self-coupling of arylboronic acids have been detected by ESI MS.[30]

Several variants of the Heck reaction (coupling of aryl compounds with alkenes) have been studied by ESI MS. As an example to illustrate this particular reaction, the phosphine-free reaction using a palladium catalyst with arene diazonium salts as the arylating reagent has been studied.[31] Thus, the reaction of the diazonium salt 4-[MeOC$_6$H$_4$N$_2$]$^+$BF$_4^-$ with [Pd$_2$(dba)$_3$] (dba = dibenzylideneacetone) in MeCN yielded the range of aryl palladium(II) species formed by addition of the diazonium salt: [4-MeOC$_6$H$_4$Pd(NCMe)$_n$]$^+$ (n = 2, 3), [4-MeOC$_6$H$_4$Pd(dba)$_2$]$^+$ and [4-MeOC$_6$H$_4$Pd-(dba) (NCMe)]$^+$. The latter of these intermediates dominated the mixture after a 90-minute reaction time and was considered to be the most likely intermediate in the reaction with an added alkene. The diazonium intermediate [4-MeOC$_6$H$_4$N$_2$Pd]$^+$ (or solvated analogues) was too short-lived to be detected. The palladium-catalysed arylation of methyl acrylate has also been monitored and a key intermediate detected.[32] Other intermediates detected in Heck-type reactions using ESI MS include ones in the palladium-catalysed intramolecular arylation of enamidines[33] and the palladium-catalysed arylation of dihydrofurans.[34]

ESI MS has also been used to characterise chiral reaction intermediates in a titanium-alkoxide mediated enantioselective sulfoxidation reaction. The reaction proceeds via a peroxometal complex, employing a titanium(IV) trialkanolamine-alkylhydroperoxide system.[35]

8.5.2 Detection and Chemistry of Reaction Intermediates in the Gas Phase

A rapidly expanding area of inorganic mass spectrometry research involves fundamental studies of reactive intermediates. Using ESI, suitable precursors can be transferred to the gas phase in the usual way. By means of CID, reactive fragments can then be generated and trapped using MS techniques described in Chapters 2 and 3. Ion-molecule reactions can then be studied in the gas phase to provide information about fundamental reaction steps that are involved in the chemistry of metal complexes. In this way, reaction intermediates can be isolated from other solution species, allowing the specific reactivity of the reactive intermediate to be studied. By using CID, thermochemical data can be obtained for parameters such as metal-ligand bond energies and energies of elementary reaction steps. Some of the processes that have been studied in this area include carbon-hydrogen bond activation, oxidation and alkene polymerisation, and the general area has been the subject of a review.[36] Some specific examples serve to illustrate the types of studies that have been done.

ESI MS has been used to study carbon-hydrogen bond activation in the gas phase; this reaction is one of great significance in organometallic chemistry and the MS studies have provided some useful insights. A study of reactions of the cations $[CpIrMe(PMe_3)]^+$ and $[Cp^*IrMe(PMe_3)]^+$ with pentane, cyclohexane and benzene was found to proceed by a dissociative mechanism, forming 16-electron iridium(III) intermediates such as **8.3**, as opposed to 18-electron iridium(V) species.[37]

8.3

The catalytic gas-phase oxidation of primary and secondary alcohols to aldehydes and ketones has been the subject of a detailed ESI MS study.[38] The dinuclear molybdenum complex $[Mo_2O_6(OCHR_2)]^-$ was the catalytically active species, which was found to participate in two catalytic cycles. Fundamental steps that were probed using kinetic measurements and by variation of the substrate alcohols (structural variations and isotopic labelling) were reaction of $[Mo_2O_6(OH)]^-$ with alcohol (with elimination of water) to give $[Mo_2O_6(OCHR_2)]^-$, oxidation of the alkoxo ligand (and liberation as the aldehyde or ketone) and regeneration of the catalyst by oxidation using CH_3NO_2; these steps are analogous to those proposed in the industrial oxidation of methanol to formaldehyde using an MoO_3 catalyst. The analogous tungsten species $[W_2O_6(OCHR_2)]^-$ underwent loss of alkene as opposed to alkoxide oxidation. The mononuclear anions $[MO_3(OH)]^-$ (M = Cr, Mo, W) were found to be inert to methanol, which indicated the importance of a binuclear complex in the catalytic cycle.

The Ziegler-Natta process for polymerisation of alkenes (which uses a zirconocene precursor such as Cp_2ZrMe_2 or Cp_2ZrCl_2 in conjunction with a strong Lewis acid e.g. methylaluminoxane) has also been studied using ESI MS techniques.[39] Reaction of $[Cp_2Zr(CH_3)_2]$ with a weak acid $[PhNMe_2H][B(C_6F_5)_4]$ in MeCN solution gave $[Cp_2Zr(CH_3)(CH_3CN)]^+$ as the dominant ion in the ESI mass spectrum. With more forcing desolvation conditions, $[Cp_2Zr(CH_3)]^+$ was formed, which was then reacted with

1-butene in the second octopole of the mass spectrometer. Insertion of 1-butene in to the Zr-CH$_3$ bond was seen, generating ions formed by multiple butene insertions up to [Cp$_2$Zr(CH$_2$CHEt)$_4$CH$_3$]$^+$.

By using an optically-transparent nanospray tip, it has been possible to identify intermediates generated by the photolysis of the cation [CpFe(C$_6$H$_6$)]$^+$.[40] This complex, which is of interest as a light-sensitive photoinitiator for polymerisation of epoxides and other monomers, was found to undergo electrospray unchanged in MeCN solution in the absence of light. However, irradiation of the tip with an argon laser (488 nm) generated [CpFe(MeCN)$_x$]$^+$ (x = 1 to 3) and [Fe(MeCN)$_y$]$^{2+}$ (y = 3 to 6), the first species formed by loss of the benzene ligand, and the second from subsequent loss of the cyclopenta-dienyl ring. When cyclohexene oxide (L) was incorporated into the photolysis solution in MeCN, species [Fe(MeCN)$_x$(L)]$^{2+}$ (x = 3 or 4) were observed, but when the solvent was changed to (poorly-coordinating) ClCH$_2$CH$_2$Cl, species [(H$_2$O)Fe(L)$_{4-12}$]$^{2+}$, together with other non-iron containing species were observed. It was reasoned that these intermediates with multiple cyclohexene oxide molecules contain a growing polymer chain attached to the metal centre.

Other applications of reactions of gas-phase species generated by ESI include reactions of rhodium phosphine complexes relevant to homogeneous catalytic hydrogenation,[41] and the reactivity of manganese oxo complexes.[42]

8.5.3 Screening of New Catalysts Using Mass Spectrometry

Traditionally, new catalysts have been discovered largely by a trial-and-error approach, where a new potential catalyst is synthesised and screened for its catalytic activity on the macroscopic scale. This is a laborious approach that has slowed down the rate of catalyst discovery. In recent years, increasing use has been made of combinatorial chemistry type approaches, where a large number ('library') of unique compounds are synthesised, either in a mixture or on polymer beads, with each individual bead holding one unique compound. Because of the ability of a mass spectrometer to act as a mass selector, potential catalysts can be synthesised in a mixture, even as a byproduct, and they do not have to be pre-purified. The selected metal species can then be screened for their activity using techniques such as ion-molecule reactions in the gas phase. While the origins of this methodology lies in the areas of pharmaceutical discovery and biology, the general techniques are increasingly being applied to areas involving inorganic chemistry, such as materials science and catalysis. When coupled to highly sensitive MS analysis techni-ques, this powerful combination allows large numbers of compounds to be synthesised and analysed quickly.[43] The applications of high throughput catalyst screening using MS techniques has been the subject of an excellent review by Chen, one of the pioneers in this area.[44]

To illustrate this general type of approach, carbene catalysts for the alkene metathesis reaction have been screened using a mass spectrometry approach.[45] In this methodology, ruthenium carbene complexes containing ancillary halide and diphosphine ligands of the type **8.4** were generated in solution from the Grubbs catalysts [RuCl$_2$(=CHR)(PCy$_3$)$_2$] and various diphosphines. The desired monomeric cations were then mass selected in the first quadrupole of the mass spectrometer, and their reactivity ascertained towards acyclic and cyclic alkenes. By variation of the various ligands and subsequent generation of a wide range of related complexes, the factors that provide for good catalytic activity were able to be defined. The high throughput nature of this methodology is nicely demonstrated by the study carrying out 180 reactions, each repeated 10 times, over a two-week period.

$$\left[\begin{array}{c} \begin{array}{c} C \begin{array}{c} PR_2 \\ PR_2 \end{array} \end{array} Ru \overset{R'}{\underset{}{\|}} - halide \end{array} \right]^+$$

8.4

In another application, a set of 21 bidentate, chiral phosphorus-nitrogen donor ligands were synthesised and converted into their rhodium complexes by reaction with [RhCl(cod)]$_2$. These complexes were then screened for catalytic activity in the hydro-silylation of 1-naphthyl methyl ketone, which identified the structural features that produced the best performing catalyst.[46] The screening was achieved using the MSEED technique (Mass Spectrometry Enantiomeric Excess Determination).[47] In this, the enantiomeric excess of chiral alcohols and amines can be determined by derivatisation of the target compound with an equimolar mixture of two 'mass-tagged' chiral derivatising agents, which differ in a substituent remote from the chiral centre. The mass of the resulting derivative is related to its absolute configuration, so that measurement of the relative amounts of the two derivatives gives a measure of the enantiomeric excess of the original material. This measurement is rapidly and simply carried out using ESI MS, on a small amount of material.

References

1. B. Hong, T. P. S. Thoms, H. J. Murfee and M. J. Lebrun, *Inorg. Chem.*, 1997, **36**, 6146.

2. J. W. Kriesel, S. König, M. A. Freitas, A. G. Marshall, J. A. Leary and T. D. Tilley, *J. Am. Chem. Soc.*, 1998, **120**, 12207.

3. T. Kemmitt and W. Henderson, *J. Chem. Soc., Perkin Trans. 1*, 1997, 729.

4. W. T. S. Huck, F. C. J. M. van Veggel and D. N. Reinhoudt, *Angew. Chem., Int. Ed. Engl.*, 1996, **35**, 1213; G. Chessa, A. Scrivanti, L. Canovese, F. Visentin, and P. Uguagliati, *Chem. Comm.*, 1999, 959.

5. C. M. Casado, B. González, I. Cuadrado, B. Alonso, M. Morán and J. Losada, *Angew. Chem., Int. Ed. Engl.*, 2000, **39**, 2135.

6. S. M. Cohen, S. Petoud and K. N. Raymond, *Chem. Eur. J.*, 2001, **7**, 272.

7. E. C. Constable, O. Eich and C. E. Housecroft, *J. Chem. Soc., Dalton Trans.*, 1999, 1363.

8. V. Huc, P. Boussaguet and P. Mazerolles, *J. Organomet. Chem.*, 1996, **521**, 253.

9. A. Marquis-Rigault, A. Dupont-Gervais, A. Van Dorsselaer and J.-M. Lehn, *Chem. Eur. J.*, 1996, **2**, 1395.

10. B. Hasenknopf, J.-M. Lehn, N. Boumediene, E. Leize and A. Van Dorsselaer, *Angew. Chem., Int. Ed. Engl.*, 1998, **37**, 3265; B. Hasenknopf, J.-M. Lehn, N. Boumediene, A. Dupont-Gervais, A. Van Dorsselaer, B. Kneisel and D. Fenske, *J. Am. Chem. Soc.*, 1997, **119**, 10956; C. Piguet, G. Hopfgartner, B. Bocquet, O. Schaad and A. F. Williams, *J. Am. Chem. Soc.*, 1994, **116**, 9092.

11. S.-S. Sun and A. J. Lees, *Inorg. Chem.*, 2001, **40**, 3154; P. J. Stang, D. H. Cao, K. Chen, G. M. Gray, D. C. Muddiman and R. D. Smith, *J. Am. Chem. Soc.*, 1997, **119**, 5163.

12. F. M. Romero, R. Ziessel, A. Dupont-Gervais and A. Van Dorsselaer, *Chem. Comm.*, 1996, 551; D. P. Funeriu, J.-M. Lehn, G. Baum and D. Fenske, *Chem. Eur. J.*, 1997, **3**, 99.

13. A. Marquis-Rigault, A. Dupont-Gervais, P. N. W. Baxter, A. Van Dorsselaer and J.-M. Lehn, *Inorg. Chem.*, 1996, **35**, 2307; E. Leize, A. Van Dorsselaer, R. Krämer and J.-M. Lehn, *Chem. Comm.*, 1993, 990.

14. M. Fujita, F. Ibukuro, H. Hagihara and K. Ogura, *Nature*, 1994, **367**, 720.

15. K. Yamaguchi, *J. Mass Spectrom.*, 2003, **38**, 473; K. Yamaguchi and S. Sakamoto, *JEOL News*, 2002, **28A**, 2.

16. S.-W. A. Fong and T. S. A. Hor, *J. Chem. Soc., Dalton Trans.*, 1999, 639.

17. Z. Li, S.-W. A. Fong, J. S. L. Yeo, W. Henderson, K. F. Mok and T. S. A. Hor, *Modern Coordination Chemistry: The contributions of Joseph Chatt*, Royal Society of Chemistry, Cambridge, 2002, 355.

18. C. E. Briant, C. J. Gardner, T. S. A. Hor, N. D. Howells and D. M. P. Mingos, *J. Chem. Soc., Dalton Trans.*, 1984, 2645.

19. S.-W. A. Fong, J. J. Vittal, W. Henderson, T. S. A. Hor, A. G. Oliver and C. E. F. Rickard, *Chem. Commun.*, 2001, 421.

20. S.-W. A. Fong, W. T. Yap, J. J. Vittal, T. S. A. Hor, W. Henderson, A. G. Oliver and C. E. F. Rickard, *J. Chem. Soc., Dalton Trans.*, 2001, 1986.

21. J. S. L. Yeo, J. J. Vittal, W. Henderson and T. S. A. Hor, *J. Chem. Soc., Dalton Trans.*, 2002, 328.

22. X Xu, S.-W. A. Fong, Z. Li, Z.-H. Loh, F. Zhao, J. J. Vittal, W. Henderson, S. B. Khoo and T. S. A. Hor, *Inorg. Chem.*, 2002, **41**, 6838.

23. S.-W. A. Fong, T. S. A. Hor, J. J. Vittal, W. Henderson and S. Cramp, *Inorg. Chim. Acta*, 2004, **357**, 1152.

24. S.-W. A. Fong, T. S. A. Hor, S. M. Devoy, B. A. Waugh, B. K. Nicholson, and W. Henderson, *Inorg. Chim. Acta, Inorg. Chim. Acta*, 2004, **357**, 2081.

25. S.-W. A. Fong, T. S. A. Hor, W. Henderson, B. K. Nicholson, S. Gardyne and S. M. Devoy, *J. Organomet. Chem.*, 2003, **679**, 24.

26. J. S. L. Yeo, J. J. Vittal, W. Henderson and T. S. A. Hor, *Inorg. Chem.*, 2002, **41**, 1194.

27. S. R. Wilson, J. Perez and A. Pasternak, *J. Am. Chem. Soc.*, 1993, **115**, 1994.

28. S. R. Wilson and Y. Wu, *Organometallics*, 1993, **12**, 1478.

29. A. O. Aliprantis and J. W. Canary, *J. Am. Chem. Soc.*, 1994, **116**, 6985.

30. M. A. Aramendía, F. Lafont, M. Moreno-Mañas, R. Pleixats and A. Roglans, *J. Org. Chem.*, 1999, **64**, 3592.

31. A. A. Sabino, A. H. L. Machado, C. R. D. Correia and M. N. Eberlin, *Angew. Chem., Int. Ed. Engl.*, 2004, **43**, 2514.

32. J. M. Brown and K. K. Hii, *Angew. Chem., Int. Ed. Engl.*, 1996, **35**, 657.

33. L. Ripa and A. Hallberg, *J. Org. Chem.*, 1996, **61**, 7147.

34. K. K. Hii, T. D. W. Claridge and J. M. Brown, *Angew. Chem., Int. Ed. Engl.*, 1997, **36**, 984.

35. M. Bonchio, G. Licini, G. Modena, S. Moro, O. Boltolini, P. Traldi and W. Nugent, *J. Chem. Soc., Chem. Commun.*, 1997, 869.

36. D. A. Plattner, *Int. J. Mass Spectrom.*, 2001, **207**, 125.

37. C. Hinderling, D. A. Plattner and P. Chen, *Angew. Chem., Int. Ed. Engl.*, 1997, **36**, 243; C. Hinderling, D. Feichtinger, D. A. Plattner and P. Chen, *J. Am. Chem. Soc.*, 1997, **119**, 10793.

38. T. Waters, R. A. J. O'Hair and A. G. Wedd, *J. Am. Chem. Soc.*, 2003, **125**, 3384.

39. D. Feichtinger, D. A. Plattner and P. Chen., *J. Am. Chem. Soc.*, 1998, **120**, 7125.

40. W. Ding, K. A. Johnson, I. J. Amster and C. Kutal, *Inorg. Chem.*, 2001, **40**, 6865.

41. Y.-M. Kim and P. Chen, *Int. J. Mass Spectrom.*, 1999, **185/186/187**, 871.

42. D. Feichtinger and D. A. Plattner, *Chem. Eur. J.*, 2001, **7**, 591.

43. A. Triolo, M. Altamura, F. Cardinall, A. Sisto and C. A. Maggi, *J. Mass Spectrom.*, 2001, **36**, 1249.

44. P. Chen, *Angew. Chem., Int. Ed. Engl.*, 2003, **42**, 2832.

45. M. A. O. Volland, C. Adlhart, C. A. Kiener, P. Chen and P. Hofmann, *Chem. Eur. J.*, 2001, **7**, 4621.

46. S. Yao, J.-C. Meng, G. Siuzdak and M. G. Finn, *J. Org. Chem.*, 2003, **68**, 2540.

47. J. Guo, J. Wu, G. Siuzdak and M. G. Finn, *Angew. Chem., Int. Ed. Engl.*, 1999, **38**, 1755.

Appendix 1 Naturally Occurring Isotopes

The table lists 287 naturally occurring isotopes, the number of protons (# p = atomic number), number of neutrons (# n), the atomic mass of each isotope (referenced to $^{12}C =$ 12 exactly), the natural abundance of each isotope, the atomic weight of each element (equal to the sum of the atomic masses of each isotope multiplied by their natural abundance), the relative intensities of each isotope where the most abundant isotope is set to 100 (MS %), and finally the isotope pattern of each element is presented in bar graph format. No data are given for isotopes that do not occur naturally, other than the atomic number, and no elements of atomic number > 92 (U) are included.

Isotope	# p	# n	Atomic mass	Natural abundance %	Atomic weight	MS %	Isotope pattern
^{1}H	1	0	1.007825035 (12)	99.9885 (70)	1.00794 (7)	100.0	
^{2}H		1	2.014101779 (24)	0.0115 (70)		0.0	
^{3}He	2	1	3.01602931 (4)	0.000137 (3)	4.002602 (2)	0.0	
^{4}He		2	4.00260324 (5)	99.999863 (3)		100.0	
^{6}Li	3	3	6.0151214 (7)	[7.59 (4)]	[6.941 (2)]	8.2	
^{7}Li		4	7.0160030 (9)	[92.41 (4)]		100.0	

Mass Spectrometry of Inorganic, Coordination and Organometallic Compounds   W. Henderson and J. S. McIndoe
© 2005 John Wiley & Sons, Ltd   ISBNs: 0-470-85015-9 (HB); 0-470-85016-7 (PB)

| ⁹Be | 4 | 5 | 9.0121822 (4) | 100 | 9.012182 (3) | 100.0 |

| ¹⁰B | 5 | 5 | 10.0129369 (3) | 19.9 (7) | 10.811 (7) | 24.8 |
| ¹¹B | | 6 | 11.0093054 (4) | 80.1 (7) | | 100.0 |

| ¹²C | 6 | 6 | 12 | 98.93 (8) | 12.0107 (8) | 100.0 |
| ¹³C | | 7 | 13.003354826 (17) | 1.07 (8) | | 1.1 |

| ¹⁴N | 7 | 7 | 14.003074002 (26) | 99.632 (7) | 14.0067 (2) | 100.0 |
| ¹⁵N | | 8 | 15.00010897 (4) | 0.368 (7) | | 0.4 |

¹⁶O	8	8	15.99491463 (5)	99.757 (16)	15.9994 (3)	100.0
¹⁷O		9	16.9991312 (4)	0.038 (1)		0.0
¹⁸O		10	17.9991603 (9)	0.205 (14)		0.2

| ¹⁹F | 9 | 10 | 18.99840322 (15) | 100 | 18.9984032 (5) | 100.0 |

²⁰Ne	10	10	19.9924356 (22)	90.48 (3)	20.1797 (6)	100.0
²¹Ne		11	20.9938428 (21)	0.27 (1)		0.3
²²Ne		12	21.9913831 (18)	9.25 (3)		10.2

| ²³Na | 11 | 12 | 22.9897677 (10) | 100 | 22.989770 (2) | 100.0 |

^{24}Mg	12	12	23.9850423 (8)	78.99 (4)	24.3050 (6)	100.0
^{25}Mg		13	24.9858374 (8)	10.00 (1)		12.7
^{26}Mg		14	25.9825937 (8)	11.01 (3)		13.9

| ^{27}Al | 13 | 14 | 26.9815386 (8) | 100 | 26.981538 (2) | 100.0 |

^{28}Si	14	14	27.9769271 (7)	92.2297 (7)	28.0855 (3)	100.0
^{29}Si		15	28.9764949 (7)	4.6832 (5)		5.1
^{30}Si		16	29.9737707 (7)	3.0872 (5)		3.4

| ^{31}P | 15 | 16 | 30.9737620 (6) | 100 | 30.973761 (2) | 100.0 |

^{32}S	16	16	31.97207070 (25)	94.93 (31)	32.065 (5)	100.0
^{33}S		17	32.97145843 (23)	0.76 (2)		0.8
^{34}S		18	33.96786665 (22)	4.29 (28)		4.5
^{36}S		20	35.96708062 (27)	0.02 (1)		0.0

| ^{35}Cl | 17 | 18 | 34.968852721 (69) | 75.78 (4) | 35.453 (2) | 100.0 |
| ^{37}Cl | | 20 | 36.96590262 (11) | 24.22 (4) | | 32.0 |

^{36}Ar	18	18	35.96754552 (29)	0.3365 (30)	39.948 (1)	0.3
^{38}Ar		20	37.9627325 (9)	0.0632 (5)		0.1
^{40}Ar		22	39.9623837 (14)	99.6003 (30)		100.0

^{39}K	19	20	38.9637074 (12)	93.2581 (44)	39.0983 (1)	100.0
^{40}K		21	39.9639992 (12)	0.0117 (1)		0.0
^{41}K		22	40.9618254 (12)	6.7302 (44)		7.2

[40]Ca	20	20	39.9625906 (13)	96.941 (156)	40.078 (4)	100.0
[42]Ca		22	41.9586176 (13)	0.647 (23)		0.7
[43]Ca		23	42.9587662 (13)	0.135 (10)		0.1
[44]Ca		24	43.9554806 (14)	2.086 (110)		2.2
[46]Ca		26	45.953689 (4)	0.004 (3)		0.0
[48]Ca		28	47.952533 (4)	0.187 (21)		0.2

[45]Sc	21	24	44.9559100 (14)	100	44.955910 (8)	100.0

[46]Ti	22	26	45.9526294 (14)	8.25 (3)	47.867 (1)	11.2
[47]Ti		27	46.9517640 (11)	7.44 (2)		10.1
[48]Ti		28	47.9479473 (11)	73.72 (3)		100.0
[49]Ti		29	48.9478711 (11)	5.41 (2)		7.3
[50]Ti		30	49.9447921 (12)	5.18 (2)		7.0

[50]V	23	27	49.9471609 (17)	0.250 (4)	50.9415 (1)	0.2
[51]V		28	50.9439617 (17)	99.750 (4)		100.0

[50]Cr	24	26	49.9460464 (17)	4.345 (13)	51.9961 (6)	5.2
[52]Cr		28	51.9405098 (17)	83.789 (18)		100.0
[53]Cr		29	52.9406513 (17)	9.501 (17)		11.3
[54]Cr		30	53.9388825 (17)	2.365 (7)		2.8

[55]Mn	25	30	54.9380471 (16)	100	54.938049 (9)	100.0

[54]Fe	26	28	53.9396127 (15)	5.845 (35)	55.845 (2)	6.4
[56]Fe		30	55.9349393 (16)	91.754 (36)		100.0
[57]Fe		31	56.9353958 (16)	2.119 (10)		2.4
[58]Fe		32	57.9332773 (16)	0.282 (4)		0.3

[59]Co	27	32	58.9331976 (16)	100	58.933200 (9)	100.0

[58]Ni	28	31	57.9353462 (16)	68.0769 (89)	58.6934 (2)	100.0
[60]Ni		33	59.9307884 (16)	26.2231 (77)		38.5
[61]Ni		34	60.9310579 (16)	1.1399 (6)		1.7
[62]Ni		35	61.9283461 (16)	3.6345 (17)		5.3
[64]Ni		37	63.9279679 (17)	0.9256 (9)		1.4

[63]Cu	29	34	62.9295989 (17)	69.17 (3)	63.546 (3)	100.0
[65]Cu		36	64.9277929 (20)	30.83 (3)		44.6

[64]Zn	30	34	63.9291448 (19)	48.63 (60)	65.409 (4)	100.0
[66]Zn		36	65.9260347 (17)	27.90 (27)		57.4
[67]Zn		37	66.9271291 (17)	4.10 (13)		8.4
[68]Zn		38	67.9248459 (18)	18.75 (51)		38.6
[70]Zn		40	69.925325 (4)	0.62 (3)		1.3

[69]Ga	31	38	68.925580 (3)	60.108 (9)	69.723 (1)	100.0
[71]Ga		40	70.9247005 (25)	39.892 (9)		66.4

[70]Ge	32	38	69.9242497 (16)	20.84 (87)	72.64 (1)	57.4
[72]Ge		40	71.9220789 (16)	27.54 (34)		75.9
[73]Ge		41	72.9234626 (16)	7.73 (5)		21.3
[74]Ge		42	73.9211774 (15)	36.28 (73)		100.0
[76]Ge		44	75.9214016 (17)	7.61 (38)		21.0

[75]As	33	42	74.9215942 (17)	100	74.92160 (2)	100.0

[74]Se	34	40	73.9224746 (16)	0.89 (4)	78.96 (3)	1.8
[76]Se		42	75.9192120 (16)	9.37 (29)		18.9
[77]Se		43	76.9199125 (16)	7.63 (16)		15.4
[78]Se		44	77.9173076 (16)	23.77 (28)		47.9
[80]Se		46	79.9165196 (19)	49.61 (41)		100.0
[82]Se		48	81.9166978 (23)	8.73 (22)		17.6

[79]Br	35	44	78.9183361 (26)	50.69 (7)	79.904 (1)	100.0
[81]Br		46	80.916289 (6)	49.31 (7)		97.3

^{78}Kr	36	42	77.920396 (9)	0.35 (1)	83.798 (2)	0.6
^{80}Kr		44	79.916380 (9)	2.28 (6)		4.0
^{82}Kr		46	81.913482 (6)	11.58 (14)		20.3
^{83}Kr		47	82.914135 (4)	11.49 (6)		20.2
^{84}Kr		48	83.911507 (4)	57.00 (4)		100.0
^{86}Kr		50	85.910616 (5)	17.30 (22)		30.4

^{85}Rb	37	48	84.911794 (3)	72.17 (2)	85.4678 (3)	100.0
^{87}Rb		50	86.909187 (3)	27.83 (2)		38.6

^{84}Sr	38	46	83.913430 (4)	0.56 (1)	87.62 (1)	0.7
^{86}Sr		48	85.9092672 (28)	9.86 (1)		11.9
^{87}Sr		49	86.9088841 (28)	7.00 (1)		8.5
^{88}Sr		50	87.9056188 (28)	82.58 (1)		100.0

^{89}Y	39	50	89.905849 (3)	100	88.90585 (2)	100.0

^{90}Zr	40	50	89.9047026 (26)	51.45 (40)	91.224 (2)	100.0
^{91}Zr		51	90.9056439 (26)	11.22 (5)		21.8
^{92}Zr		52	91.9050386 (26)	17.15 (8)		33.3
^{94}Zr		54	93.9063148 (28)	17.38 (28)		33.8
^{96}Zr		56	95.908275 (4)	2.80 (9)		5.4

^{93}Nb	41	52	92.9063772 (27)	100	92.90638 (2)	100.0

^{92}Mo	42	50	91.906809 (4)	14.84 (35)	95.94 (2)	61.5
^{94}Mo		52	93.9050853 (26)	9.25 (12)		38.3
^{95}Mo		53	94.9058411 (22)	15.92 (13)		66.0
^{96}Mo		54	95.9046785 (22)	16.68 (2)		69.1
^{97}Mo		55	96.9060205 (22)	9.55 (8)		39.6
^{98}Mo		56	97.9054073 (22)	24.13 (31)		100.0
^{100}Mo		58	99.907477 (6)	9.63 (23)		39.9

Tc	43	–	–	–	–	–

[96]Ru	44	52	95.907599 (8)	5.54 (14)	101.07 (2)	17.6	
[98]Ru		54	97.905287 (7)	1.87 (3)		5.9	
[99]Ru		55	98.9059389 (23)	12.76 (14)		40.4	
[100]Ru		56	99.9042192 (24)	12.60 (7)		39.9	
[101]Ru		57	100.9055819 (24)	17.06 (2)		54.1	
[102]Ru		58	101.9043485 (25)	31.55 (14)		100.0	
[104]Ru		60	103.905424 (6)	18.62 (27)		59.0	

[103]Rh	45	58	102.905500 (4)	100	102.90550 (2)	100.0	

[102]Pd	46	56	101.905634 (5)	1.02 (1)	106.42 (1)	3.7	
[104]Pd		58	103.904029 (6)	11.14 (8)		40.8	
[105]Pd		59	104.905079 (6)	22.33 (8)		81.7	
[106]Pd		60	105.903478 (6)	27.33 (3)		100.0	
[108]Pd		62	107.903895 (4)	26.46 (9)		96.8	
[110]Pd		64	109.905167 (20)	11.72 (9)		42.9	

[107]Ag	47	60	106.905092 (6)	51.839 (7)	107.8682 (2)	100.0	
[109]Ag		62	108.904756 (4)	48.161 (7)		92.9	

[106]Cd	48	58	105.906461 (7)	1.25 (6)	112.411 (8)	4.4	
[108]Cd		60	107.904176 (6)	0.89 (3)		3.1	
[110]Cd		62	109.903005 (4)	12.49 (18)		43.5	
[111]Cd		63	110.904182 (3)	12.80 (12)		44.6	
[112]Cd		64	111.902757 (3)	24.13 (21)		84.0	
[113]Cd		65	112.904400 (3)	12.22 (12)		42.5	
[114]Cd		66	113.903357 (3)	28.73 (42)		100.0	
[116]Cd		68	115.904755 (4)	7.49 (18)		26.1	

[113]In	49	64	112.904061 (4)	4.29 (5)	114.818 (3)	4.5	
[115]In		66	114.903882 (4)	95.71 (5)		100.0	

[112]Sn	50	62	111.904826 (5)	0.97 (1)	118.710 (7)	3.0	
[114]Sn		64	113.902784 (4)	0.66 (1)		2.0	
[115]Sn		65	114.903348 (3)	0.34 (1)		1.0	
[116]Sn		66	115.901747 (3)	14.54 (9)		44.6	
[117]Sn		67	116.902956 (3)	7.68 (7)		23.6	
[118]Sn		68	117.901609 (3)	24.22 (9)		74.3	
[119]Sn		69	118.903311 (3)	8.59 (4)		26.4	
[120]Sn		70	119.9021991 (29)	32.58 (9)		100.0	
[122]Sn		72	121.9034404 (30)	4.63 (3)		14.2	
[124]Sn		74	123.9052743 (17)	5.79 (5)		17.8	

^{121}Sb	51	70	120.9038212 (29)	57.21 (5)	121.760 (1)	100.0
^{123}Sb		72	122.9042160 (24)	42.79 (5)		74.8

^{120}Te	52	68	119.904048 (21)	0.09 (1)	127.60 (3)	0.3
^{122}Te		70	121.903050 (3)	2.55 (12)		7.5
^{123}Te		71	122.9042710 (22)	0.89 (3)		2.6
^{124}Te		72	123.9028180 (18)	4.74 (14)		13.9
^{125}Te		73	124.9044285 (25)	7.07 (15)		20.7
^{126}Te		74	125.9033095 (25)	18.84 (25)		55.3
^{128}Te		76	127.904463 (4)	31.74 (8)		93.1
^{130}Te		78	129.906229 (5)	34.08 (62)		100.0

^{127}I	53	74	126.904473 (5)	100	126.90447 (3)	100.0

^{124}Xe	54	66	123.9058942 (22)	0.09 (1)	131.293 (6)	0.3
^{126}Xe		68	125.904281 (8)	0.09 (1)		0.3
^{128}Xe		70	127.9035312 (17)	1.92 (3)		7.1
^{129}Xe		71	128.9047801 (21)	26.44 (24)		98.3
^{130}Xe		72	129.9035094 (17)	4.08 (2)		15.2
^{131}Xe		73	130.905072 (5)	21.18 (3)		78.8
^{132}Xe		74	131.904144 (5)	26.89 (6)		100.0
^{134}Xe		76	133.905395 (8)	10.44 (10)		38.8
^{136}Xe		78	135.907214 (8)	8.87 (16)		33.0

^{133}Cs	55	78	132.905429 (7)	100	132.90545 (2)	100.0

^{130}Ba	56	74	129.906282 (8)	0.106 (1)	137.327 (7)	0.2
^{132}Ba		76	131.905042 (9)	0.101 (1)		0.1
^{134}Ba		78	133.904486 (7)	2.417 (18)		33.7
^{135}Ba		79	134.905665 (7)	6.592 (12)		9.2
^{136}Ba		80	135.904553 (7)	7.854 (24)		10.9
^{137}Ba		81	136.905812 (6)	11.232 (24)		15.7
^{138}Ba		82	137.905232 (6)	71.698 (42)		100.0

^{138}La	57	81	137.907105 (6)	0.090 (1)	138.9055 (2)	0.1
^{139}La		82	138.906347 (5)	99.910 (1)		100.0

^{136}Ce	58	78	135.907140 (50)	0.185 (2)	140.116 (1)	0.2
^{138}Ce		80	137.905985 (12)	0.251 (2)		0.3
^{140}Ce		82	139.905433 (4)	88.450 (51)		100.0
^{142}Ce		84	141.909241 (4)	11.114 (51)		12.6

^{141}Pr	59	82	140.907647 (4)	100	140.90765 (2)	100.0	

^{142}Nd	60	82	141.907719 (4)	27.2 (5)	144.24 (3)	100.0
^{143}Nd		83	142.909810 (4)	12.2 (2)		44.9
^{144}Nd		84	143.910083 (4)	23.8 (3)		97.5
^{145}Nd		85	144.912570 (4)	8.3 (1)		30.5
^{146}Nd		86	145.913113 (4)	17.2 (3)		63.2
^{148}Nd		88	147.916889 (4)	5.7 (1)		21.0
^{150}Nd		90	149.920887 (4)	5.6 (2)		20.6

Pm	–	–	–	–	–	–

^{144}Sm	62	82	143.911998 (4)	3.07 (7)	150.36 (3)	11.5
^{147}Sm		85	146.914894 (4)	14.99 (18)		56.0
^{148}Sm		86	147.914819 (4)	11.24 (10)		42.0
^{149}Sm		87	148.917180 (4)	13.82 (7)		51.7
^{150}Sm		88	149.917273 (4)	7.38 (1)		27.6
^{152}Sm		90	151.919728 (4)	26.75 (16)		100.0
^{154}Sm		92	153.922205 (4)	22.75 (29)		85.0

^{151}Eu	63	88	150.919702 (8)	47.81 (3)	151.964 (1)	91.6
^{153}Eu		90	152.921225 (4)	52.19 (3)		100.0

^{152}Gd	64	88	151.919786 (4)	0.20 (1)	157.25 (3)	0.8
^{154}Gd		90	153.920861 (4)	2.18 (3)		8.8
^{155}Gd		91	154.922618 (4)	14.80 (12)		59.6
^{156}Gd		92	155.922118 (4)	20.47 (9)		82.4
^{157}Gd		93	156.923956 (4)	15.65 (2)		63.0
^{158}Gd		94	157.924019 (4)	24.84 (7)		100.0
^{160}Gd		96	159.927049 (4)	21.86 (19)		88.0

^{159}Tb	65	94	158.925342 (4)	100	158.92534 (2)	100.0

^{156}Dy	66	90	155.924277 (8)	0.06 (1)	162.500 (1)	0.2
^{158}Dy		92	157.924403 (5)	0.10 (1)		0.4
^{160}Dy		94	159.925193 (4)	2.34 (8)		8.3
^{161}Dy		95	160.926930 (4)	18.91 (24)		67.1
^{162}Dy		96	161.926795 (4)	25.51 (26)		90.5
^{163}Dy		97	162.928728 (4)	24.90 (16)		88.4
^{164}Dy		98	163.929171 (4)	28.18 (37)		100.0

^{165}Ho	67	98	164.930319 (4)	100	164.93032 (2)	100.0

^{162}Er	68	94	161.928775 (4)	0.14 (1)	167.259 (3)	0.4
^{164}Er		96	163.929198 (4)	1.61 (3)		4.8
^{166}Er		98	165.930290 (4)	33.61 (35)		100.0
^{167}Er		99	166.932046 (4)	22.93 (17)		68.2
^{168}Er		100	167.932368 (4)	26.78 (26)		79.7
^{170}Er		102	169.935461 (4)	14.93 (27)		44.4

^{169}Tm	69	100	168.934212 (4)	100	168.93421 (2)	100.0

^{168}Yb	70	98	167.933894 (5)	0.13 (1)	173.04 (3)	0.4
^{170}Yb		100	169.934759 (4)	3.04 (15)		9.6
^{171}Yb		101	170.936323 (3)	14.28 (57)		44.9
^{172}Yb		102	171.936378 (3)	21.83 (67)		68.6
^{173}Yb		103	172.938208 (3)	16.13 (27)		50.7
^{174}Yb		104	173.938859 (3)	31.83 (92)		100.0
^{176}Yb		106	175.942564 (4)	12.76 (41)		40.1

^{175}Lu	71	104	174.940770 (3)	97.41 (2)	174.967 (1)	100.0
^{176}Lu		105	175.942679 (3)	2.59 (2)		2.7

^{174}Hf	72	102	173.940044 (4)	0.16 (1)	178.49 (2)	0.5
^{176}Hf		104	175.941406 (4)	5.26 (7)		15.0
^{177}Hf		105	176.943217 (3)	18.60 (9)		53.0
^{178}Hf		106	177.943696 (3)	27.28 (7)		77.8
^{179}Hf		107	178.9458122 (29)	13.62 (2)		38.8
^{180}Hf		108	179.9465457 (30)	35.08 (16)		100.0

^{180}Ta	73	107	179.947462 (4)	0.012 (2)	180.9479 (1)	0.0
^{181}Ta		108	180.947992 (3)	99.988 (2)		100.0

^{180}W	74	106	179.946701 (5)	0.12 (1)	183.84 (1)	0.4
^{182}W		108	181.948202 (3)	26.50 (16)		86.5
^{183}W		109	182.950220 (3)	14.31 (4)		46.7
^{184}W		110	183.950928 (3)	30.64 (2)		100.0
^{186}W		112	185.954357 (4)	28.43 (19)		92.8

^{185}Re	75	110	184.952951 (3)	37.40 (2)	186.207 (1)	59.7
^{187}Re		112	186.955744 (3)	62.60 (2)		100.0

^{184}Os	76	108	183.952488 (4)	0.02 (1)	190.23 (3)	0.0
^{186}Os		110	185.953830 (4)	1.59 (3)		3.9
^{187}Os		111	186.955741 (3)	1.96 (2)		4.8
^{188}Os		112	187.955830 (3)	13.24 (8)		32.5
^{189}Os		113	188.958137 (4)	16.15 (5)		39.6
^{190}Os		114	189.958436 (4)	26.26 (2)		64.4
^{192}Os		116	191.961467 (4)	40.78 (19)		100.0

^{191}Ir	77	114	190.960584 (4)	37.3 (2)	192.217 (3)	59.2
^{193}Ir		116	192.962917 (4)	62.7 (2)		100.0

^{190}Pt	78	112	189.959917 (7)	0.014 (1)	195.078 (2)	0.0
^{192}Pt		114	191.961019 (5)	0.782 (7)		2.3
^{194}Pt		116	193.962655 (4)	32.967 (99)		97.5
^{195}Pt		117	194.964766 (4)	33.832 (10)		100.0
^{196}Pt		118	195.964926 (4)	25.242 (41)		74.6
^{198}Pt		120	197.967869 (6)	7.163 (55)		21.2

^{197}Au	79	118	196.966543 (4)	100	196.96655 (2)	100.0

^{196}Hg	80	116	195.965807 (5)	0.15 (1)	200.59 (2)	0.4
^{198}Hg		118	197.966743 (4)	9.97 (20)		33.4
^{199}Hg		119	198.968254 (4)	16.87 (22)		56.5
^{200}Hg		120	199.968300 (4)	23.10 (19)		77.4
^{201}Hg		121	200.970277 (4)	13.18 (9)		44.1
^{202}Hg		122	201.970617 (4)	29.86 (26)		100.0
^{204}Hg		124	203.973467 (5)	6.87 (15)		23.0

^{203}Tl	81	122	202.972320 (5)	29.524 (14)	204.3833 (2)	41.9
^{205}Tl		124	204.974401 (5)	70.476 (14)		100.0

^{204}Pb	82	122	203.973020 (5)	1.4 (1)	207.2 (1)	2.7
^{206}Pb		124	205.974440 (4)	24.1 (1)		46.0
^{207}Pb		125	206.975872 (4)	22.1 (1)		42.2
^{208}Pb		126	207.976627 (4)	52.4 (1)		100.0

^{209}Bi	83	126	208.980374 (5)	100	208.98038 (2)	100.0

Po	84	–	–	–		–
At	85	–	–	–		–
Rn	86	–	–	–		–
Fr	87	–	–	–		–
Ra	88	–	–	–		–
Ac	89	–	–	–		–
^{232}Th	90	142	232.0380508 (23)	100	232.0381 (1)	100.0
Pa	91		–	–		–
^{234}U	92	142	234.0409468 (24)	[0.0055 (2)]	238.02891 (3)	0.0
^{235}U		143	235.0439242 (24)	[0.7200 (51)]		0.7
^{238}U		146	238.0507847 (23)	[99.2745 (106)]		100.0

Appendix 2 Periodic Table of the Elements (1 to 102), Shaded According to Number of Naturally Occurring Isotopes

Mass Spectrometry of Inorganic, Coordination and Organometallic Compounds W. Henderson and J. S. McIndoe
© 2005 John Wiley & Sons, Ltd   ISBNs: 0-470-85015-9 (HB); 0-470-85016-7 (PB)

Periodic Table

Legend (key):

Atomic number
Symbol
Element name
Atomic weight

Tc — no naturally occurring isotopes
Nb — 1 isotope
Ta — 2 isotopes
Cr — 3–5 isotopes
Mo — 6+ isotopes

Group	1	2	3	4	5	6	7	8	9	10	11	12	13	14	15	16	17	18
	1 **H** Hydrogen 1.0079																	2 **He** Helium 4.0026
	3 **Li** Lithium 6.941	4 **Be** Beryllium 1.0079											5 **B** Boron 10.811	6 **C** Carbon 12.011	7 **N** Nitrogen 14.007	8 **O** Oxygen 15.999	9 **F** Fluorine 18.998	10 **Ne** Neon 20.180
	11 **Na** Sodium 22.990	12 **Mg** Magnesium 24.305											13 **Al** Aluminium 26.982	14 **Si** Silicon 28.086	15 **P** Phosphorus 30.974	16 **S** Sulfur 32.065	17 **Cl** Chlorine 35.453	18 **Ar** Argon 39.948
	19 **K** Potassium 39.098	20 **Ca** Calcium 40.078	21 **Sc** Scandium 44.956	22 **Ti** Titanium 47.867	23 **V** Vanadium 50.942	24 **Cr** Chromium 51.996	25 **Mn** Manganese 54.938	26 **Fe** Iron 55.845	27 **Co** Cobalt 58.933	28 **Ni** Nickel 58.693	29 **Cu** Copper 63.546	30 **Zn** Zinc 65.39	31 **Ga** Gallium 69.723	32 **Ge** Germanium 72.61	33 **As** Arsenic 74.922	34 **Se** Selenium 78.96	35 **Br** Bromine 79.904	36 **Kr** Krypton 83.80
	37 **Rb** Rubidium 85.468	38 **Sr** Strontium 87.62	39 **Y** Yttrium 88.906	40 **Zr** Zirconium 91.224	41 **Nb** Niobium 92.906	42 **Mo** Molybdenum 95.94	43 Tc Technetium	44 **Ru** Ruthenium 101.07	45 **Rh** Rhodium 102.91	46 **Pd** Palladium 106.42	47 **Ag** Silver 107.87	48 **Cd** Cadmium 112.41	49 **In** Indium 114.82	50 **Sn** Tin 118.71	51 **Sb** Antimony 118.71	52 **Te** Tellurium 127.60	53 **I** Iodine 126.90	54 **Xe** Xenon 131.29
	55 **Cs** Caesium 132.91	56 **Ba** Barium 137.33	71 **Lu** Lutetium 174.97	72 **Hf** Hafnium 178.49	73 **Ta** Tantalum 180.95	74 **W** Tungsten 183.84	75 **Re** Rhenium 186.21	76 **Os** Osmium 190.23	77 **Ir** Iridium 192.22	78 **Pt** Platinum 195.08	79 **Au** Gold 196.97	80 **Hg** Mercury 200.59	81 **Tl** Thallium 204.38	82 **Pb** Lead 207.2	83 **Bi** Bismuth 208.98	84 *Po*	85 *At*	86 *Rn*
	87 *Fr*	88 *Ra*	57–70															

lanthanides (57–70):

57 **La** Lanthanum 138.91	58 **Ce** Cerium 140.12	59 **Pr** Praseodymium 140.91	60 **Nd** Neodymium 144.24	61 *Pm* Promethium	62 **Sm** Samarium 150.36	63 **Eu** Europium 151.96	64 **Gd** Gadolinium 157.25	65 **Tb** Terbium 158.93	66 **Dy** Dysprosium 162.50	67 **Ho** Holmium 164.93	68 **Er** Erbium 167.26	69 **Tm** Thulium 168.93	70 **Yb** Ytterbium 173.04

actinides (89–102):

Ac	90 **Th** Thorium 232.04	*Pa*	92 **U** Uranium 238.03	*Np* Neptunium	*Pu*	*Am*	*Cm*	*Bk*	*Cf*	*Es*	*Fm*	*Md*	*No*

Appendix 3   Alphabetical List of Elements (1 to 102)

Element	Symbol	#	Element	Symbol	#	Element	Symbol	#	Element	Symbol	#
Actinium	*Ac*	*89*	Erbium	Er	68	Molybdenum	Mo	42	Scandium	Sc	21
Aluminium	Al	13	Europium	Eu	63	Neodymium	Nd	60	Selenium	Se	34
Americium	*Am*	*95*	*Fermium*	*Fm*	*100*	Neon	Ne	10	Silicon	Si	14
Antimony	Sb	51	Fluorine	F	9	*Neptunium*	*Np*	*93*	Silver	Ag	47
Argon	Ar	18	*Francium*	*Fr*	*87*	Nickel	Ni	28	Sodium	Na	11
Arsenic	As	33	Gadolinium	Gd	64	Niobium	Nb	41	Strontium	Sr	38
Astatine	*At*	*85*	Gallium	Ga	31	Nitrogen	N	7	Sulfur	S	16
Barium	Ba	56	Germanium	Ge	32	*Nobelium*	*No*	*102*	Tantalum	Ta	73
Berkelium	*Bk*	*97*	Gold	Au	79	Osmium	Os	76	*Technetium*	*Tc*	*43*
Beryllium	Be	4	Hafnium	Hf	72	Oxygen	O	8	Tellurium	Te	52
Bismuth	Bi	83	Helium	He	2	Palladium	Pd	46	Terbium	Tb	65
Boron	B	5	Holmium	Ho	67	Phosphorus	P	15	Thallium	Tl	81
Bromine	Br	35	Hydrogen	H	1	Platinum	Pt	78	Thorium	Th	90
Cadmium	Cd	48	Indium	In	49	*Plutonium*	*Pu*	*94*	Thulium	Tm	69
Calcium	Ca	20	Iodine	I	53	Polonium	Po	84	Tin	Sn	50
Californium	*Cf*	*98*	Iridium	Ir	77	Potassium	K	19	Titanium	Ti	22
Carbon	C	6	Iron	Fe	26	Praseodymium	Pr	59	Tungsten	W	74
Cerium	Ce	58	Krypton	Kr	36	*Promethium*	*Pm*	*61*	Uranium	U	92
Cesium	Cs	55	Lanthanum	La	57	*Protactinium*	*Pa*	*91*	Vanadium	V	23
Chlorine	Cl	17	Lead	Pb	82	Radium	Ra	88	Xenon	Xe	54
Chromium	Cr	24	Lithium	Li	3	Radon	Rn	86	Ytterbium	Yb	70
Cobalt	Co	27	Lutetium	Lu	71	Rhenium	Re	75	Yttrium	Y	39
Copper	Cu	29	Magnesium	Mg	12	Rhodium	Rh	45	Zinc	Zn	30
Curium	Cm	96	Manganese	Mn	25	Rubidium	Rb	37	Zirconium	Zr	40
Dysprosium	Dy	66	*Mendelevium*	*Md*	*101*	Ruthenium	Ru	44			
Einsteinium	*Es*	*99*	Mercury	Hg	80	Samarium	Sm	62			

Elements without naturally occurring isotopes appear *italicised.*

Mass Spectrometry of Inorganic, Coordination and Organometallic Compounds   W. Henderson and J. S. McIndoe
© 2005 John Wiley & Sons, Ltd   ISBNs: 0-470-85015-9 (HB); 0-470-85016-7 (PB)

Appendix 4   Glossary of Terms

Accurate mass

A mass determination sufficiently accurate (to within 5 *ppm*, at least) to enable the elemental composition of the ion to be predicted. Most useful for ions with $m/z < 500$.

amu

Atomic mass unit. See *Da*.

APCI

Atmospheric pressure chemical ionisation. Reagent ions formed in a corona discharge perform *chemical ionisation* at atmospheric pressure.

APPI

Atmospheric pressure photoionisation. The corona discharge in APCI is replaced by a UV lamp, photons from which ionise a suitable solvent (typically acetone or benzene) which then react with the analyte.

Atm

Atmospheric pressure, defined as 101 325 Pa.

Average mass

The mass of an ion of a given formula calculated using the atomic weight of each element, e.g. $C = 12.01104$, $H = 1.007976$, $N = 14.00672$.

B

The magnetic field, in Tesla. Also an abbreviation for a *magnetic sector*.

Bar

10^5 Pa. Unit of pressure approximating to one atmosphere. Low pressures often listed in millibar, 1 mbar = 100 Pa.

Mass Spectrometry of Inorganic, Coordination and Organometallic Compounds   W. Henderson and J. S. McIndoe
© 2005 John Wiley & Sons, Ltd   ISBNs: 0-470-85015-9 (HB); 0-470-85016-7 (PB)

Base peak

The most intense peak in a spectrum. By convention, its relative intensity is set at 100% and all other peaks in the spectrum are normalised to this value.

CI

Chemical Ionisation. Reagent ions, generated *in situ*, are used to ionise molecules through transfer of a proton, electron or other charged species from one species to another.

CID

Collision-Induced Dissociation. The fragmentation of an ion by collision with a neutral gas atom (e.g. argon or xenon) or molecule (e.g. nitrogen). Synonymous with CAD (collision-activated decomposition).

Centroid

Computer processing of a mass spectrum replaces each peak profile with a single line, typically with the intention of further computational treatment (library searching, calibration, etc.).

Cluster ion

An ion incorporating more than one molecule in association with a charged species, for example $[(Ph_3PO)_nNa]^+$ ($n = 2$ to 4).

Collision cell (chamber)

A volume containing a relatively high pressure of an inert gas (typically helium, argon or xenon), within which *CID* takes place. The cell usually employs an ion focusing device such as an rf-only *quadrupole* or *hexapole*. The collision voltage may be adjusted to induce more or less fragmentation.

Cone voltage

A voltage applied in the intermediate-pressure region between an atmospheric pressure source and the mass analyser to energise the ions and cause fragmentation through *CID* processes between ions and residual desolvation gas and solvent molecules.

Configuration

The way in which the analysers in a multianalyser instrument are arranged. The shorthand notation for this uses the abbreviations *B*, *E*, *FTICR*, *Q*, *QIT* and *TOF* in the order in which they are encountered by the ion beam, e.g. QTOF (*hybrid*; *quadrupole* followed by *time-of-flight*), EB (*double-focusing sector* instrument).

Continuum mode

Each recorded spectrum is saved separately (as opposed to *MCA*).

Cryogenic pump

Used to maintain exceptionally high vacuums, such as in FTICR instruments. The final stage of pumping, cryogenic pumps condense gases through cooling to very low temperatures (4.2 K, liquid He).

Da

Dalton. Unit for atomic mass, defined as 12 Da = 12 amu = mass of ^{12}C.

Da/e

Synonymous with *m/z*.

Daughter ion

See *product ion*.

Delayed extraction

A method of focusing ions prior to acceleration down the flight tube in *MALDI-TOF*.

Detection limit

The smallest amount of sample that provides a signal distinguishable from the background noise. Experimental conditions must be specifying in detail.

DI

Desorption Ionisation. Ionisation directly from a solid or liquid surface, as in MALDI, FAB/LSIMS, PD, FD.

Diffusion pump

A high speed jet of oil vapour collides with residual gas molecules and sweeps them in the direction of the roughing pump (which is required to reduce the exhaust pressure to a tolerable limit). The oil is recycled by condensation. Capable of reducing pressure to the region of 10^{-7} mbar.

Direct probe

A means of introducing a solid or liquid sample directly into an ion source.

Double focusing

A high *resolution* instrument using both a *magnetic sector* and an *electrostatic analyser* to separate ions. Forward (BE) and reverse (EB) geometries are possible.

e

Fundamental unit of charge, 1.6×10^{-19} coulombs.

E

An *electrostatic analyser*. Also the voltage of the same.

EDESI

Energy-dependent *ESI*. A presentation method for *CID* experiments plotting the data in 3D, *m/z* vs. fragmentation energy with contours on a map representing ion intensity.

EI

Electron Ionisation (synonymous with 'electron impact'). An energetic beam of electrons interacts with and ionises gas-phase molecules.

Electron multiplier

Used in detectors to enhance the current associated with an arriving ion. Electrons are accelerated into the surface of an electrode, which in turn release more secondary electrons which are accelerated into a second electrode, and so on. An avalanche effect ensues, generating a detectable current.

ES

See *ESI*.

ESI

Electrospray Ionisation. Ions are transported into the gas phase by spraying a solution through a capillary at high potential. Solvent is evaporated by heating or by a warm bath gas.

eV

Electron volt. The kinetic energy acquired by an electron upon acceleration through a potential difference of 1 V $= 1.602177 \times 10^{-19}$ J.

Even-electron ion

An ion containing no unpaired electrons, i.e. $[M]^{\pm}$.

FAB

Fast Atom Bombardment. A fast-moving beam of inert gas atoms desorbs ions dissolved in a liquid *matrix*. Also see *LSIMS*.

Faraday cup

A hollow cup-shaped detector, robust but not especially sensitive.

FI/FD

Field Ionisation/Field Desorption. Electrons tunnel from a gaseous (FI) or solid (FD) sample to an emitter under the influence of an extremely high electric field gradient.

Forward geometry

EB *double-focusing* mass spectrometer.

Fragmentation

Decomposition of an ion, usually to another ion and a neutral molecule or radical.

FTICR

Fourier Transform Ion Cyclotron Resonance. *ICR* modified so that all ions are excited and detected simultaneously, and the complex waveform of overlapping signals is converted into frequency and amplitude values (and hence into a mass spectrum) by means of a mathematical operation called a Fourier transform.

FWHM

Full Width at Half Maximum. The full width of a single peak measured at half the height of the peak. Used to calculate *resolution*, by dividing the FWHM by the m/z ratio of that peak.

GC/MS

Gas Chromatography/Mass Spectrometry. A mass spectrometer with an *EI* or *CI* source is used as a detector for a gas chromatography column. A very powerful technique for the analysis of complex, volatile mixtures.

h

Hexapole ion guide. Six parallel rods arranged in an hexagonal array used to focus ions by means of an applied rf-field.

Hybrid

A mass spectrometer with *MS/MS* capabilities that includes mass analysers from more than one of the *FTICR*, *Q*, *QIT*, *TOF* and *sector* categories, e.g. Q-TOF.

ICP-MS

Inductively coupled plasma mass spectrometry. A sample is entrained in a flow of gas and carried into a high temperature plasma, where it is converted into monatomic ions.

ICR

Ion cyclotron resonance. Ions are trapped in a tight orbit by a strong magnetic field, and their orbiting frequency is characteristic of their m/z ratio. Exciting the ions into orbits near the walls of the trapping cell induces detectable image currents. Also see *FTICR*.

Inlet

The means by which a sample is introduced to the mass spectrometer.

In-source fragmentation

CID in the region between an atmospheric pressure source and the mass analyser; see *cone voltage*.

In space, in time

Terms used to differentiate *MS/MS* carried out in spatially separated mass analysers (in space) or the same location but separated chronologically (in time). *MS/MS* in space is commonly performed by *QqQ* and *hybrid* instruments; *MS/MS* in time by *QIT* and *FTICR* instruments.

Intensity

The intensity of a peak (0 – 100) relative to the *base peak* (by convention, 100).

Ion trap

See *QIT*.

Ion tunnel

A strongly focusing ion guide consisting of ring electrodes of alternating polarity; these are replacing *rf-only* quadrupoles and hexapoles in some instruments.

Ionisation

The process that transforms a neutral species into an ion.

IRMPD

Infrared multi-photon dissociation. A means of ion fragmentation using IR radiation in an *FTICR-MS*.

Isobaric Ions

Ions of the same nominal mass but with different atomic compositions, e.g. CO^+ and $C_2H_4^+$.

ITMS

Ion trap mass spectrometry. See *QIT*.

LC/MS

Liquid chromatography/mass spectrometry. A mass spectrometer with an *ESI* or *APCI* source is used as a detector for a high performance liquid chromatography (HPLC) column. A very powerful technique for the analysis of complex mixtures of soluble polar compounds.

LDI

Laser Desorption/Ionisation. A pulsed laser is used to ablate material from a surface, the irradiation causing ionisation through a thermal process. Also see *MALDI*.

Linked scan

Double focusing sector instruments can have their magnetic (B) and electric (E) fields scanned in a fixed ratio (e.g. B/E, B^2/E) to allow observations of fragmentation in the *field-free* regions of a double-focusing sector instrument.

LSIMS

Liquid Secondary Ion Mass Spectrometry. An energetic beam of ions (typically Cs^+) strikes a liquid *matrix* in which the analyte is dissolved, ejecting secondary (analyte) ions into the gas phase. Closely related to *FAB*.

Magnetic sector

A mass analyser that applies a magnetic field perpendicular to the direction the ions enter, which focuses all ions with the same momentum and *m/z*.

MALDI

Matrix Assisted Laser Desorption/Ionisation. A solid *matrix* cocrystallised with a small amount of analyte is ablated by a laser. The matrix absorbs most of the energy and is vaporised to form an energetic plume in which the analyte is ionised.

Mass analysis

Separation of ions according to their *m/z* ratios.

Mass spectrometer

A device which produces gas-phase ions, separates them according to *m/z* and counts them, so generating a mass spectrum.

Mass spectrum

A plot of *m/z* vs. intensity, usually normalized such that the *base peak* = 100%. Mass spectra may be displayed as a line graph or as a histogram ('stick plot', where each peak is represented by a *centroid*).

Matrix

A material used to support the analyte and absorb most of the incident energy in a desorption technique e.g. glycerol in *LSIMS*, nicotinic acid in *MALDI*, nitrocellulose in *PD*.

MCA

Multi Channel Analysis. Spectra are summed together as they are recorded, generating a single spectrum with improved *signal-to-noise ratio*.

Metastable ion

An ion that decomposes after leaving the source but before arriving at the detector.

MCP

Microchannel Plate. A detector consisting of a closely packed array of electron multipliers that can detect ions arriving over a wide area. Most frequently employed in *TOF* instruments.

mm Hg

one atmosphere of pressure will force a column of mercury to a height of 760 mm, so 1 mm Hg = 1/760 atm = 133.32 Pa. Equivalent to Torr.

Molecular ion

An ion formed by the removal or addition of an electron from a molecule. Also see *pseudomolecular ion*.

Monoisotopic mass

The mass of an ion of a given formula calculated using the exact weight of the most abundant isotope of each element, e.g. ^{12}C = 12 exactly, ^{1}H = 1.007825, ^{14}N = 14.003074.

MS/MS

The measurement of the *m/z* ratio of ions before and after reaction in the mass spectrometer. This is generally carried out by using one mass analyser to select ions of a particular *m/z* (*precursor* or *parent* ions) which are fragmented (usually by *CID*) and a second mass analyser is used to detect the ions resulting from the reaction (the *product* or *daughter* ions).

MS^{n}

Trapping mass spectrometers (*QIT*, *FTICR*) perform MS experiments *in time*, allowing a *precursor ion* to be selected and fragmented to form *product ions*, one of which may be selected in turn to become a precursor itself, etc. Many stages of such an experiment may be performed (hence MS^{3}, MS^{4}, MS^{5}, etc.).

m/z

The mass-to-charge ratio. A dimensionless quantity obtained by dividing the mass of an ion by the number of charges it carries. Alternatively, *Th* (Thomson).

Nanospray

Nanoflow Electrospray. A type of *electrospray ionisation* in which very low flow rates (nL min^{-1}) are achieved through use of a conducting microcapillary, greatly improving sensitivity and lowering sample consumption.

NICI

Negative Ion Chemical Ionisation. Any form of *CI* that results in a negatively charged ion, but most commonly ionisation takes place through capture of an electron (hence also Electron Capture Ionisation).

Nominal mass

The mass of an ion of a given formula calculated using the integer mass number of the most abundant isotope of each element, e.g. C = 12, H = 1, N = 14.

oa-TOF

Orthogonal *TOF*. Time-of-flight mass analysis where the ions are accelerated into the drift region in a direction at right angles to the original direction of motion; most frequently used in conjunction with continuous ion sources.

Odd-electron ion

A *radical ion*, $[M]^{\bullet\pm}$.

Parent ion

See *precursor ion*.

PD

Plasma Desorption, also known as ^{252}Cf fission ionisation. Extremely energetic fission fragments are used to desorb and ionise molecules from a surface.

PI

Photoionisation. Ionisation *via* irradiation with photons.

ppm

Part per million. In MS terms, the accuracy of a mass determination. If an experimental *m/z* value is 1000.005 and the theoretical value is 1000.000, the two values are correct to within $0.005/1000 = 5 \times 10^{-6} = 5$ ppm.

Precursor ion

A charged species that undergoes fragmentation. It does not have to be the *molecular ion*; a fragment may be selected as a precursor ion if it is then induced to undergo decomposition.

Product ion

The charged product of the reaction (usually, but not necessarily, fragmentation) of a specific *precursor ion*.

PSD

Post Source Decay. Fragmentation of an ion while in the drift region of a TOF mass analyser; observed only if fitted with a *reflectron*.

Pseudomolecular ion

Ions formed by association of a molecule with a charged species e.g. $[M + H]^+$, $[M + NH_4]^+$, $[M + Na]^+$ or by deprotonation i.e. $[M - H]^-$.

q

Abbreviation for an *rf-only quadrupole* ion focusing device.

Q

Abbreviation for a *quadrupole* mass analyser.

QIT

Quadrupole Ion Trap. A mass analyser consisting of a ring electrode and two end cap electrodes, within which a three-dimensional quadrupole field is established to trap ions in a complicated trajectory. Ions may be ejected sequentially in order of m/z value to generate a mass spectrum, and MS^n studies may be performed *in time*.

QqQ

Abbreviation for a *triple quadrupole* instrument, in which two *quadrupole* mass analysers are separated by a *collision cell*, enabling *MS/MS* studies to be performed. Also QhQ, where the collision cell is a *hexapole*.

QTOF

A popular *hybrid instrument* consisting of a *quadrupole* mass analyser, a *collision cell* and a *TOF* mass analyser.

Quadrupole

Four parallel cyclindrical metal rods arranged in a square array act as a mass analyser by the application of appropriate AC and DC voltages. Ions undergo a complicated trajectory along the central axis and only those with a certain m/z value survive the journey without discharging on one of the rods. If operated using the AC voltage alone (*rf-only* mode), a quadrupole passes and focuses ions of all m/z.

Quasimolecular ion

See *pseudomolecular ion*.

Quistor

See *QIT*.

Radical ion

An ion with an unpaired electron, e.g. $[M]^{\bullet+}$, a radical cation. Radical cations are the *molecular ions* in *EI*; sometimes called 'odd-electron' ions.

Reflectron

An ion mirror, used to reflect ions traveling along a flight tube back in the opposite direction, which has the effect of greatly improving the *resolution* of a *TOF* mass analyser.

Resolution

A key measure of the performance of a mass spectrometer. Measured in different ways, most simply from a single peak using the *FWHM* measurement, but more traditionally by assessing the ability of the mass spectrometer to discriminate between two ions close to one another in *m/z* ratio. If two peaks Δm apart in mass overlap such that the valley between them reaches 10% the intensity of the main peaks, the resolution is given by $\Delta m/m$. A 50% valley definition is sometimes used for quadrupole instruments.

Resolving power

The ability to differentiate between ions close in mass. See *resolution*.

Reverse Geometry

BE *double-focusing* mass spectrometer.

rf-only quadrupole

Radio-frequency only *quadrupole*. A type of ion guide in which an AC current is applied to the rods, passing (and focusing) ions of all *m/z* values. rf-only hexapoles, octapoles (and indeed, N-poles where N is an even number) are also known.

Roughing pump

A mechanical pump which physically sweeps air from a system, often using a rotary device. They have large gas handling capacities, but cannot achieve high vacuum (only about 10^{-3} mbar), and are used to remove the bulk of the air from a system to reduce the exhaust pressure of pumps capable of higher vacuum (*diffusion, turbomolecular, cryogenic pumps*).

Savitsky-Golay

A *smoothing* algorithm. A computationally efficient way of plotting a least-squares fit of a polynomial function from the points of a single data set.

Sensitivity

Ion current generated from a given amount of sample, from a well-described set of operating conditions.

SID

Surface Induced Decomposition. Energetic ions impinging on an inert surface will fragment to generate product ions, which can be detected by a second mass analyser to enable *MS/MS* studies.

SIMS

Secondary Ion Mass Spectrometry. A fast-moving ion beam is used to abrade a surface, removing and ionising material. Conditions may be set in such a way as to obtain a depth profile.

Single Focusing

A low *resolution* instrument that separates ions using a single *magnetic sector* only.

Smoothing

Cosmetic improvement of data by application of a suitable algorithm, e.g. the moving average or *Savitsky-Golay* methods. Often applied prior to converting a *mass spectrum* into *centroid* data.

SNR

Signal-to-noise ratio. The degree to which a signal is differentiated from random noise; the higher the better.

Soft Ionisation

Any ionisation method that produces ions with low *internal energy*, manifested by strong *molecular ion* (or *pseudomolecular ion*) signals and little or no fragmentation, e.g. *MALDI* and *ESI*.

Source

The region immediately before the mass analyser, where gas-phase ions are formed.

Tandem Mass Spectrometry

A now less-frequently used term for *MS/MS*.

TDC

Time-to-digital converter. Used for measuring the times of arrival at a detector in a *TOF* instrument.

Th

Thomson. $= m/z$. Unit named after the 'father of mass spectrometry', Sir J. J. Thomson.

Thermospray

Ionisation source that involves passing a solution of the analyte through a heated capillary.

TIC

Total Ion Current. The sum of all the current carried by the separate ions contributing to the mass spectrum.

TOF

Time-of-Flight mass analyser. Ions are accelerated in a pulse and allowed to pass along a *flight tube*. Their arrival time at the other end is a function of their *mass-to-charge ratio*.

Torr

See *mm Hg*.

Triple quad

See *QqQ*.

Turbomolecular pump

A rapidly spinning turbine rotor which pushes gas from the inlet towards the exhaust. A *roughing pump* is also employed to reduce the exhaust pressure.

Unit mass resolution

Resolution sufficiently high to discriminate between peaks 1 *m/z* apart. Only meaningful when the mass range is also specified, e.g. 'unit mass resolution at 500 *m/z*'.

V

Voltage; if unspecified, the voltage used to accelerate ions into the mass analyser.

Vacuum system

The evacuated volume of a mass spectrometer and all the pumps (see *roughing*, *turbomolecular*, *diffusion*, and *cryogenic pumps*) and other devices associated with creating, monitoring and maintaining the low pressure (= high vacuum).

z

The charge number, i.e. a doubly charged ion, $[M]^{2\pm}$, has $z = 2$.

Appendix 5 Useful Sources of Information

The Encyclopedia of Mass Spectrometry, M. L. Gross and R. M. Caprioli. (Eds-in-chief), Elsevier, Oxford. This will be a 10-volume work published over a 4-year period, 2003-6. As at June 2004, Volume 1 (Theory and Ion chemistry) has been published. Of most interest to inorganic chemists will be Volume 4 (Fundamentals of and Applications to Organic (and Organometallic) Compounds, Volume 5 (Elemental, Isotopic and Inorganic Analysis by Mass Spectrometry), and Volume 6 (Molecular Ionization Methods).

Mass Spectrometry – A Textbook, J. H. Gross, Springer, 2004. A welcome all-new addition to the list of general mass spectrometry student textbooks. Problems and solutions from the book are freely available online at *http : //www.ms-textbook.com/*.

Mass Spectrometry Basics, C. G. Herbert and R. A. W. Johnstone, CRC Press, Boca Raton, 2003. Consists of short chapters mostly concerning the various components of a mass spectrometer – inlets, ionisation sources, mass analysers, detectors, data systems. Some notable holes in coverage (nothing on ion traps, for example) and with no citations to the primary literature, this reference book is nonetheless an important resource to anyone trying to get to grips with the terminology and instrumentation used in modern mass spectrometry. Appendices include the most comprehensive glossary of terms available and an excellent and comprehensive bibliography listing mass spectrometry textbooks covering the full range of topics.

Mass Spectrometry: Principles and Applications, E. de Hoffmann and V. Stroobant, John Wiley & Sons Ltd., Chichester, UK, 2001. A complete overview of the principles, theories and key applications of modern mass spectrometry. Many examples and illustrations, as well as problem sets for students.

Applications of Inorganic Mass Spectrometry, J. R. Laeter, Wiley, New York, 2001. Covers mass spectrometry applications in metrology, cosmochemistry, geosciences, nuclear, environmental, materials and planetary science, as well as an introduction to the techniques used in these fields.

Mass Spectrometry of Inorganic, Coordination and Organometallic Compounds W. Henderson and J. S. McIndoe
© 2005 John Wiley & Sons, Ltd ISBNs: 0-470-85015-9 (HB); 0-470-85016-7 (PB)

Inorganic Spectroscopic Methods, A. K. Brisdon, Oxford University Press, Oxford, UK, 1998. This book, primarily aimed at the undergraduate student, contains a brief introduction to mass spectrometry that is useful for the complete beginner.

Electrospray Ionization Mass Spectrometry. Fundamentals, Instrumentation and Applications, R. B. Cole (Ed.) John Wiley & Sons, Ltd, Chichester, UK, 1997. This is a comprehensive book containing 15 chapters, organised into four parts covering: fundamental aspects of electrospray ionisation; electrospray coupling to mass analysers; interfacing of solution-based separation techniques to electrospray; and applications of electrospray ionisation.

Web References

The world-wide web is a huge and rapidly growing repository for information on all aspects of mass spectrometry. Impossible as it is to even scratch the surface with a list of links, we provide just one: Base Peak (*http : //base-peak.wiley.com/*). Self-described as 'the web's leading Mass Spectrometry resource', Base Peak is an enormously useful reference, providing free access to mass spectrometry research papers, discussion forums, conferences, books, links, educational materials etc. A good place to start a web-based search for information.

Index

Mass Spectrometry of Inorganic, Coordination and Organometallic Compounds W. Henderson and
J. S. McIndoe
© 2005 John Wiley & Sons, Ltd ISBNs: 0-470-85015-9 (HB); 0-470-85016-7 (PB)

www.ingramcontent.com/pod-product-compliance
Lightning Source LLC
Chambersburg PA
CBHW080941260125
20788CB00015BA/169